SIGNAL PROCESSING, SPEECH AND MUSIC

STUDIES ON NEW MUSIC RESEARCH

Serie Editors:

Marc Leman, Institute for Psychoacoustics and Electronic
Music, University of Ghent, Belgium
Paul Berg, Royal Conservatory, The Hague, The Netherlands

SIGNAL PROCESSING, SPEECH AND MUSIC

STAN TEMPELAARS

Routledge
Taylor & Francis Group

LONDON AND NEW YORK

First published 1996 by
SWETS & ZEITLINGER PUBLISHERS

Published 2014 by Routledge
2 Park Square, Milton Park, Abingdon, Oxfordshire OX14 4RN
711 Third Avenue, New York, NY, 10017, USA

First issued in paperback 2016

Routledge is an imprint of the Taylor & Francis Group, an informa business

Library of Congress Cataloging-in-Publication Data

Tempelaars, Stan, 1938-
    Signal processing, speech, and music / Stan Tempelaars.
        p.   cm. -- (Studies on new music research)
    Includes bibliographical references and index.
    1. Signal processing - - Digital techniques. 2. Music- -Data
processing. 3. Speech processing systems.   I. Title.   II. Series.
TK5102.9.T43 1996
621.382'8 --dc21   96-46383
                                      CIP

Cover design: Ivar Hamelink

ISBN 13: 978-1-138-98192-8 (pbk)
ISBN 13: 978-90-265-1481-4 (hbk)

Publisher's Note
The publisher has gone to great lengths to ensure the quality of this reprint but points out
that some imperfections in the original may be apparent

PREFACE

In the academic year 1967/1968 the Institute of Sonology at Utrecht University organized a course that turned out to be the first of a long series. Originally, this course was intended for composers who wanted to start activities in the field of electronic music, but gradually, it became a course in that scientific/artistic discipline which we now call 'sonology'. My contribution to the program was initially called 'electro-acoustics' and later 'signal processing'. This change is also characteristic for a shift of the emphasis from the treatment of practical systems to a more theoretical and fundamental approach.

This book is written in the first place as a text book to be used on courses in sonology. Therefore, in some places in the text it can be detected that the book was originally intended for students with a musical background. To make it possible to use it in a course for students in phonetics as well, I have added a few topics from the field of speech signal processing, particularly in chapter 7.

My experience has taught me that for some students it is necessary to update or refresh their knowledge of mathematics, which is often fragmentary or incomplete. For that purpose, I have added a short crash course in mathematics in chapter 2 (a good introduction can be found in Batschelet (1975) and Lax et al.(1976)). To allow students with limited mathematical capacity to work with this book I have decided to avoid the use of complex numbers. This has made some derivations and proofs more cumbersome, but I hope that it has also improved the accessibility. The derivations and proof are, furthermore, quite complete, mainly for the benefit of those who want to know how certain results are reached.

The subject of this book is 'acoustic communication' and it deals with the two most important acoustical communication systems for human beings: speech and music. As with all communication systems the goal here is to transmit a message, a quantity of information, from one person ('source') to another ('receiver'). In an acoustical communication system this occurs by means of air vibrations. Such a message linked to a physical carrier we call a 'signal'. If we wish to study this communication process, we have three possible entries: at the source we can investigate how the signal is produced; with the receiver we can delve into the perception of the signal; and we can direct ourselves to the sound wave itself. These three areas of investigation are naturally closely related but each of them has its own difficulties and possibilities. The phase in between, that of the acoustical signal, has the advantage that we are dealing with pure physical processes that can be exactly measured and analysed with physical equipment. This book deals with this phase of the communication process, the theoretical and practical aspects of sound signals and of systems for processing (in the most general sense of the word) these signals.

The theoretical treatment is based on signal and system theory. In this theory, as much as possible is abstracted from the physical aspects: signals and systems are treated as mathematical functions and operators. Then the practical aspects of signals and systems are considered. Only since the invention of the microphone and the loudspeaker has it really been possible to analyse and manipulate sound signals. The acoustical signal itself, the sound vibration, is elusive and intractable. How difficult the investigation of the air vibrations is, is shown by von Helmholtz's famous book, "On the Sensation of Tone" which appeared in 1863 and which, by the way, is still worth studying. Thanks to the microphone, we can convert the fluctuations of the air-pressure into a proportional ('analog') variation of an electrical voltage. That this causes no degradation of the signal can be checked by using a loudspeaker to monitor the signal from the microphone. For the investigation and manipulation of the electrical version of acoustical signals a whole arsenal of measuring, analysis and registration equipment has been developed. This book discusses such electrical (or electronic) systems. Those who are interested in acoustical and electro-acoustical systems are referred to other textbooks.

The plan of this book is as follows: after the introductory chapter 1 that deals with acoustical communication, chapter 2 considers the principal concepts of the 'function' and the 'signal function', in which we shall have a look at the mathematical tools for the theoretical models of signals and systems. Various important mathematical functions (exponential, logarithmic, trigonometric) and operations with functions, such as differentiation and integration, will be discussed. In chapter 3 the second theme, namely 'systems', is introduced via a particular system that is simultaneously simple and important: the harmonic oscillator. In chapter 4 we go back to signal functions. Now we learn about various possible representations and transformations, among others the analog and digital representation and the transformation from the time domain into the frequency domain and back (the Fourier transform). The next chapter, no. 5, is again devoted to systems, in particular to the theory of linear systems. Chapters 6 and 7 deal with practical aspects. In chapter 6 various linear and nonlinear systems are discussed while in chapter 7 analysis and synthesis systems are considered. Chapter 8, the final chapter, is an epilogue in which we return to the subject of chapter 1, acoustical communication.

The amount of subjects is quite large for a single course. Therefore, I strongly advise the teacher who wants to use this book to organize two courses. From my own teaching experience, I know that a good solution is to discuss in the first course chapter 1, chapter 2 (sections 2.1, 2.2 and 2.6), chapter 4 (sections 4.1, 4.2.A, 4.3, 4.4.A, 4.4.D and chapter 5, while skipping all applications of calculus. In the second course, the remaining chapters and sections could then be discussed.

I thank all who have helped me with suggestions. I thank Paul Goodman who translated the first version, Peter Pabon for allowing me to use his presentation of the Fast Fourier Transform, and Lies Zondag for the final corrections.

Stan Tempelaars
Summer 1996

CONTENTS

CHAPTER 1

Acoustical Communication

Acoustical communication is possible thanks to the fact that we have an organ that is sensitive to pressure fluctuations in the medium (air) in which we live. A similar or analogous organ is to be found also in more primitive beings and here the function of this organ is clear: it *warns* the organism of danger and makes it aware of the presence of food etc. This is the primary function of our hearing organ. The sound signals here are irregular and often pulse-like (the breaking of a twig, the falling of a stone, the noise of water etc.). This fact is very important for a proper insight into the functioning of the ear. For perception experiments, for example, we should not restrict ourselves to regular stimuli such as sinusoidal vibrations. This holds especially for directional hearing because this ability is directly connected with the warning function of the hearing organ. Also the strong tendency people have to identify unknown sounds (for example electronically generated sounds) can be related to this function.

The 'pressure sensitivity' is in fact a sensitivity to pressure *changes* where the pressure fluctuation must be greater than a particular threshold level to excite the nerves. This threshold level is not constant but depends among other things on the 'speed' of the vibration (still better: its 'frequency', but this term is introduced later in the next chapter). When the pressure fluctuations are too slow, they no longer integrate into a sound impression. I will deal with this aspect further on in this chapter. At the upper boundary there is much variation. The limiting factor here is the mass of the vibrating parts of the hearing organ. With fast vibrations most of the sound energy is converted into heat in the middle ear and does not reach the sensory cells in the inner ear. A high upper threshold is therefore to be found in small animals (with sound vibrations in water and thus with sea creatures the situation is different). The upper threshold is also dependent on age, sex and other factors. Between these two limits there is still a wide range of audible sounds that are also available for the second function of the hearing organ, the communicative function, which is the consequence of the ability of the higher animals to produce sound vibrations themselves. This led to certain applications such as the 'radar' system of bats, but much more important was the possibility of communication between members of a species in connection with reproduction, collection of food, mutual defence etc. Here a curious phenomenon occurs: with our muscles we indeed can produce air pressure variations, for example, by waving our hands, but as these fluctuations are rather slow, the produced sound waves have wavelengths that are relatively long (between tens and hundreds of meters) compared to the dimensions of the source of the vibrations, the human body. This is important because to radiate a signal efficiently the source dimensions should be of the same order of magnitude as the wavelength. A violin for example radiates wavelengths between 0.22 and 1.7 m, a double bass between 1 and 8 m. With our muscles we can only produce sounds effectively via clapping the hands, drumming, hitting, stamping etc., i.e. by interrupting a movement and not by the movement itself. The repertory of such sounds however is far too limited to function as the basis for a high-level communication system. Thus the human body is a very inefficient source of the sound vibrations that are required for acoustical communication.

The solution to this problem has been the development of a special organ in many animals and in man to make acoustical communication possible. In man this is the voice. With this organ air vibrations with the necessary speed characteristics (easily perceivable, efficiently transmittable) can be produced. Simplified the principle of voice production is as follows: with the muscles of the thorax and diaphragm the air in the lungs is compressed. This air can only escape via the trachea by pressing apart the vocal chords that normally shut off the trachea. After a very short time the vocal chords are closed again by the aerodynamic effect of the escaping air. This process is repeated so quickly that 60 to 300 times a second a small portion of air escapes. These periodic air pushes produce a sound vibration with the speed properties required for a sound to be audible. These vibrations come thus into existence with the help of muscle energy but not as 'muscle vibrations'. Information is coded into this signal via (slow) changes in the parameters of this fast vibration, because

- a slow vibration is not audible
- a slow change in a fast vibration is audible.

Here a comparison can be made with the technical process of **modulation** used in radio and television transmissions. Only many years after the invention of this technique the resemblance to the process of speech (and music) production was recognized. In speech and music we see 'natural' examples of modulation that, in radio technique, is applied for the same purpose, namely that of improving the signal transmission. After amplification an electrical signal can be broadcast as an electromagnetic fluctuation with the help of an antenna. Due to the enormous speed of the electromagnetic waves, the wavelength would be so large that for an efficient radiation the antenna would have to be of an absolutely impossible size, and, what is more, signals coming from different transmitters would be difficult to distinguish from each other. Therefore, modulation is applied: in the transmitter a fast, stationary vibration, the so-called 'carrier wave' is generated. When a signal is to be broadcast, the maximal deviation of this vibration is made to correspond with this signal. Here we speak of 'amplitude modulation' (AM). (In radio technique other forms of modulation are employed as well). This process can be seen in the figure below (fig.1.1):

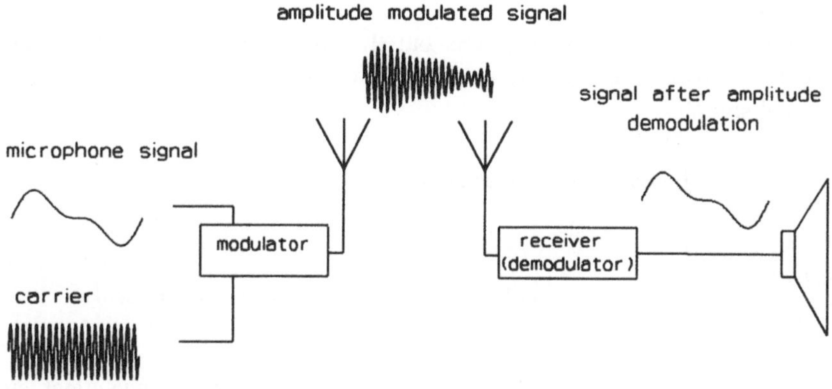

amplitude modulated signal

signal after amplitude
demodulation

microphone signal

modulator

receiver
(demodulator)

carrier

Figure 1.1 Amplitude modulation in radio transmission.

Due to the much higher speed of the vibration the antenna size can be kept to a reasonable level and by tuning the receiver (in which, by means of a process that is called 'demodulation', the original signal can be derived from the modulated one) to the carrier wave the signal can be distinguished from other signals with different carrier waves. Furthermore due to modulation, signals are less 'vulnerable' which means that they are not so easily distorted by interferences. This holds as well for speech as for musical signals. Along with making the transmission efficient, modulation makes the signal more robust; it remains detectable also under difficult circumstances. The modulation is much more complicated than with its technical equivalent because various parameters of the vibration are modulated simultaneously. In speech the system looks as follows (fig.1.2):

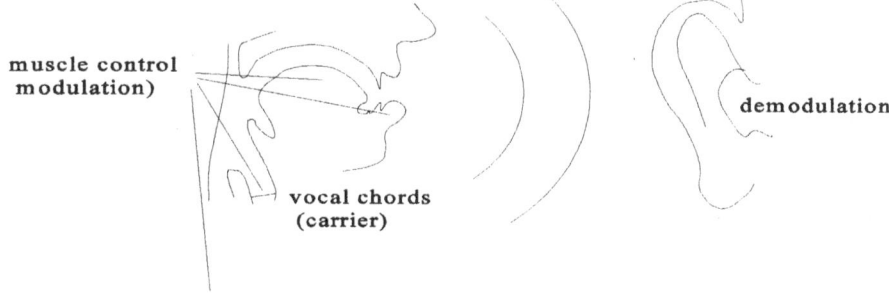

muscle control
modulation)

demodulation

vocal chords
(carrier)

Figure 1.2 Modulation in speech communication.

The carrier wave vibration is produced by means of the vocal chords and simultaneously there and higher in the vocal tract the carrier wave is modulated via (changing) the number of vibrations, the excursion and the 'form' of the vibration, the articulation etc.

In music the same thing happens but the role of the various parameters is different. It is therefore possible to give the following scheme for the production of musical signals:

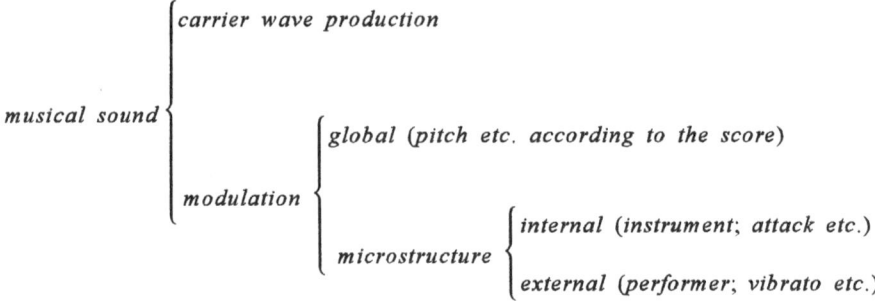

The analogy with a radio broadcast holds as well for the receiver. From the above it is clear that the perceptual process is a form of 'demodulation' in which the signal information is extracted from the modulated carrier wave. Having fulfilled its task, the supporting vibration is no longer required. That the carrier wave is not essential can be shown by means of the technique of Linear Predictive Coding (LPC) which will be discussed in detail in chapter seven. With LPC speech can be transmitted very efficiently by transmitting only the modulation parameters and not the carrier wave. Demodulation, i.e., separating carrier wave and modulating signal, is only possible when the frequency of the latter is below that of the former signal. In speech and musical signals this condition is fulfilled: the maximal 'muscle frequency' is ca. 20 Hz, which is also the lowest audible tone and thus the minimal carrier frequency.

Choosing the acoustic signal (in its electrical form) as an object for investigation thus means that we are dealing with modulated signals. We should always take this aspect into account when studying the process of signal transmission or the more general problem of acoustical communication. We shall be dealing with signals, the information carriers, and with systems that transmit and/or process these signals. Signals and systems are tightly coupled. Although we can construct an abstract mathematical model of the signal, the *signal function* (to be discussed in chapter 2) the real signal is always linked via its physical carrier to the system of which that physical carrier is a part (a space, an amplifier and so on).

Taking into account the modulation aspect, we can draw the conclusion that the communication chain looks as follows (fig.1.3):

Figure 1.3 The communication chain.

The signals and systems that will be our main subject are located between the output of the modulator and the input of the demodulator. It is of interest to check whether a more comprehensive approach is possible and to include aspects of the modulation and demodulation process as well. We will discuss this question in the final chapters.

For further reading, see Mayr (1980), Plomp(1984), Corliss (1990) and Pierce (1983).

CHAPTER 2

Functions

2.1 Registrations and signal functions

One possibility resulting from the conversion of the acoustical signal into an electrical one is that of making a registration of it. Examples of registration equipment are the *pen recorder* and the *oscilloscope*. Furthermore a computer equipped with converters (see chapter 4) and graphical output is often used for this purpose. The registrations shown here were made in this manner.

The pen recorder consists of an amplifier which increases the power of the electrical fluctuations and an electromechanical converter which converts an electrical voltage into a proportional deviation of a pen. Coupling the recorder with a microphone has the effect that the pen follows the air pressure fluctuations. If simultaneously we run a strip of paper at a constant speed under the pen, the result is a 'registration' of the pressure fluctuations with time (see fig.2.1.1).

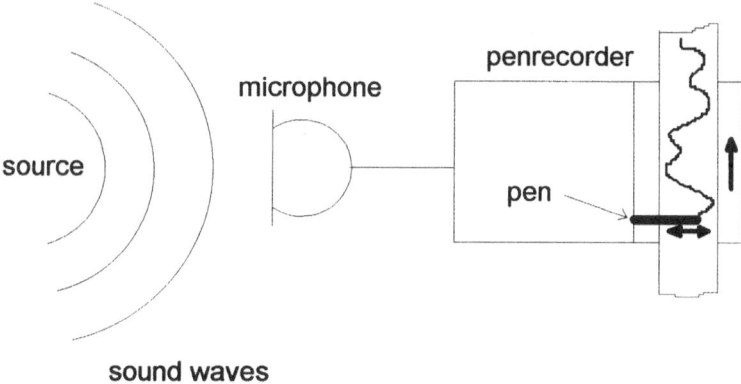

Figure 2.1.1 Registration of a signal with a penrecorder.

On the next page some registrations of this type are shown: in fig.2.1.2 the registration of a speech sound, in fig.2.1.3 that of a clarinet tone and in fig.2.1.4 that of a violin sound.

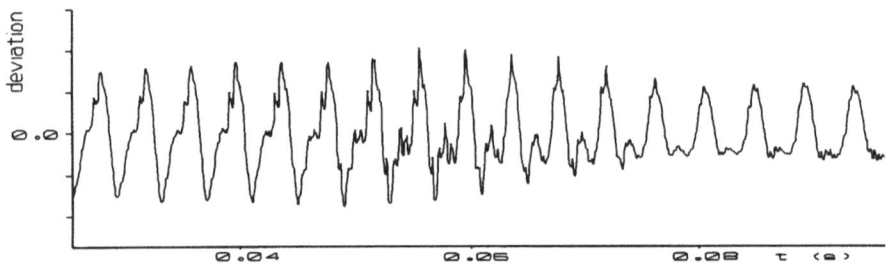

Figure 2.1.2 Registration of a speech signal.

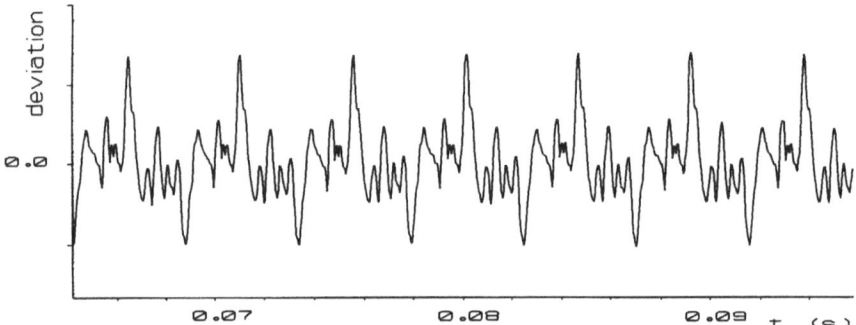

Figure 2.1.3 Registration of a clarinet tone.

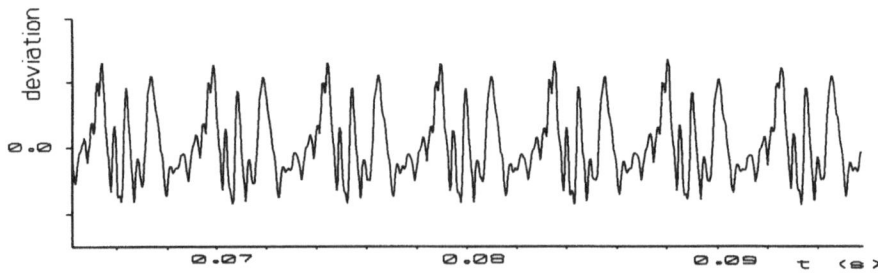

Figure 2.1.4 Registration of a violin tone.

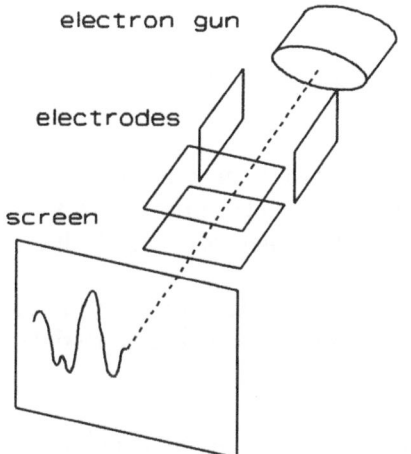

electron gun

electrodes

screen

Figure 2.1.5 Principle of an oscilloscope.

An oscilloscope (see diagram in fig.2.1.5) differs in two ways from a recorder:

1 - Instead of a pen an electron beam in a vacuum tube (as in a TV tube) is used. The screen lights up in the places where it is hit by the electrons. The amplified signal is connected to electrodes that move the beam in a vertical direction up and down.

2 - Instead of simultaneously moving the screen at a constant speed in a horizontal direction (as the paper in the pen recorder) the beam is, at a constant speed, moved from left to right by means of a second set of electrodes.

Through the combination of these two movements the registration appears on the screen as a light track. A limitation of the oscilloscope in comparison with the recorder is, of course, the length of the screen. While the paper strip can, in principle, be many metres long, the screen of the oscilloscope can only be some 20 centimetres at the most. This problem is solved by returning the beam when it reaches the end of the screen very quickly back to the beginning again. Obviously this limits its possible applications. Advantages of the oscilloscope are the high speed with which the electron beam can be moved across the screen and the fact that there is no consumption of materials such as ink and paper as in the case of the pen recorder. However, its registration is not permanent.

The registration makes clear that with every time point t a certain deviation y of the vibration corresponds. Mathematically the signal exists of a series of succeeding (y,t)-pairs. In mathematics such a relation is called a *function* and in this case a *signal function*. Because the deviation value is connected with the time point t we call t the independent variable and y the dependent variable. This hierarchy also appears in the usual mathematical notation for a function, which is not (y,t) but $y = f(t)$. It is characteristic for a function that with every value of the independent variable t there is only one value of the corresponding dependent variable y. There can, however, be more than one dependent variable. This is for example the case with a signal function that describes a moving image. Here we have to do with a plane in which the light intensity is dependent on place (two coordinates) and on time while we must specify the intensity for at least three wavelengths (colours).

The possible t- and y-values are given by respectively the start and the end time-point of the registration and by the maximal and minimal air pressure values. Within these boundaries all possible values can occur as is the case with points on a line. We say the t- and y-values are *real numbers*. Because we will be dealing with various sorts of numbers a short summary of the different number types we will encounter, is presented first.

- *natural numbers* (1, 2, 3, 4, 5,...).

These are the oldest type of numbers, necessary for counting. It is possible to add and subtract but certain subtractions are not possible, for example 3 - 5 = ?. To make this subtraction possible the concept of numbers is enlarged with

- *integer numbers* (...,-2, -1, 0, 1, 2, 3, ...).

With these it is possible to add, subtract, multiply but not always to divide, for example 3/5 = ? Such divisions of integer numbers are indeed possible after the introduction of

- *rational numbers*

These are numbers that can be written as p/q (q≠0), where p and q are integers. To this category also belong the finite decimal fractions, such as 4.57, because 4.57 = 457/100.

The awareness that there are numbers which can be approximated quite closely by rational numbers but are not themselves rational (cannot be written as the quotient of two integers) as with the number that gives the relation between the circumference and the diameter of a circle and is indicated with 'π' ('pi') led to the introduction of

- *real numbers*

which are usually introduced with the help of points on a line, the number line.

After the real numbers came the introduction of the *complex numbers*. They are very useful and very common in signal theory. Still, they will not be made use of in this book in order not to set the mathematical requirements too high.

Now we will return to the signal functions. The fact that a signal function consists of pairs of real numbers, the (y,t)-pairs, means that no matter how short the signal function is, the amount of (y,t)-pairs is infinite (just as any line contains an infinite number of points). This is no problem as long as the equipment we use does not work with numbers but with physical quantities that are proportional to the signal function (electrical voltage, mechanical deviation, strength of magnetic field etc.). The numbers which express the values of these quantities are also real and continuous numbers. We speak here of an *analog* representation, of an analog signal, and allude in this way to this parallel. The designation 'time-continuous' is used as well, which indicates only that the time values are expressed in real numbers.

When one tries to introduce such a signal function into a computer, that does work with (integer) numbers, one is confronted with the problem that the computer can only work with a finite number supply. Before discussing the means of resolving this problem we can ask ourselves what would be the reason for the application of computers. Although this book as a whole is an answer to this question we can briefly name the following three reasons:

1. Analog signal transmission is very efficient, which means that the available transmission capacity is almost completely used. This may sound positive but that is not entirely the case. For example, a transmission system's non-utilized capacity can be used to send more information than is strictly necessary and with the excess information anything lost or damaged during transportation can be reconstructed. This is analogous to the situation in which a person who wants to be as sure as possible that an important message will reach its destination, sends three identical letters instead of one. We call such a system with extra capacity *redundant* and computer systems happen to be very redundant. In information theoretical terms they are not economical but 'robust' because the degeneration of information is prevented. This aspect is of importance for the technique of sound registration and reproduction and there the change from analog to digital (= computer) techniques is in full swing.

2. A second advantage that is important for the scientific investigation of sound signals is that in a computer signals can be subjected to many refined analysis methods for which there are no analog alternatives. Several of these techniques will be discussed later in this book.

3. Conversely signals can be produced via calculations in the computer and this synthesis possibility is the third important aspect of digital sound technique that offers ever more practical applications (synthetic speech, computer music).

The solution to how to introduce an analog signal into a computer is rather like that which allows the making of films. Here a time-continuous process is simulated with a finite series of single frames taken at a short distance in time of each other.

As for sound signal functions the following procedure occurs:

a. To solve the problem of the infinite supply of number pairs a selection process is required. From the infinite series of t-values (time points) which are included in the infinite series of (y,t)-pairs a finite selection of bounded t-values is made. This selection process is called 'sampling'. The time points of this finite, countable set are indicated with t_k. The integer number k is the index, the 'number' of the time point. Usually the time points are at a fixed distance Δt of each other and then $t_k = k \cdot \Delta t$. The index t takes over the role of the independent variable.

b. The selected y-values may be numbers with very long to infinitely long series of decimals. Such numbers cannot be handled by a computer. This asks for another intervention: each y-value corresponding to a certain t_k (called a 'sample') is rounded off to a finite decimal number $\hat{y}$ with for example no more than five positions after the decimal point. The value $y = \pi$ can thus not occur anymore (it has an infinite series of decimals) but it is thus rounded off to $\hat{y} = 3.14159$. In which way this selection and rounding off can be done with a minimal loss of information will be discussed later.

The set of deviation values y has become finite and countable as well by this operation. Actually we have stepped over from real numbers to integers (for instead with $\hat{y} = 3.14159$ we can as well work with $\hat{y} = 314159$). The infinite series of (y,t)-pairs is in this way replaced by a finite and countable series of $(\hat{y},t_k)$-pairs.

The theory that has been developed parallel to that for analog signals and that normally is indicated with DSP (Digital Signal Processing, see Oppenheim et al. 1975), however, works with discrete time values and real function values, thus with (y,t_k)-pairs because this allows a simple connection with the existing theory. The rounding-off effects are considered separately. The difference between both signal functions is apparent from the terms *time-continuous* and *time-discrete* which are more accurate than the popular expressions *analog* and *digital*. The same terms are used for systems working with these signals.

The relation between these two sorts of functions will be made clearer if we find a technique to specify (signal) functions. For this the following two possibilities exist:

1. The first method is to give the total list of number pairs. In principle with time-discrete functions this is feasible, but with time-continuous functions it is impossible to specify the complete list and so we should do with a pseudo list in the form of a graphic representation of the relation between y and t. The graphical registrations shown above can be considered as such 'pseudo lists'.
2. The second method is the use of a function rule, a 'receipt', formula or prescription that states how from a given value of the independent variable (in our case 't') the corresponding y-value can be calculated.

Two examples:

$$\text{(time−continuous)} \quad y(t) = \tfrac{1}{2}t^3 + 3t^2 - 6$$

$$\text{(time−discrete)} \quad y_k = \tfrac{1}{2}k^3 + 3k^2 - 6$$

In practice it has become usual to identify the function with the function rule.

Thus we speak thus for example of "the function $y(t) = \sqrt{t^2 - 1}$ "

Knowing the function rule is very attractive because all the numerical relations are taken together in a very concise way, and the rule itself can be subjected to further mathematical operations.

For an arbitrary signal function the function rule is, of course, not known. However we shall learn a technique that will allow us to derive a function rule for a given arbitrary signal function that is specified as a list. The opposite, namely that of deriving a list or graph from the function rule, is usually very simple. In fig.2.1.6 and fig.2.1.7 two graphs are given which correspond to the above examples.

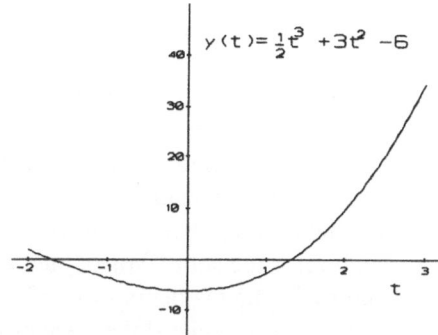

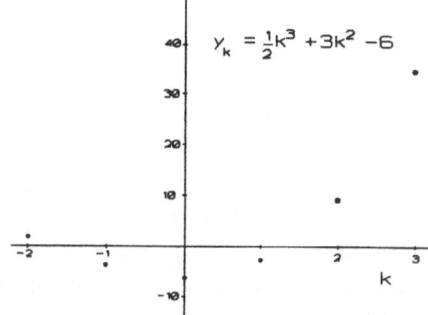

Figure 2.1.6
A time-continuous function.

Figure 2.1.7
A time-discrete function.

The behaviour of systems can also be described by means of functions. Such a function is then a mathematical model of that system. If the mathematical description of the parts of the system is known it is often possible to derive that of the whole system from this. The results calculated by means of the theoretically derived function rule are then compared with the experimental results and thus the correctness of the theoretical model can be checked.

Let us again have a look at the above registrations and see if, without further measuring apparatus, we can say something about what is perceived with these sound signals. We are then confronted with the problem discussed in Chapter 1, namely that we are dealing with a modulated signal, and can only say something about its perception if we know something about the demodulation process that takes place. We thus leave the area of signal functions and enter that of the perception. In the first place we can learn something from the registration about the global time structure (start, end, duration, segmentation) of the signal. This is

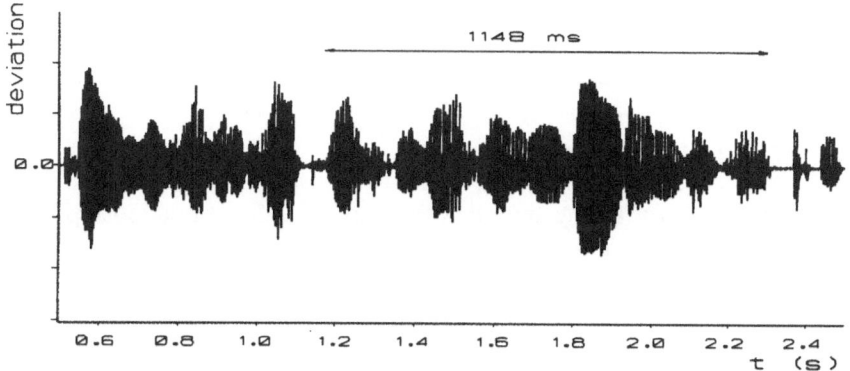

Figure 2.1.8 Slow registration of a speech signal.

most easily done with 'slow' registrations (low paper speed) as that given in fig.2.1.8. where we can see without any problem that the duration of the marked signal fragment is 1148 ms.

As we study registrations of signal functions it becomes apparent that there are regular and irregular signal functions. With the regular ones we see a particular repeating vibration pattern. This appears to happen especially with musical sounds and with vowels in speech sounds. We know the explanation for this: we are dealing with a modulated carrier wave. Helmholtz already knew this distinction. He spoke of *periodic* vibrations which form the set of 'tones' and of *non-periodic* vibrations which form the set of 'noises'. The mathematical definition of a periodic function is as follows: a function $y(t)$ is periodic if there exists a value T so that $y(t + T) = y(t)$ for all values of t. The smallest T for which this holds is the *period*. Obviously true periodic signal functions do not exist because they would last indefinitely long. Furthermore they would not be modulated and thus would contain no information. Helmholtz' subdivision should not be applied strictly; speech and musical sounds are at the most quasi-periodic and often (and even sometimes chiefly) contain non-periodic components.

Representations of such quasi-periodic signals are to be seen in the above diagrams 2.1.2, 2.1.3 and 2.1.4. In fig.2.1.9 an electronically produced vibration can be seen which very well approximates the ideal of a pure periodic vibration while fig.2.1.10 shows a non-periodic signal. With (quasi-)periodic vibrations the duration of the period can be measured. In fig.2.1.9. it is for example 5 ms.

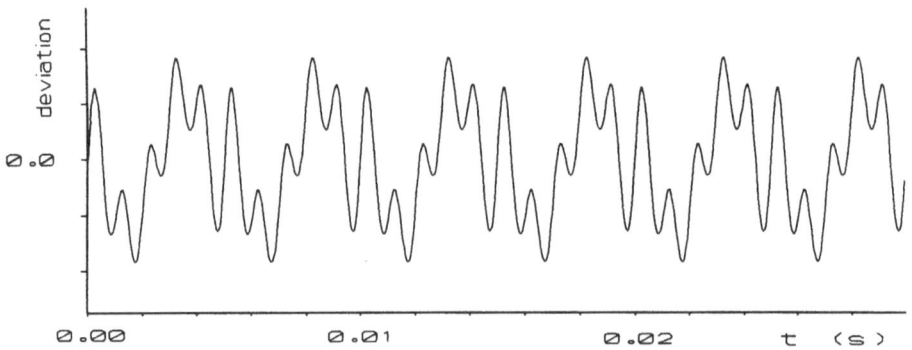

Figure 2.1.9 Registration of an electronic sound.

It has been known for a long time that there exists a direct relationship between the duration of a period and the observed pitch:

period duration (ms)	pitch
3.822	c (middle c of a piano)
3.608	c#
3.405	d

etc.

The determination of the period duration of a quasi-periodic signal can be quite problematic. I will come back to this in Chapter 7.

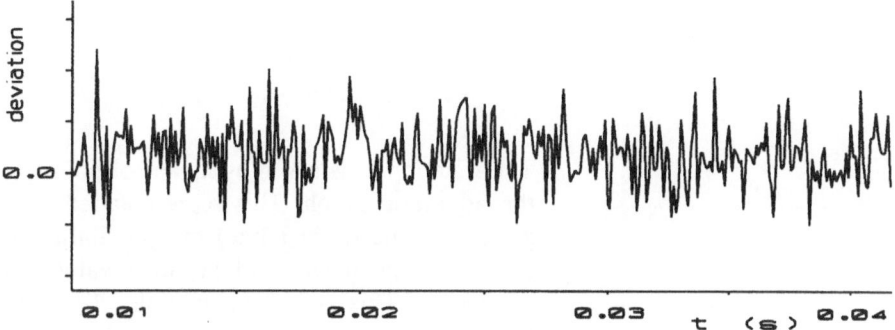

Figure 2.1.10 Registration of a non-periodic signal.

Another relation between registration and perception concerns the loudness. How loud a signal is, depends upon the extent of the vibration deviations. This relationship is actually much more complicated than that between the period and the pitch. In the first place we must know what the energy of a signal function is (this is handled in the following section), but this is only the first step. To treat this subject in detail is actually outside the scope of this book. The same holds for the relation between the shape of the vibration and the timbre. We shall see further on how it is possible to analyse the shape of the vibration, but we shall not delve into the perception of the timbre.

To discuss signal functions or systems in more detail we must become acquainted with some mathematical tools. Consequently we must first pay some attention to a few important mathematical functions. The following short course in mathematics can only be superficial and incomplete. Many textbooks are available for those who want more information; see for example Batschelet (1975), Ross, Lax et al.(1976), Smirnov (1964) and Szabo (1974).

2.2 Exponential and logarithmic functions

A. *The definition of the exponential function*

In mathematics the exponential notation is introduced as an abbreviation for the operation of multiplying a number of times a particular number by itself:

$$2^n = 2 \cdot 2 \cdot 2 \cdot 2 \cdot 2 \cdot 2 \dots\dots(n \text{ times})$$

Thus:

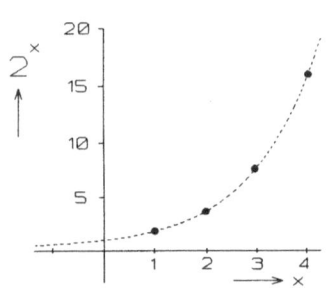

Figure 2.2.1
Exponential function.

$$2^1 = 2$$
$$2^2 = 2 \cdot 2$$
$$2^3 = 2 \cdot 2 \cdot 2$$
$$2^4 = 2 \cdot 2 \cdot 2 \cdot 2$$
$$\text{etc.}$$

The function 2^n is given by means of points in the adjoining graph. This representation suggests (via the dashed line) the possibility of allowing non-integer and negative values as exponents. This is indeed possible. To show this we proceed from the two most important calculating rules for exponents:

$$g^n \cdot g^m = g^{n+m} \quad \text{e.g. } 2^2 \cdot 2^3 = (2 \cdot 2)(2 \cdot 2 \cdot 2) = 2^5 \tag{2.1}$$
$$(g^n)^m = g^{n \cdot m} \quad \text{e.g. } (2^2)^3 = (2 \cdot 2)(2 \cdot 2)(2 \cdot 2) = 2^6 \tag{2.2}$$

The reduction of an exponent by 1 is the same as dividing once with the base number g. This leads to an interpretation for $y = g^x$. Starting with $10^3 = 1000$ we get $10^2 = 1000/10 = 100 \rightarrow 10^1 = 100/10 = 1 \rightarrow 10^0 = 10/10 = 1 \rightarrow 10^{-1} = 1/10 = 0.1$.

Thus: $g^{-n} = \dfrac{1}{g^n}$.

In the same intuitive manner we can arrive at the interpretation of $y = g^x$ for

non-integer values of x: $16^{\frac{1}{4}} = (2^4)^{\frac{1}{4}} = 2^{4 \cdot \frac{1}{4}} = 2$.

Thus $16^{\frac{1}{4}} = 2$, or in words: $16^{\frac{1}{4}}$ is the number that, when multiplied four times by itself, gives the value 16. We usually write for this: $\sqrt[4]{16}$ (the 'root to the fourth power of 16'). With this g^x is defined for all rational numbers:

$$g^{\frac{p}{q}} = \sqrt[q]{g^p}$$

and this definition is easily extended to hold for all real numbers. The calculation of such exponential expressions can be done with the help of an electronic calculator or computer.

B. *Applications of exponential functions*

1. The energy of a signal function

By definition the energy of a signal function is equal to the square of the function values. This means that the energy is always a positive quantity, the fluctuations of which follow those of the signal function. Often though the average energy over a particular time interval is more interesting. How do we calculate the average value? If the number of function values is finite (thatis if we are dealing with a time-discrete function) it can be found in the usual way: mean = sum/total number. Thus if we indicate the mean energy between time points t_1 and t_N with the letter E with a bar on top of it we have:

$$\overline{E} = \frac{\sum_{k=1}^{N} y^2(t_k)}{N} \tag{2.3}$$

The energy level of a signal is usually characterized by the *square root* of the mean energy and this quantity is named the Root Mean Square or RMS value of the signal function. Thus, y_{RMS} is defined by:

$$y_{RMS} = \sqrt{\overline{E}} = \sqrt{\overline{y^2}} \tag{2.4}$$

From the RMS-value the mean energy can easily be derived by squaring y_{RMS}. The advantage of taking the square root is that the RMS value has the same physical unit (Volt, Pascal, . . .) as the signal function itself. The three operations (root/mean/square) are easily recognizable in the definition. While for the calculation of the average energy (or RMS value) of a time-discrete function formula (2.3) can be used, for the same calculation with a time-continuous function we need an 'integral'. Therefore, this calculation must be postponed till section 2.5.A.

2. The notation of numbers
When writing very large and very small numbers it is advantageous to make use of exponential notation with, as a basis, $g = 10$. For example:

$67540000 = 6.754 \cdot 10000000 = 6.754 \cdot 10^7$
$0.000000198 = 1.98/10000000 = 1.98/10^7 = 1.98 \cdot 10^{-7}$

The advantages of this notation are:
   the numbers are shorter
   a smaller risk to make mistakes
   uniform format (advantage with computer input and output)
A disadvantage is that the visual recognition of the magnitude of the quantity disappears.

For the designation of physical quantities often only exponents of 10 are used which are themselves multiples of 3, because these multiples have been given names:

10^{12} = 'tera' 10^9 = 'giga' 10^6 = 'mega' 10^3 = 'kilo'
10^{-12} = 'pico' 10^{-9} = 'nano' 10^{-6} = 'micro' 10^{-3} = 'milli'
e.g. 600240 m $= 6.0024 \cdot 10^5$ m $= 600.24 \cdot 10^3$ m $= 600.24$ km

3. Exponential decay

In certain musical instruments and in the production of vowels as well, we observe that the excursions of the vibration after the start become gradually smaller. This is easily seen in a registration when working with a low paper speed as in fig.2.2.2. If this decay is the result of an energy loss leading to a reduction of the deviation in each successive period of the vibration with a constant factor b (with $b < 1$), it can be written with the help of an exponential function, for the series:

$$y(t=0) \ = \ A$$
$$y(t=T) \ = \ b \cdot A$$
$$y(t=2T) \ = \ b \cdot (b \cdot A) \ = \ b^2 A$$
$$\text{etc.}$$

is described by the function $y(t) = A \cdot b^{t/T}$ (t=0, T, $2T$, $3T$,..). If we admit 'all' t-values ($t \geq 0$) this function describes the smooth curve that connects the maxima of the vibration, often called the 'envelope' of the vibration.

On theoretical grounds it is advantageous with such exponential expressions to use always the same base number and to include the differences in the exponent. As a uniform basis the (real) number 'e' (= 2.7182818..) has been chosen (see section 2.3.E) and the factor $b^{t/T}$ is replaced by e^{-pt}. The required value of p can be calculated with the help of the logarithm. This subject is treated in the next section and there we shall see how this can be done. With this new factor the function rule for $E_y(t)$, the exponential envelope of signal function y is

$$E_y(t) \ = \ A \cdot e^{-pt}$$

The decay speed for this exponential function depends upon p (and thus on b and T). In fig.2.2.3 a few exponential curves with various p-values are shown.

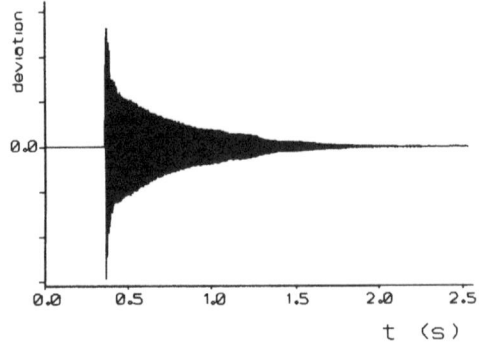

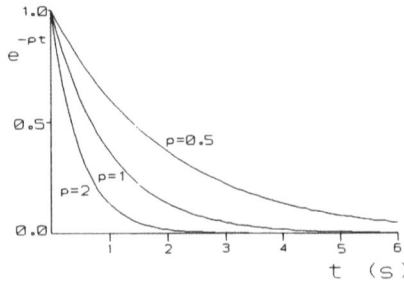

Figure 2.2.2
Registration of a harp tone.

Figure 2.2.3
Exponential functions.

4. Well-tempered tuning

Many musical instruments have fixed pitches. If one works with such instruments it is necessary to introduce a pitch 'grid'. In western music the well-tempered tuning system, in which the twelve semitone intervals in the octave have all the same frequency proportion, has been applied for a long time.

The concept 'frequency' (number of vibrations per second) and the relation it has to the duration of the period (and thus the pitch) has not yet been dealt with. Accepting for the time being that an interval is characterized by the proportion of the frequencies of the two vibrations, it follows that, if we divide the octave (proportion 2 : 1) into 12 equal intervals, the frequency factor 'r' of this interval must be such that $r^{12} = 2$. To confirm this we write a chromatic scale and indicate the frequencies of the successive tones with $f_0, f_1, ...f_{12}$. It now holds that:

$$\frac{f_1}{f_0} = r \quad \text{or} \quad f_1 = r \cdot f_0$$

$$\frac{f_2}{f_1} = r \quad \text{or} \quad f_2 = r \cdot f_1 = r \cdot (r \cdot f_0) = r^2 f_0$$

$$\frac{f_3}{f_2} = r \quad \text{or} \quad f_3 = r \cdot f_2 = r \cdot (r^2 f_0) = r^3 f_0$$

etc.

Continuing in this way, we get $f_{12} = r^{12} \cdot f_0$. But we also know that f_{12} is the octave of f_0, in other words: $f_{12} = 2 \cdot f_0$. The combination of these two results yields:

$$r^{12} = 2, \; r = 2^{\frac{1}{12}} = \sqrt[12]{2} = 1.059463..$$

A disadvantage of this tuning method is that the important intervals of the fifth and third are not pure. The fifth should have the frequency relation 3:2, but we find:

$$r^7 = (1.0594631)^7 = 1.4983071,$$

and regarding the third we find instead of 1.25 (proportion 5:4): $r^4 = 1.2599211$.

C. *The logarithm*

Above we saw what the expression g^x means. Now we ask the question whether it is possible with two numbers g and a to find an exponent x so that $g^x = a$. Such an exponent is always to be found if g and a are positive, but instead of $g^? = a$ the following notation for this unknown exponent x has become usual:

$$x = {}^g\!\log a \; (\text{'the g-}logarithm\text{ of a'})$$

This expression is thus totally equivalent to $g^x = a$

The definition of the logarithm can also be presented in the following two ways:

$$g^{\,^g\log a} = a \quad \text{and} \quad {}^g\log g^{\,b} = b \quad \text{(check this)}$$

Two important rules for dealing with logarithms are

(a) $^g\log a \cdot b = {}^g\log a + {}^g\log b$ (2.5)

Proof:

Given

$$\left.\begin{array}{l} {}^g\log a \cdot b = z \rightarrow g^z = a \cdot b \\ {}^g\log a = x \rightarrow g^x = a \\ {}^g\log b = y \rightarrow g^y = b \end{array}\right\} \quad \begin{array}{l} g^z = a \cdot b = g^x \cdot g^y = g^{x+y} \\ z = x + y \rightarrow {}^g\log a \cdot b = {}^g\log a + {}^g\log b \end{array}$$

This is thus in fact just another version of rule (2.1)

Example: $^{10}\log 100000 = 5 = {}^{10}\log 100 \cdot 10000 = {}^{10}\log 100 + {}^{10}\log 1000 = 2 + 3$

(b) $^g\log a^x = x \cdot {}^g\log a$ (2.6)

For integer values of x this rule comes down to a repeated application of the previous rule. However it holds also for non-integer values of x and is in its turn the counterpart of rule (2.2). For the proof we go on from the 'alternative definition' for the logarithm:

$$a = g^{\,^g\log a} \quad \text{thus} \quad a^x = \left(g^{\,^g\log a}\right)^x = g^{x \cdot {}^g\log a}$$

Now we take from this the g-logarithm:

$$^g\log a^x = {}^g\log g^{x \cdot {}^g\log a} = x \cdot {}^g\log a$$

While g may take any value, in practice $g = 10$ ('Brigg's logarithm') is used for calculations and $g = e$ ('natural logarithm', also written as 'ln' instead of ^{e}log) for theoretical applications. Most calculators can calculate both.

D. *An important application of the logarithm: the decibel*

Not long ago logarithms (and tools based on them such as slide rules) were very important for carrying out calculations. This is no longer the case thanks to modern calculators and computers. For theoretical derivations, however, the logarithm is still very important. For example the p-factor of an exponentially decaying envelope can now be calculated, because from $b^{1/T} = e^{-p}$ follows:

$$\frac{1}{T}\ln b = -p \ .$$

In general the logarithm is a convenient tool in calculations with proportions instead of differences as with the intensity of sensory perception. If for example a loudspeaker produces a particular signal with a power (energy per second) W_1 and a second loudspeaker does the same with twice the power, thus $W_2 = 2W_1$, then naturally the second signal sounds louder than the first. If we now were to add a third loudspeaker that emitted the same signal and we regulate the power in such a way that the difference in loudness between loudspeaker 3 and 2 is equal to that between loudspeaker 2 and 1, it turns out that the amount of power we must give to the third loudspeaker must be $W_3 = 4W_1$, in other words equal *proportions* of power are perceived as equal *differences* in loudness. So we must constantly calculate proportions. Imagine for example that $W_2/W_1 = 2.56$ and $W_3/W_2 = 4.69$ then

$$\frac{W_3}{W_1} = \frac{W_3}{W_2} \cdot \frac{W_2}{W_1} = 4.69 \cdot 2.56 = 12.0064$$

Because it is easier to work with differences than with proportions it has been decided not to work with the proportions themselves but with their (10-)logarithm. This is because in this way multiplications can be reduced to additions, and divisions to subtractions (with rules (2.5) and (2.6)):

$$\log \frac{W_2}{W_1} = 0.40844 \qquad \log \frac{W_3}{W_2} = 0.67177$$

$$\log \frac{W_3}{W_1} = \log \frac{W_3}{W_2} \cdot \frac{W_2}{W_1} = \log \frac{W_3}{W_2} + \log \frac{W_2}{W_1} = 1.07941$$

From this it is possible to calculate W_3/W_1 if necessary. Log W_2/W_1 is called the number of Bel level *difference* (!) between W_2 and W_1.

In practice, to acquire neater numbers we, multiply the numbers thus found by 10 and then speak of decibels (dB) (just as we say 15 decimeters instead of 1.5 metres):

$$10 \cdot \log \frac{W_2}{W_1} = \text{the number of } decibel \text{ level difference}$$

If $W_2 = 2W_1$ then the level difference is $\Delta L \approx 3$ dB
,, $W_2 = 10W_1$,, $\Delta L \approx 10$ dB
,, $W_2 = 100W_1$,, $\Delta L \approx 20$ dB

We have seen that the energy of a signal function is defined as the square of the deviation. If we apply this to the power (energy per second) and put this in the definition of the decibel we find for the number of decibels:

$$10\log\frac{W_2}{W_1} = 10\log\frac{y_2^{\,2}}{y_1^{\,2}} = 10\log\left(\frac{y_2}{y_1}\right)^2 = \mathbf{20\log}\frac{Y_2}{y_1} \tag{2.7}$$

Again the _mean_ energy of the signals is used in these expressions, and in the second version of (2.7) thus the RMS value.

A well-known and straightforward application of the decibel is the specification of the quality of a transmission or recording system via the _signal-to-noise ratio_ (SNR), the level difference in dB between the signal and the noise that in every system is inevitably added to the signal. The history of sound recording can be tracked by the continuous improvement of the SNR from a few dB in the Edison phonograph to more than 90 dB in the compact disc.

Apart from comparing the mean energies (or RMS values) per second of two vibrations the dB is also used as an absolute measure for comparison with a reference vibration. In acoustics the (arbitrarily chosen) reference vibration has an average pressure fluctuation p_0 that enables it to be just perceived. The value of p_0 is internationally defined as $2 \cdot 10^{-5}$ N/m² (the unit of pressure being the Newton (N) per square meter).

$$20\log\frac{p_1}{2\cdot 10^{-5}} = \text{the number of dB SPL}$$

Some SPL values (SPL = _Sound Pressure Level_):

50 dB	conversation
80 dB	train
110 dB	plane
120 dB	threshold of pain (energy 10^{12} times that at threshold!)

In electro-acoustics the reference level is often the normal modulation level of a tape recorderwhich means that we usually work with negative dB-values. In telecommunications the reference is one milliWatt and the levels are expressed in dBm, the 'm' referring to this reference.

Most dB-meters are voltmeters equipped with a pressure-sensitive microphone and an amplifier. The voltage fluctuations are measured but the scale is calibrated in dBs. An extra complication in the relationship existing between loudness and energy is the fact that the sensitivity of the ear is pitch-dependent. Therefore, dB-meters are equipped with filters to simulate this behaviour. There are three standard filter curves, designated by A, B and C (for various sound levels), but in practice we usually work with the A-curve. The measurement values are then expressed in dBA and called the _Sound Level_.

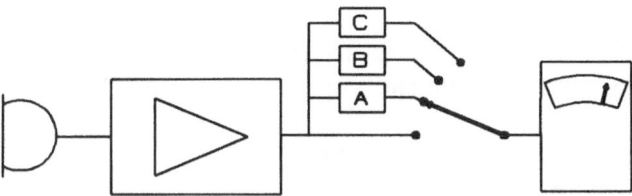

Figure 2.2.4 Principle of a sound level meter.

Working with dBs has advantage with multiplicative operations, but disadvantages with additive operations as is evident from the following example: if one has a loudspeaker with a level of 60 dB SPL and if then a second loudspeaker with the same level is switched on, the energy is doubled and the level increased by 3 dB: '60 dB + 60 dB = 63 dB'. If the two levels are different the level of the sum signal can be calculated with the following rule (2.8) which indicates the relationship between the *dB*-level c_{dB} resulting from adding signal *a* with level a_{dB} and an independent ('incoherent') signal *b* with level b_{dB} ($c_{dB} = a_{dB} + b_{dB}$):

$$c_{dB} = a_{dB} + 10 \log \left(1 + 10^{\frac{b_{dB} - a_{dB}}{10}} \right) \tag{2.8}$$

2.3 Differentiating functions

A. *The differential quotient*

The modern scientific approach incorporates the theoretic analysis of physical systems and the experimental verification of this analysis. This method was developed during the Renaissance period, simultaneously with the enormous revival of mathematics. Galileo, for example, discovered experimentally that the length of the trajectory of a falling object increases by the square of the elapsed time, but he was not yet able to prove this relation theoretically. This only became possible in an elegant manner after the discovery of *differential* and *integral* calculus by Newton (and independently by Leibniz) and the famous law of forces, also formulated by Newton:

force = mass · acceleration (in short: $f = m \cdot a$)

This law implies that a falling object, which thus is under the influence of a constant gravity force, falls with a constant acceleration and the question arises how to determine the displacement with this information. That there exists a relation between acceleration and displacement is clear and the description of this relation, which is also very important for the theoretic analysis of vibrations, requires the mathematical concept of the 'derivative' of a function. This will be our interest at this point.

The derivative of a function relates to the steepness of its graphic representation. In every point of a ('normal') function this steepness can be found
- by the eye (crudely)
- by ruler and protractor (slightly more accurate)
- if the function rule is known, by calculus (very accurate)

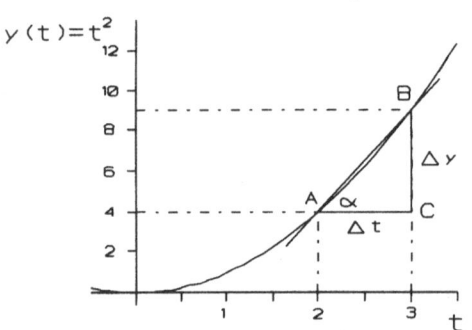

Figure 2.3.1 Steepness of a function.

An example of the third approach: the function $y(t) = t^2$ is drawn in fig.2.3.1. Imagine that we want to calculate the steepness of this curve in point A ($t=2$). To do so we first determine the steepness of the line (the so-called 'chord') that connects A to the point B ($t=3$) situated further on. The steepness of this chord can be given by the angle α, but also by the ratio of the lengths of line sections BC and AC, in other words BC/AC. In the next section we will see that there is a simple relation between α and BC/AC and how the value of α can be calculated from BC/AC (e.g. if AC/BC=1, α=45°). From now on I will refer to BC/AC as 'the' steepness. For the chord AB we find

$$\text{steepness} = \frac{BC}{AC} = \frac{9-4}{3-2} = 5$$

From this is derived $\alpha = 78.69°$. Written in somewhat more complicated terms:

$$\text{steepness} = \frac{y(3) - y(2)}{3-2} = \frac{y(2+1) - y(2)}{2+1-2}$$

By indicating the distance AC (=1) by Δt, this becomes:

$$\text{steepness} = \frac{y(2+\Delta t) - y(2)}{2+\Delta t - 2} = \frac{y(2+\Delta t) - y(2)}{\Delta t}$$

The steepness in point A itself (which is obviously smaller) is found by shifting the position of B along the curve towards A, in other terms, by reducing Δt in the t-coordinate $2 + \Delta t$. So for a series of t-values we find:

Δt	steepness	$\alpha(°)$
1.0	5.00	78.69
0.5	4.50	77.47
0.2	4.20	76.61
0.1	4.10	76.29
0.0001	4.0001	75.96

We are also able to calculate the steepness in A:

$$y(2+\Delta t) = (2+\Delta t)^2 = 4+4\Delta t+\Delta t^2$$

$$y(2) \qquad\qquad = 4$$

$$\underline{\qquad\qquad\qquad\qquad\qquad\qquad}$$

$$y(2+\Delta t) - y(2) \qquad = 4\Delta t + \Delta t^2$$

$$\text{steepness} = \frac{4\Delta t + \Delta t^2}{\Delta t} = 4 + \Delta t$$

If Δt approaches 0 the steepness approaches 4 and α approaches 75.9637..°.
We may determine the steepness in different points by repeating the above reasoning with 't' as coordinate of A. Then:

$$\text{steepness of chord} = \frac{y(t+\Delta t)-y(t)}{\Delta t} = \frac{(t+\Delta t)^2 - t^2}{\Delta t}$$

$$= \frac{t^2+2t\cdot\Delta t+\Delta t^2 - t^2}{\Delta t} = \frac{2t\cdot\Delta t+\Delta t^2}{\Delta t}$$

$$= 2t + \Delta t$$

We find the steepness at point t by substituting 0 for Δt: $2t$. Now we can calculate the steepness for every value of t, e.g.

t	$y(t)$	steepness	$\alpha(t)°$
2.0	4.0	4.0	75.96
1.0	1.0	2.0	63.43
0.5	0.25	1.0	45.00
0.02	0.0004	0.04	2.29
10.0	100.0	20.0	87.14
100.0	10000.0	200.0	89.71

So for every t there is a function value $y(t)$ as well as a value for the steepness and an angle α; stated otherwise, the value of steepness and α are functions of t as well. Symbolically the steepness function belonging to the function $y(t)$ is indicated by $y'(t)$. It is shortly called the *derivative* of y. (The angle function $\alpha(t)$ is no longer important.) Determining the derivative is called the *differentiation* of the function y. In the following formula the relation between y and y' is the

differential quotient: $y'(t) \ = \ \lim\limits_{\Delta t \to 0} \dfrac{y(t + \Delta t) - y(t)}{\Delta t}$ (2.9)

Abbreviating the difference $y(t + \Delta t) - y(t)$ to Δy (CB in fig.2.3.1) we get

$$y'(t) \ = \ \lim\limits_{\Delta t \to 0} \frac{\Delta y}{\Delta t}$$

and this is in its turn abbreviated to $y'(t) \ = \ \dfrac{dy}{dt}$

B. *Displacement, speed and acceleration*

Let us consider a practical application of this principle. Assume an object moving rectilinearly. The displacement at time t is represented by $s(t)$, displacement as a function of time. The mean velocity between two points in time is the distance covered in that time divided by the time:

$$\text{mean velocity} \ = \ \frac{\text{distance}}{\text{time}} \ = \ \frac{s(t_2) - s(t_1)}{t_2 - t_1} \ = \ \frac{s(t_1 + \Delta t) - s(t_1)}{\Delta t}$$

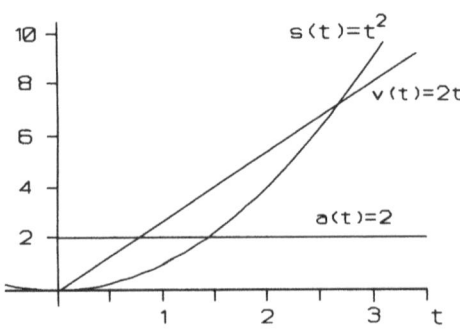

Figure 2.3.2
Displacement, speed and acceleration.

E.g. $t_1 = $ 8:00 h, $\Delta t = $ ½ h, $s(t=8:00) = 0$ (starting point), $s(t=8:30) = 40$ km. The mean velocity is $40/½ = 80$ km/h. Taking the curve in fig.2.3.2 to give a graphical representation of the relation between s and t, the steepness of the drawn chord is equal to this mean velocity.

If 'velocity' corresponds to 'steepness', we can talk about 'the' velocity for a certain value of t, $v(t)$, and this velocity function equals the steepness function of s: $v(t) = s'(t)$. The velocity is the derivative of the displacement. In the same way we may define the concept of acceleration. The mean acceleration a between t_1 and t_2 is equal to the change of velocity per time unit:

$$a \;=\; \frac{v(t_2) - v(t_1)}{t_2 - t_1} \;=\; \frac{v(t_1 + \Delta t) - v(t)}{\Delta t}$$

The acceleration at a certain moment t is equal to the steepness of the velocity function. Summarizing:

$$v(t) \;=\; s'(t) \;=\; \frac{d\,s(t)}{dt} \quad \text{and} \quad a(t) \;=\; v'(t) \;=\; \frac{d\,v(t)}{dt} \tag{2.10}$$

This is combined to $\quad a(t) \;=\; \dfrac{d\,\dfrac{ds}{dt}}{dt} \;=\; \dfrac{d^2 s(t)}{dt^2} \quad (\text{short: } \dfrac{d^2 s}{dt^2}) \tag{2.11}$

In fig.2.3.2 not only the displacement function $s(t) = t^2$ is displayed but also the velocity function $v(t) = 2t$ and the acceleration function $a(t) = 2$. Because the graph of $v(t)$ is an oblique straight line (thus having a constant steepness), the derivative is a constant. It can be proven in the same manner as in the previous example that the value of the constant is 2, but in practice it is easier to make use of calculation rules for differentiation. These rules directly give the derivative for particular functions or types of functions, so that the calculation of it via rule (2.9) can be avoided. Here follow a few of the most important rules.

C. *Rules for differentiation (1)*
-    If $y(t) = t^x$, then $y'(t) = x \cdot t^{x-1}$ (2.12)

Proof:
$$y'(t) \;=\; \lim_{\Delta t \to 0} \frac{(t + \Delta t)^x - t^x}{\Delta t} \;=$$

$$=\; \lim_{\Delta t \to 0} \frac{(t^x + x \cdot t^{x-1} \cdot \Delta t + \ldots\ldots) - t^x}{\Delta t} \;=\; \lim_{\Delta t \to 0} \frac{x \cdot t^{x-1} \cdot \Delta t + \ldots\ldots}{\Delta t}$$

The points represent products of powers of t with powers of Δt; we divide by Δt and then make Δt infinite small so that $y'(t) = x \cdot t^{x-1}$

Example: $\qquad\qquad\qquad y(t) = t^3 \;\rightarrow\; y'(t) = 3t^2$

-    If $y(t) = C \cdot g(t)$, then $y'(t) = C \cdot g'(t)$ (2.13)

Proof:
$$y'(t) \;=\; \lim_{\Delta t \to 0} \frac{C \cdot g(t + \Delta t) - C \cdot g(t)}{\Delta t} \;=\; C \cdot \lim_{\Delta t \to 0} \frac{g(t + \Delta t) - g(t)}{\Delta t} \;=\; C \cdot g'(t)$$

Example: $\qquad\qquad\qquad y(t) = 3t^2 \;\rightarrow\; y'(t) = 3 \cdot 2t = 6t$

-     If $y(t) = g(t) + h(t)$, then $y'(t) = g'(t) + h'(t)$ (2.14)

Proof:

$$y'(t) = \lim_{\Delta t \to 0} \frac{g(t + \Delta t) + h(t + \Delta t) - g(t) - h(t)}{\Delta t}$$

$$= \lim_{\Delta t \to 0} \frac{g(t + \Delta t) - g(t)}{\Delta t} + \lim_{\Delta t \to 0} \frac{h(t + \Delta t) - h(t)}{\Delta t} = g'(t) + h'(t)$$

Example: $y(t) = 3t^2 + 2t$ ➜ $y'(t) = 6t + 2$

D. *Differentiation without differential quotient*

Additional rules will be discussed later. First I would like to pay attention to the
question how a function can be differentiated without knowing its function rule,
as, for example, in the case of an arbitrary signal function. Let us first consider
the case of a time-continuous electric signal $V(t)$, a fluctuating electrical voltage.
We can make use of a differentiating network, which is a simple system consisting of a resistor and a capacitor (fig.2.3.5). To understand the working of this
network it is necessary to know the physical characteristics of these components.

1. Differentiating network.

RESISTOR.
If a conducting connection is made between two
points between which an electrical voltage V
(unit: volt) exists, then an electrical current i
(unit: ampere) will flow which is proportional to
V. The proportion constant is designated by R and
is called the *resistance* of the connection. In formula: $V = i \cdot R$, the famous law of Ohm. If with a
voltage of 1 volt a current of 1 ampere flows then
it is said that the resistance is 1Ω ('1 ohm'). It is
also possible to give conductors a particular resistance value. Then such a conductor is itself
called a 'resistor'. One speaks, for example, of a
resistor of 220 ohm.

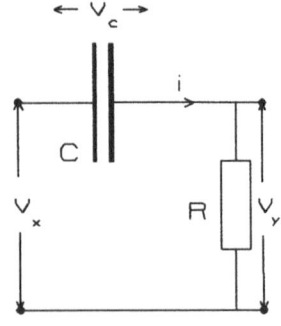

Figure 2.3.3
Differentiating network.

CAPACITOR.
This consists of two flat conductors at a very short distance from each other.
Something of this form is to be recognized in the capacitor symbol (see

fig.2.3.5). Positive and negative charges in these conductors are held together by electrostatic forces and so a capacitor behaves as a charge reservoir. It is possible to charge a capacitor by bringing the two poles in contact with a voltage source, for example a battery. The quantity of charge Q is proportional to the voltage applied. The proportion constant is given by C and is called the *capacity* of that particular capacitor. The unit is the farad (symbol: F; named after Faraday), and the capacity is 1 F if, with a voltage of 1 volt, there is a charge of 1 Coulomb. In formula: $Q = C \cdot V$. When a conductive connection is made between the poles of a capacitor, the charge flows away. The strength of this electrical current is equal to the speed of the change of the charge per unit of time:

$$i = \frac{dQ}{dt} = C \frac{dV}{dt} \tag{2.15}$$

Now for the differentiating network: if at the input a voltage $V_x(t)$ is connected, a current $i(t)$ begins to flow. For this holds:

$$i = C \frac{dV_c}{dt} = C \frac{d(V_x - V_y)}{dt}$$

Furthermore $V_y = i \cdot R$;   together: $V_y = RC \frac{dV_x}{dt} - RC \frac{dV_y}{dt}$

With the condition $RC \frac{dV_y}{dt} \ll V_y$ (RC small enough)

this becomes $V_y \approx RC \frac{dV_x}{dt}$

Thus the output voltage is the derivative of the input voltage. Fig.2.3.4 shows the effect on a few signal functions of a differentiating network. As the above conditions are not always met, deviations from the theory occur.

2. Differentiating numerically.

Time-discrete functions cannot be differentiated because Δt cannot be made infinitely small. But because Δt can become very small the terms 'differentiation' and 'derivative' are used in reference to these function with the, in fact, arbitrary value $\Delta t = 1$. This leads to the simple expression for the derivative:
$$y'(k) = y(k + 1) - y(k)$$

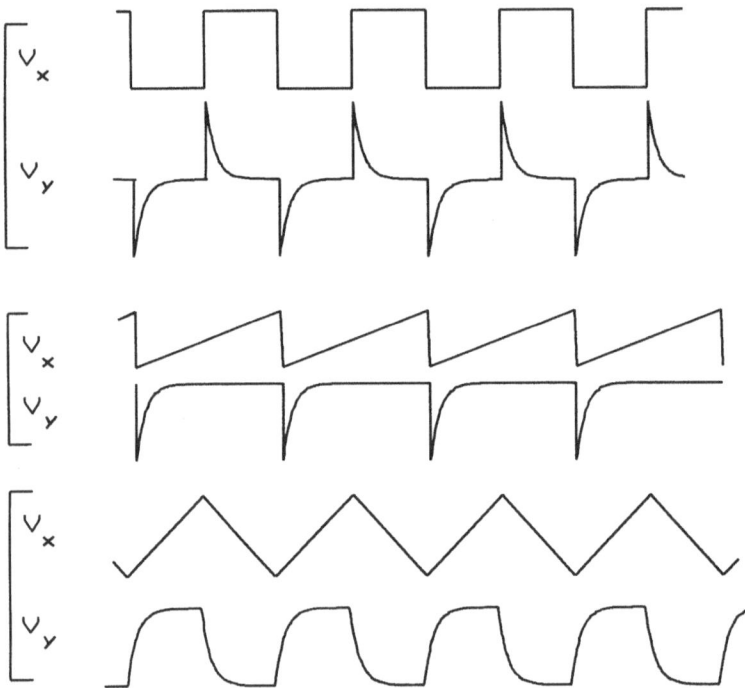

Figure 2.3.4 Differentiating signal functions.

E. *Rules for differentiation (2)*

Let us return to the calculating rules for the differentiation of time-continuous functions. We have already encountered the number e ($= 2.718...$). This number was 'discovered' by the German mathematician Euler (hence the 'e') who showed that the exponential function $y(t) = e^t$ has the characteristic property to remain unchanged by differentiation. This function is its own derivative; the steepness function is equal to the function itself as can be checked by inspecting fig.2.3.5a. This gives us a new rule:

-     If $y(t) = e^t$, then $y'(t) = e^t$ (2.16)

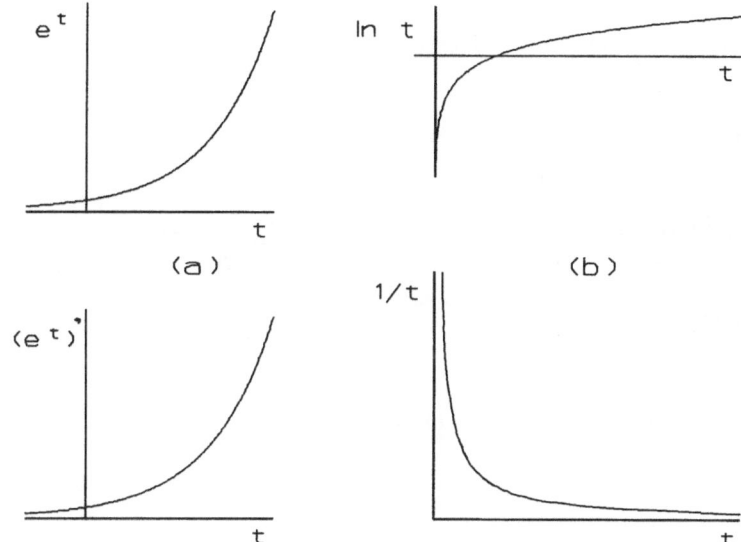

Figure 2.3.5
(a) Exponential function with derivative (b) logarithmic function with derivative.

It was furthermore shown that the function e^t can be written as an infinite series:

$$y(t) \ = \ e^{\,t} \ = \ 1 \ + \ t \ + \ \frac{t^2}{2!} \ + \ \frac{t^3}{3!} \ + \ ... \tag{2.17}$$

The symbol $n!$, pronounced 'n factorial', is an abbreviation for
$$n \cdot (n-1) \cdot (n-2) \cdot (n-3) \cdot \, ... \, . \cdot 3 \cdot 2 \cdot 1$$
With (2.17) we can check (2.16) by differentiating the series term by term:

$$y'(t) = 0 + 1 + \frac{2t}{2!} + \frac{3t^2}{3!} + \frac{4t^3}{4!} + ... = 1 + t + \frac{t^2}{2!} + \frac{t^3}{3!} + \frac{t^4}{4!} + ... = e^{\,t}$$

Now without proof the rule that says that the derivative of the function $y(t) = {}^{\mathrm{e}}\!\log t \ (= \ln t)$ is equal to $y'(t) = 1/t$. This can be checked in fig.2.3.5b via the steepness relations between the graphs.

-    If $y(t) \ = \ \ln t$, then $y'(t) \ = \ \dfrac{1}{t}$ (2.18)

A few other important calculation rules:
-    If $y(t) \ = \ x(t) \cdot z(t)$ then $y'(t) \ = \ x'(t) \cdot z(t) \ + \ x(t) \cdot z'(t)$ (2.19)

Proof:

$$y'(t) = \lim_{\Delta t \to 0} \frac{x(t+\Delta t)z(t+\Delta t) - x(t)z(t) - x(t)z(t+\Delta t) + x(t)z(t+\Delta t)}{\Delta t}$$

$$= \lim_{\Delta t \to 0} \left(\frac{[x(t+\Delta t) - x(t)]z(t+\Delta t)}{\Delta t} + \frac{[z(t+\Delta t) - z(t)]x(t)}{\Delta t} \right)$$

$$= z(t) \cdot \lim_{\Delta t \to 0} \frac{x(t+\Delta t) - x(t)}{\Delta t} + x(t) \cdot \lim_{\Delta t \to 0} \frac{z(t+\Delta t) - z(t)}{\Delta t}$$

$$= x'(t) \cdot z(t) + x(t) \cdot z'(t)$$

Example: $y(t) = 3t^2 \cdot \ln t \to y'(t) = 6t \cdot \ln t + 3t^2 \cdot \dfrac{1}{t} = 6t \cdot \ln t + 3t$

-   If $y(t) = \dfrac{1}{x(t)}$ then $y'(t) = -\dfrac{x'(t)}{x^2(t)}$ (2.20)

Proof:

$$y'(t) = \lim_{\Delta t \to 0} \frac{\dfrac{1}{x(t+\Delta t)} - \dfrac{1}{x(t)}}{\Delta t} = \lim_{\Delta t \to 0} \frac{1}{\Delta t} \cdot \frac{x(t) - x(t+\Delta t)}{x(t+\Delta t) \cdot x(t)}$$

$$= -\frac{1}{x^2(t)} \cdot \lim_{\Delta t \to 0} \frac{x(t+\Delta t) - x(t)}{\Delta t} = \frac{x'(t)}{x^2(t)}$$

Example: $y(t) = \dfrac{1}{\ln t} \to y'(t) = -\dfrac{\dfrac{1}{t}}{(\ln t)^2}$

-   If $y(t) = \dfrac{x(t)}{z(t)}$, then $y'(t) = \dfrac{x'(t)z(t) - x(t)z}{z^2(t)}$ (2.21)

The proof of this rule follows directly from rules (2.19) and (2.20).

Example: $y(t) = \dfrac{\ln t}{t} \Rightarrow y'(t) = \dfrac{\dfrac{1}{t} \cdot t - \ln t}{t^2} = \dfrac{1 - \ln t}{t^2}$

The following rule, also called the 'chain rule', concerns the differentiation of functions that have been combined as for example the function

$$y(t) = \sqrt{t^2 - 1}$$

which can be considered as a combination of the function $x(t) = \sqrt{z(t)}$ and the function $z(t) = t^2 - 1$. The rule shows:

-    If $y(t) = x(z(t))$, then $y'(t) = x'(t) \cdot z'(t)$ (2.22)

The full proof will not be given here, only an indication of how it goes:

$$y'(t) = \lim \frac{\Delta y}{\Delta t} = \lim \frac{x(z + \Delta z) - x(z)}{\Delta x} \cdot \frac{\Delta x}{\Delta t}$$

$$= \lim \frac{x(z + \Delta z) - x(z)}{\Delta z} \cdot \frac{z(t + \Delta t) - z(t)}{\Delta t} = x'(z) \cdot z'(t)$$

Example: $y(t) = \sqrt{t^2 - 1} \rightarrow y'(t) = \frac{1}{2}(t^2 - 1)^{-\frac{1}{2}} \cdot 2t = \dfrac{t}{\sqrt{t^2 - 1}}$

Finally one more rule: sometimes in calculations use is made of the following approximations:

$$y'(t) \approx \frac{y(t + \Delta t) - y(t)}{\Delta t} \quad \text{or} \quad y(t + \Delta t) \approx y(t) + y'(t) \cdot \Delta t \tag{2.23}$$

2.4 Equations

A. *Algebraic equations*

Now we may approach the problem of the motion of a falling object as follows: an object having mass m is under influence of a constant force G, the gravitation force. Using Newton's law of forces we find:

$$G = m \cdot a = m \cdot s''(t)$$

and we may wonder whether $s(t)$ can be derived from this. An expression like this, in which unknown values or functions appear, is called an *equation*. An equation implies a task: Find the unknown, find the solution. The most general form of a 'common' (or algebraic) equation in which one or more variable values occur is: *y(t) = 0*

E.g.: $3t - 5 = 0$ solution: $t = 5/3$

$3t^2 - 2t - 5 = 0$ solutions: $t = 5/3$ and $t = -1$

In the second example we use the formula for the general solution of the quadratic equation $at^2 + bt + c = 0$ which is $t_{1,2} = \dfrac{-b \pm \sqrt{b^2 - 4ac}}{2a}$

By solving the equation $y(t) = 0$ we determine the zero crossings of the function, i.e. those values of t where the graph crosses the zero axis. With a given function $y(t)$ we may also solve the equation $y'(t) = 0$, and if we do, we determine the t-values for which the steepness of the function equals zero, in other words we determine those points where the graph is horizontal: the maxima and minima.

Example: The equation $y(t) = t^3 + 2t^2 - 3t = 0$
             has three solutions: $t = -3, t = 0$ and $t = 1$.
             The equation $y'(t) = 3t^2 + 4t - 3 = 0$
             has the two solutions: $t_1 = 0.535$ and $t_2 = -1.869$.

As shown in fig.2.4.1 the function reaches a minimum and then a maximum for these two values. The values of these can be found by substituting the t-values in $y(t)$. From this we find -0.879 and 6.065. This possibility of finding the extremes of a function will be used in the following sections.

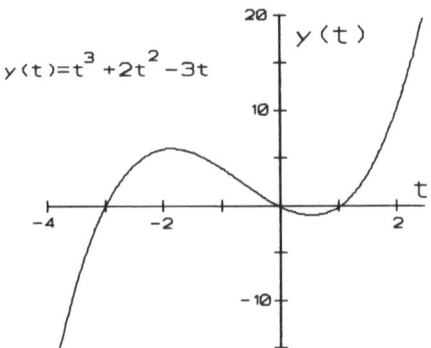

Figure 2.4.1 Maxima and minima.

B. *Difference and differential equations*

A second category of equations is formed by the differential and difference equations, where an unknown *function* is to be found. When the function is (time-)continuous we speak of a *differential equation*, with (time-)discrete functions of a *difference equation*. The example at the beginning of this section represents a differential equation:

$$G = m \cdot s''(t) \quad \text{or} \quad s''(t) = \frac{G}{m}$$

Using the calculation rules for differentiating it is easy to see that the velocity $v(t)$ follows from this: $v(t) = s'(t) = \frac{G}{m}t + C_1$

because the differentiation of this expression yields the forgoing version.
In the same way we find for $s(t)$ itself:

$$s(t) = \frac{G}{2m} t^2 + C_1 t + C_2$$

The constants C_1 and C_2 can only be determined if more information about the
system is available, e.g. that the falling motion started from a standstill ($v(0)=0$)
and that the distance moved will be measured from the start position ($s(0) = 0$),
the so-called initial conditions. The first term tells us that $v(t = 0) = C_1$ and as
$v(t=0) = 0$: $C_1 = 0$. Equally: $s(t = 0) = C_2$.

Because it is known that $s(t=0) = 0$ it follows that C_2 is also equal to 0 (note:
other initial conditions give other values for C_1 and C_2!). Thus the final result is

$$s(t) = \frac{G}{2m} t^2$$

and in this way the quadratic rule of Galileo has been derived theoretically.

The solution to a differential equation thus leads to a function. We shall use
this method later to derive signal functions from, for example, a simplified
mathematical model of a system such as a vibrating string. I shall, however,
spend no time to explain how solutions to differential equations are found;
instead I will simply give the solution as it is always possible to check it via
substitution. As has already been mentioned in the previous section, time-discrete
functions cannot be differentiated because the time-distance between function
values cannot be smaller than the sampling interval Δt. Yet there does exist a
technique that is comparable to that of differential equations. Let us proceed from
the 'normal' differential equation $y'(t) = b$.

If we write out y' this becomes:
$$\lim_{\Delta t \to 0} \frac{y(t + \Delta t) - y(t)}{\Delta t} = b$$

This equation has as solution: $\qquad y(t) = b \cdot t + C$
which is checked by differentiating this expression: $\qquad y'(t) = b$

If the function y is time-discrete, which means that we must substitute the discrete variable t_k for the continuous variable t, then Δt can no more become infinitely small. Because Δt is constant, we can, to simplify matters, assume $\Delta t = 1$.

The above equation then becomes $\qquad y(t_{k+1}) - y(t_k) = b$
or shorter $\qquad y_{k+1} - y_k = b$

Such an equation in which a relation between time-discrete time-values occurs is
called a difference equation. Here the task is also to find the solution and again

the correctness of the solution can be checked by means of substitution. Here the

solution is: $\qquad y_k = b \cdot k + C$

check:
$$y_{k+1} = b(k+1) + C = b \cdot k + b + C$$
$$y_k \qquad\qquad\qquad = b \cdot k + C$$

$$\overline{\qquad\qquad\qquad\qquad}$$

$$y_{k+1} - y_k \qquad = b$$

Here is a practical example of working with difference equations:

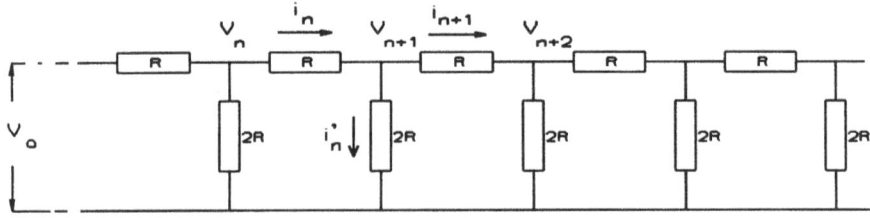

Figure 2.4.2 R-2R ladder network.

To the resistance network shown here (a so-called R-$2R$ ladder network) applied among other thing in DA-converters) a voltage of V_0 volt is connected. The question is which voltage will be found on each node. As no charge can be stored at a junction of connections, it holds that the algebraic sum of the currents is equal to 0 (first Kirchhoff law). For the given currents i_n, i_n' en i_{n+1} this means
$$i_n = i_n' + i_{n+1}$$
If we apply to all three Ohm's law then this becomes
$$\frac{V_n - V_{n+1}}{R} = \frac{V_{n+1}}{2R} + \frac{V_{n+1} - V_{n+2}}{R}$$

Multiply left and right by R:
$$V_n - V_{n+1} = \frac{1}{2}V_{n+1} + V_{n+1} - V_{n+2}$$

or
$$V_n - \frac{5}{2}V_{n+1} + V_{n+2} = 0$$

This is a difference equation, in which the discrete function V_n is unknown. I assert the solution is:

$$V_n = V_0 \cdot 2^{-n} \quad \text{(so } V_1 = \frac{1}{2}V_0, \quad V_2 = \frac{1}{4}V_0 \quad \text{etc.)}$$

To check this via substitution we need

$$V_{n+1} = V_0 \cdot 2^{-(n+1)} = V_0 \cdot 2^{-n} \cdot 2^{-1} = \frac{1}{2}V_0 \cdot 2^{-n}$$

$$V_{n+2} = V_0 \cdot 2^{-(n+2)} = V_0 \cdot 2^{-n} \cdot 2^{-2} = \frac{1}{4}V_0 \cdot 2^{-n}$$

Substitution: $\qquad V_0 \cdot 2^{-n} - \frac{5}{2} \cdot \frac{1}{2}V_0 \cdot 2^{-n} + \frac{1}{4}V_0 \cdot 2^{-n} = 0$

The *R-2R* ladder network can thus be used to generate a series of voltages that decreases by a factor 2.

2.5 The integration of functions

A. *Integral and mean value*

The solution to a differential equation is a matter of finding a function of which the derivative is known. We call the first function the *antiderivative* of the second. Thus if *y'(t) is the derivative of y(t)* we may also say that *y(t) is the antiderivative of y'(t)*. As said before I shall not deal with the problem how to find an antiderivative, but content myself with mentioning that it is possible to set up rules that are in part a 'mirror image' of the rules for differentiation; for example:

$$\text{the antiderivative of the function } t^x \text{ is} \quad \frac{1}{x+1}t^{x+1} + C$$

The antiderivative of the functions t^3 is thus equal to $\frac{1}{4}t^4 + C$. The constant C disappears when differentiated. For a very important application of antiderivatives we go on from rule (2.23):

$$y(t + \Delta t) - y(t) \approx y'(t) \cdot \Delta t$$

Another way to write the same rule, starting from the function $y(t)$ and its antiderivative, $Y(t)$, is $\qquad Y(t + \Delta t) - Y(t) = y(t) \cdot \Delta t$

Imagine that y and Y are known (for example like the y shown in fig.2.5.1) and we apply this rule to the point $t = a$:

$$Y(a + \Delta t) - Y(a) = y(a) \cdot \Delta t = \text{the area of strip 1}$$

If we apply this rule a number of times we arrive at

$$
\begin{aligned}
Y(a + \Delta t) - Y(a) &= y(a) \cdot \Delta t &&= \textit{area } 1 \\
Y(a + 2\Delta t) - Y(a + \Delta t) &= y(a + \Delta t) \cdot \Delta t &&= \textit{area } 2 \\
Y(a + 3\Delta t) - Y(a + 2\Delta t) &= y(a + 2\Delta t) \cdot \Delta t &&= \textit{area } 3 \\
&\quad= \\
Y(a + n\Delta t) - Y(a + (n-1)\Delta t) &= y(a + (n-1)\Delta t) \cdot \Delta t &&= \textit{area } n
\end{aligned}
$$

+

$$
Y(b) - Y(a) = y(a) \cdot \Delta t + y(a + \Delta t) \cdot \Delta t + \ldots = \textit{sum of areas} =
$$

$$
= \sum_{p=1}^{n} y(a_p) \Delta t = \text{total area below the curve} \tag{2.24}
$$

$$
(N.B. \quad a + n\Delta t = b, \quad a_p = a + (p-1) \cdot \Delta t)
$$

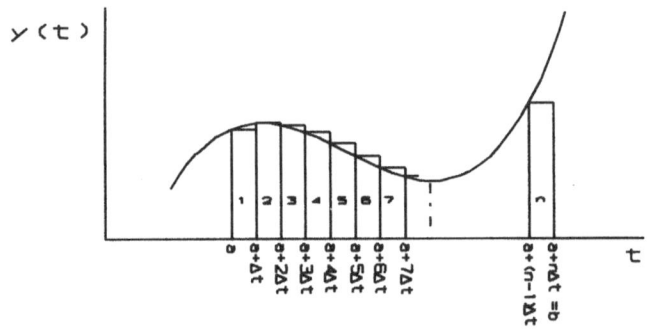

Figure 2.5.1 Area of a function.

The difference between the sum and the area under the curve will decrease if we reduce the width of the strips ($\Delta t \to 0$) and increase their number until finally we have an infinite number of infinite narrow strips. It is usual here to replace the symbol $\sum$ by the 'integral' symbol $\int$. More accurate, we make the following

substitution: $\qquad \sum_{p=1}^{n} y(a_p)\Delta t \to \int_{a}^{b} y(t)\, dt \qquad$ Thus: the area under the curve

between a and b is equal to $\qquad \int_{a}^{b} y(t)\, dt = Y(b) - Y(a)$

Instead of $Y(b)-Y(a)$ we also write: $\quad Y(t)\Big|_a^b$

Example (see fig.2.4.4): the area under the curve (the function $y = t^2$) between the boundaries 1 and 3 is noted symbolically in this way:

$$\int_1^3 t^2\, dt$$

This area is thus equal to:

$$\frac{1}{3}t^3\Big|_1^3 = \frac{1}{3}3^3 - \frac{1}{3}1^3 = 8.67$$

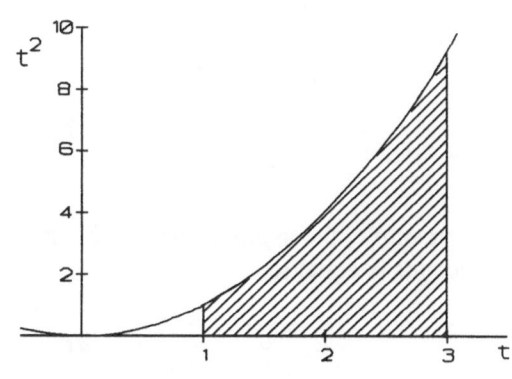

Figure 2.5.2 Area.

With this method we can calculate the area under a continuous curve, if its function rule is known. This makes it possible, among other things, to determine the *average value* of a function over a particular interval. Imagine that we should calculate the average value between a and b of the function shown in fig.2.5.3.

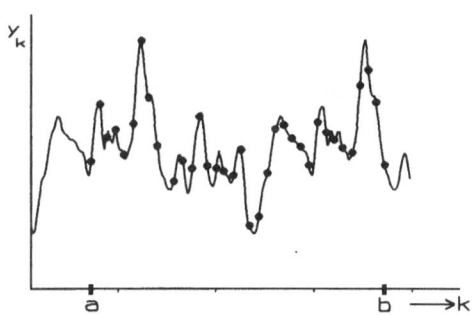

Figure 2.5.3 Mean value.

If it concerns a time-discrete function then that is not a problem, because the calculation proceeds in the familiar way: all function values are added and then divided by their number N:

$$\bar{y} = \frac{\sum\limits_{k=1}^{N} y_k}{N} \qquad (2.25)$$

To find the expression for the average value of a time-continuous function we reduce the time distance between the samples continuously, and thus increase the number of samples. In the expression for y this means that the numerator and denominator both increase. Their quotient is then undefined. Let us therefore first multiply the numerator and denominator by Δt:

$$\bar{y} = \frac{\Delta t \cdot \sum\limits_{k=1}^{N} y_k}{\Delta t \cdot N} = \frac{\sum\limits_{k=1}^{N} y_k \cdot \Delta t}{N \cdot \Delta t}$$

The denominator is equal to $N \cdot \Delta t = b - a$ and if we decrease the value of Δt and increase that of N the denominator does not change. The numerator becomes an integral and we arrive at:

$$\bar{y} = \frac{\int_a^b y(t)\,dt}{b - a} = \frac{1}{b - a}(Y(b) - Y(a)) \tag{2.26}$$

So, for example, the average energy of a signal function $y(t)$ over a time duration of T seconds is equal to:

$$\bar{E} = \overline{y^2} = \frac{1}{T}\int_0^T y^2(t)\,dt \tag{2.27}$$

According to rule (2.4) we have for the RMS value of $y(t)$:

$$y_{RMS} = \sqrt{\bar{E}} = \sqrt{\frac{1}{T}\int_0^T y^2(t)\,dt} \tag{2.28}$$

B. *Rules, integrating network, integrating numerically*

The upper limit of an integral is sometimes less than the lower limit. This does not have to present a problem:

$$\int_a^b y(t)\,dt = Y(b) - Y(a) = -[Y(a) - Y(b)] = -\int_b^a y(t)\,dt \tag{2.29}$$

If the function to be integrated is equal to the sum of two other functions, $y(t) = x(t) + z(t)$, it holds for the integral:

$$\int_a^b [x(t) + z(t)]\,dt = X(b) + Z(b) - X(a) - Z(a) =$$
$$= X(b) - X(a) + Z(b) - Z(a) = \int_a^b x(t)\,dt + \int_a^b z(t)\,dt \tag{2.30}$$

The third rule is:

$$\int_{a}^{b} y(t)\, dt + \int_{b}^{c} y(t)\, dt \ = \ Y(b) - Y(a) + Y(c) - Y(b)$$

$$= \ Y(c) - Y(a) \ = \ \int_{a}^{c} y(t)\, dt \qquad\qquad (2.31)$$

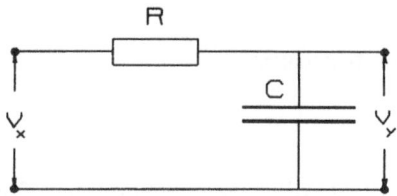

Figure 2.5.4 Integrating network.

As with differentiation it is sometimes possible to integrate a function without knowing the function rule. Time-continuous functions can be input in an integrating network, in electrical form, time-discrete functions can be integrated numerically. Fig.2.5.4 shows an integrating network, the counterpart of fig.2.3.3.

We find for V_x and V_y:

$$V_x \ = \ V_y + iR \ = \ V_y + C\frac{dV_y}{dt}R$$

If *RC* is large enough, this becomes:

$$\frac{dV_y}{dt} \ = \ \frac{1}{RC}V_x \ \text{ or } \ V_y \ = \ \frac{1}{RC}\int V_x\, dt$$

The 'numerical integration' of discrete functions is very simple, because the integration boils down to a summation ($\Delta t = 1$); the calculation is in fact the addition of function values.

C. *The RMS value of an asymmetrical signal*

In problem 2.3 the task is to calculate the RMS value of some simple signals. The solution is given in the appendix. Here I should like to discuss the problem of the RMS value of a signal with a non-zero mean value, an asymmetrical signal. We can split such a signal in two parts, a constant value y_c and a symmetrical time function $y_s(t)$: $y(t) = y_c + y_s(t)$.
Let us square $y(t)$ and then determine the mean value:

$$\overline{y(t)^2} \ = \ \overline{[y_c + y_s(t)]^2} \ = \ \overline{y_c^2} + \overline{2y_c y_s(t)} + \overline{y_s(t)^2}$$

with $\overline{2y_c y_s(t)} = 2y_c\overline{y_s(t)} = 0$ we find $\overline{y(t)^2} = y_c^2 + \overline{y_s(t)^2}$.

In some cases the constant term is not relevant. If we are, for example, interested in the loudness of a sound signal $y(t)$ the constant term represents the inaudible part of the signal that can be neglected. In that case we use only $y_s(t)^2$ and find for the RMS value:

$$y_{RMS} = \sqrt{\overline{y_s(t)^2}} = \sqrt{\overline{y(t)^2} - y_c^2} \qquad (2.32)$$

If, on the other hand, $y(t)$ is a voltage connected to an electrical heater, the constant term cannot be ignored as both the constant and the time-variant parts of $y(t)$ contribute to the production of heat.

2.6 Sinusoidal vibrations and trigonometric functions

A. *Sine function and sinusoidal movement*

We shall now, after having studied 'natural' vibrations via registrations, focus our attention on 'artificial' vibrations, that is to say a vibration that does occur in nature but is especially important theoretically. One of the few cases where this vibration can be perceived as such is in the movement of a bicycle pedal (or a crank), and therefore this somewhat exotic example will be worked out further. We observe the up and down movements of the bicycle pedal from behind (by preference pedals with reflectors that can be seen in the dark). It involves thus a one-dimensional movement like that of a pen in a pen recorder. We can indicate the displacement at a particular moment t with $y(t)$. Imagine now that we know the displacement at a certain moment ($t = 0$), and ask ourselves whether the displacement at a later time point (for example $t = 1.09$) can be derived

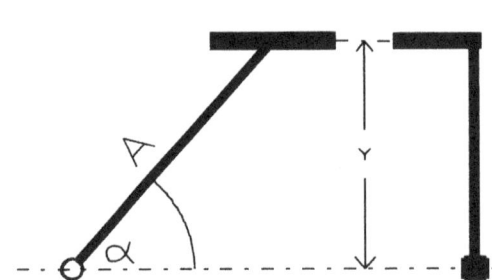

Figure 2.6.1 Bicycle step experiment.

from this. This would be quite easy when a registration of the movement were available. This is not the case but it is possible to derive this registration theoretically. For this we must of course know something about the movement of the pedal, such as:
- the pedal speed, or better: the duration of one rotation. Imagine that we perceive that for 10 rotations 6 seconds are necessary, then the duration of one rotation is $6/10 = 0.6$ sec.
- The start position of the pedal. Imagine that at the start of our experiment (the arbitrary time point $t = 0$) the angle α amounts to $80°$.
- The pedal length A; let us assume A has a value of 20 cm.

We can now calculate the value of α at the moment chosen by us ($t = 1.09$ s); because for a rotation of 360° 0.6 seconds are required, and if the start position is 80°, then at the moment $t = 1.09$ α will have the following value:

$$\alpha(t = 1.09) = 360° \frac{1.09}{0.6} + 80° = 734° = 14°$$

(We may subtract 360° or multiples of it from the total angle; this does not change the position of the pedal.) It is easy to see from this calculation that the general formula to calculate α for a moment t is the following:

$$\alpha(t) = 360° \frac{t}{0.6} + 80°$$

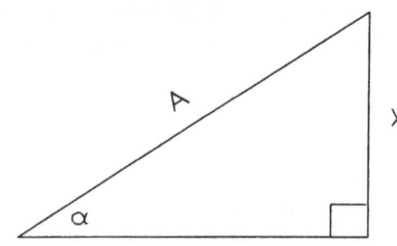

Figure 2.6.2 Definition sine.

We are, however, interested in the value of y, in other words, we should determine what the relation between α and y is. For this we gather the relevant facts in a triangle. The question how it is possible to calculate side y from one side and the angle α is one of the oldest questions in mathematics. It has been known for a long time that in the figure at the side the proportion y/A is independent of the place where the vertical side is drawn. If we shift this side to the right then both y and A increase, but the proportion y/A does not change. That proportion depends solely upon angle α; with a particular value of α a particular value of y/A is connected, and with a particular value of y/A a corresponding value of α is connected. This proportion has therefore been given a name:

y/A is called the *sine* of α (abbreviated: sin α).

We can also speak of a sine *function* because with every value of α one value of sin α corresponds. The sine value of a particular angle can nowadays be determined very simply with a calculator. Type in the angle value, press the key SIN and the calculator gives the sine value, for example

α	sine
20°	0.3420
30°	0.5000
40°	0.6428
14°	0.2419

The last number in the list puts us in the position to calculate the value of y, for with $\alpha = 14°$ corresponds $y/A = 0.2419$, and because $A = 20$ we find: $y = 4.838$.

The relation $y/20 = \sin \alpha$ and the above expression for α allow us to calculate y at any moment t, thus to find $y(t)$:

$$y(t) = 20 \sin \alpha (t) = 20 \sin \left(360° \frac{t}{0.6} + 80°\right)$$

For example:

t	$\alpha(t)$	$y(t)$
0.00	80°	19.696
1.09	14°	4.838
2.33	38°	12.313

Because the vertical pedal movement is described by a sine function the movement is called *sinusoidal*. We could now draw y as a function of t and in this way find the registration of the pedal movement, but for this the definition of the sine must still be expanded, because with the definition based upon the above triangle we will have a problem when α does not lie between $0°$ and $90°$.

Imagine, for example, that we wish to calculate $y(t)$ for $t = 1.28$ s; we then find for $\alpha(t)$: $128°$ and although a calculator gives as sine value '0.7880', the meaning of this is not clear. The expanded definition of the sine function is arrived at as follows: place the angle α in a circle (as shown in fig.2.6.3) that has a radius equal to 1. One leg of the angle coincides with the horizontal axis, the other intersects the circle in P. From P we draw a horizontal line. This intersects the vertical axis in P'. If we now calculate $\sin \alpha$ we find:

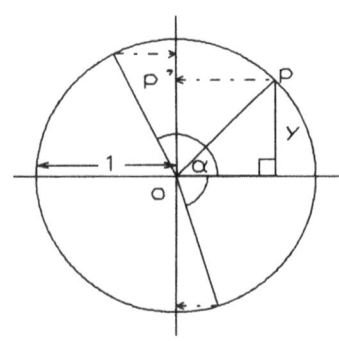

$$\sin \alpha \; = \frac{y}{1} \; = OP'$$
$$= \text{the length of the projection on the vertical axis}$$

Figure 2.6.3 Definition sine.

This definition can also be used for values of α that are larger than $90°$, and also for negative angles because they are expanded in opposing directions (as with the hands of a clock). In the figure an angle of $130°$ and one of $-75°$ are drawn. A calculator gives the following sine values for these two angle: 0.7660 and -0.9659. We can now calculate the displacement y for all t-values and bring them in a graph. The result is shown in fig.2.6.4. Please note: the sinusoidal movement or vibration is one-dimensional!

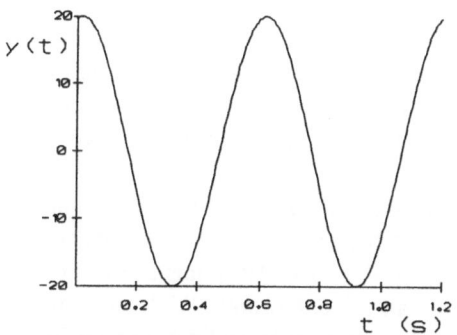

Figure 2.6.4 Bicycle step movement.

The registration which shows the relationship between *y*- and *t*-values is two-dimensional, and has the well-known sine shape here. (This remark holds, moreover, for all the vibration registrations discussed above.) We can generalize the formula for the pedal movement, so that *y* can be calculated for other pedal lengths than 20 cm, other rotation times than 0.6 s and other start angles than 80°. Clearly, we can directly substitute the other values. If we refer to the rotation time as *T*, the pedal length as *A* and the start angle as φ the general formula becomes:

$$y(t) = A \cdot \sin\left(360° \frac{t}{T} + \phi\right) \qquad (2.33)$$

Generally we make use of more neutral terms for *T*, *A* and φ. We call

T : the *period* (or period duration)
A : the *amplitude*
φ : the *initial phase angle* (the total angle α is called the *phase angle*.)

Instead of working with the period duration *T* which we determine by dividing the time in which a number of rotations take place by that number (thus *T* = time/number) one can also work with the number of rotations per time unit. This quantity, equal to number/time, and thus equal to 1/*T*, is called the *frequency f* and expressed not in 'per seconds' but in the unit called 'hertz' (abbreviated: 'Hz') which means the same thing. When we substitute in our formula 1/*T* by *f* we get:

$$y(t) = A \cdot \sin(360° ft + \phi) \qquad (2.34)$$

Until now we have expressed the size of angles in degrees. There is, however, another important angle unit, the *radian* which is also used very often. Therefore, first something about these two units.

An angle of one degree (1°) is defined as the 1/360-th part of a circle. This is an old unit of which the definition is rather arbitrary, but for measuring angles as in astronomy and navigation it is still very much in use. In this system a right angle has 90 degrees. The measuring instrument for determining the size of an angle in degrees is the protractor.

Much more recent is the radian. The definition of this unit is based upon a certain geometric characteristic that shows some similarity to that upon which the sine definition is based. In fig.2.6.5 an angle α can be seen, and an arc with the point of the angle as the centre. One can prove that the proportion of the arc length *a* to the radius *r*, thus *a/r*, does not depend on the length of the radius, but only upon the angle α (to which *a/r* is proportional).

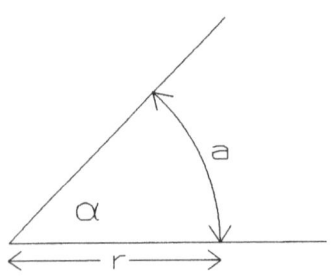

This means that the proportion *a/r* can be used as a measure for angle α. That angle for which this proportion has the value 1 is the unit angle, one radian. With an arbitrary angle the size of it in radians can be found by drawing an arc, measuring its length and the length of the radius, and dividing one by the other. It is easier to proceed from an angle that encompasses a complete circle. The arc (in this case the circumference of the circle) of such an angle is $2\pi r$ and the radius is *r*. The number of radians thus amounts to

Figure 2.6.5 Definition radian.

$$2\pi r/r = 2\pi \approx 6.2831853..$$

Because a complete circle encompasses 360° we have the conversion factor between both angle units:

1 radian = 57.29578° (= 57° 17' 44.8") and 1 degree = 0.0174533 rad

Many calculators have these conversion factors built-in which is convenient when it is necessary to convert from one unit to the other. This happens quite often because in practical measurements the degree is used, while in theoretical calculations the radian is preferred. (Calculators also offer the unit shown as GRAD. This is a variant of the degree by which a right angle has 100 units instead of 90. For our applications it is not important.) If we wish to use radians in formula (2.34) for the sinusoidal vibration we have only to replace 360° by 2π, and of course we must express the initial phase angle in radians as well:

$$y(t) = A \sin(2\pi f t + \phi) \qquad (2.35)$$

$2\pi f$ is often abbreviated to ω (the circular or angular frequency):

$$y(t) = A \sin(\omega t + \phi) \qquad (2.36)$$

B. *Time-discrete sine functions*

With time-discrete signal functions we are dealing with 'samples', signal function values that exist only at certain time points t_k. Usually these time points lie at constant intervals from each other and as they are normally determined by a pulse generator, the 'clock', one uses also the sampling frequency $f_s = 1/\Delta t$ instead of the sample time Δt. For the time points t_k it thus holds that $t_k = k \cdot \Delta t = k/f_s$.

We can create a time-discrete sine function starting from a time-continuous one by substituting the discrete variable t_k for the continuous t:

$$y(t_k) \ = \ y(k) \ = \ y_k \ = \ A \cdot \sin (360°fk\Delta t \ + \ \phi)$$

$$= \ A \cdot \sin (360°f\frac{k}{f_s} \ + \ \phi)$$

$$= \ A \cdot \sin (\frac{2\pi f}{f_s}k \ + \ \phi)$$

$$(2.37)$$

Here $2\pi f/f_s$ is often abbreviated, for example to γ: $y_k \ = \ A \sin (\gamma \, k \ + \ \phi)$.
E.g. $A = 1$, $f = 285$ Hz, $f_s = 8200$ Hz ($\gamma = 0.2184$), $\phi = 80° = 1.3963$ r. See the list below and the graph in fig.2.6.6.

k	y_k
0	0.985
1	0.999
2	0.966
3	0.887
4	0.765
5	0.608
6	0.421

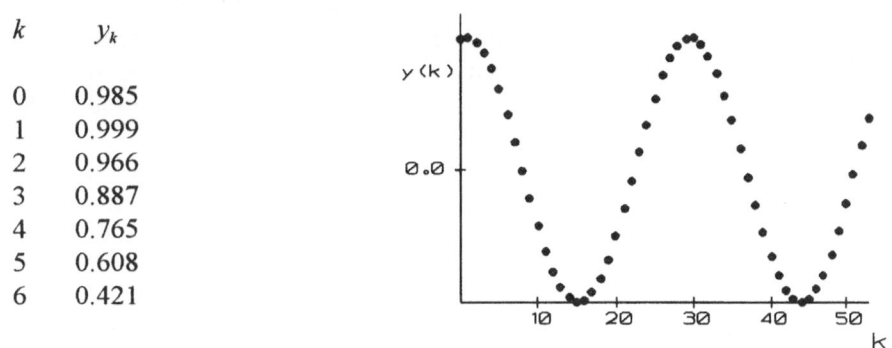

Figure 2.6.6 Time-discrete sinusoidal signal.

As we shall see with time-discrete signal functions the signal frequency should always be lower than half the value of the clock frequency. This thus means:

$$f < \tfrac{1}{2} f_s \quad \text{or} \quad \frac{2\pi f}{f_s} < \pi \quad \text{thus } \gamma < \pi$$

This is an important difference from time-continuous sine functions; for ω there is no upper limit, in contrast with γ. Another remarkable difference between a time-continuous and a time-discrete sinusoidal function with the same T-value is that the former is periodic and the latter is only then periodic when an integer number of sample intervals Δt fits into one sine period T which means that f_s is a multiple of f. Only then the list of samples of the next period is equal to that of the previous one. That is for example the case when $T = 10\Delta t$; now 10 sample periods Δt fit into T. If $t = 9.5 \ \Delta t$ then $2T = 19$. Now the sample lists are identical with a period $2T$, or the fundamental frequency $\frac{1}{2}f$. In general the following can be said about this:

Assume: $T = \dfrac{p}{q}\Delta t$ (p and q integer) , then: $q \cdot T = p \cdot \Delta t$ or $q \cdot f_s = p \cdot f$

In other words, the fundamental period is $q \cdot T$, the fundamental frequency is f/q.
The conclusion is that the fundamental frequency is equal to the GCD (Greatest Common Divisor) of f and f_s.

C. *The importance of sinusoidal vibrations*

Sinusoidal vibrations take a central place in signal theory for the following three reasons:
1. There are systems that vibrate (almost) sinusoidal. Such a system is called a *harmonic oscillator*. The following chapter is devoted to this. A tuning fork that is not struck too hard is an example of a 'natural' harmonic oscillator.
2. It is possible to lay a link between non-sinusoidal vibrations and sinusoidal vibrations. This important reduction of arbitrary vibrations to sinusoidal vibrations is called *Fourier analysis* and is dealt with in chapter 4.
3. With an important group of systems that are called *linear systems*, sinusoidal vibrations are given a sort of preference treatment in the sense that the sinusoidal shape in these systems is not affected. In chapter 5 attention is paid to this 'sine in/sine out'-principle.

D. *Trigonometric functions*

Because sinusoidal signal functions are so important in the theory of vibrations, it is necessary to be familiar with the mathematical characteristics of the sine functions. The following summary gives the most important rules for our applications (see also Szabo et al. 1974).
1. Definitions

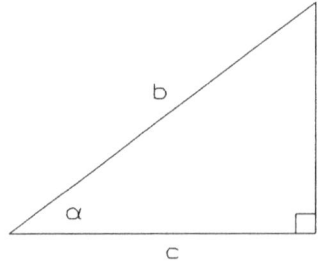

Figure 2.6.7
Rectangular triangle.

In fig.2.6.7 you see again the triangle with the original definition of the sine:

$$\sin \alpha = \frac{a}{b} \qquad (2.38)$$

If this proportion is known then all other possible proportions between the sides of the triangle are set. In principle we can limit ourselves thus to the sine. Still, it is sometimes easy to give a name to some of the other proportions as well. This has been done among others with the two proportions

$\dfrac{c}{b}$ the cosine of α: $\cos \alpha$ (2.39)

$\dfrac{a}{c}$ the tangent of α: $\tan \alpha$ (2.40)

For broader definitions of cos and tan use is made, just as for the sine, of a circle with radius 1. Have a look at the figure to the right and compare the definitions with the previous one.

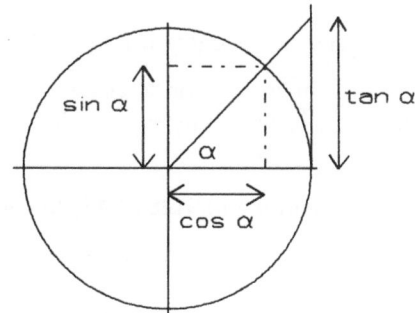

2. Special cases
For certain values of α the values for sine, cos and tan can be derived from the geometric characteristics. Check for yourself:

Figure 2.6.8
Definition sine, cosine and tangent.

$\sin 0° = 0$ $\sin 30° = 0.5$
$\tan 45° = \tan \frac{1}{4}\pi = 1$
$\cos 0° = 1$ $\cos 60° = 0.5$

3. Relations
With the help of the circle definition and by use of certain symmetries it is possible to derive conversion formulae between sin, cos and tan:
a. $\sin -\alpha = -\sin \alpha; \ \cos -\alpha = \cos \alpha$ (2.42)
b. $\sin \alpha = -\cos(\alpha+90°) = \cos(\alpha-90°)$
c. $\cos \alpha = \sin(\alpha+90°) = -\sin(\alpha-90°)$
d. $\sin \alpha = -\sin(\alpha+180°); \ \cos \alpha = -\cos(\alpha+180°)$

$$\sin^2 \alpha + \cos^2 \alpha = 1$$ (2.43)

This follows from Pythagoras's theorem (see fig.2.6.7):

$$\frac{a^2}{b^2} + \frac{c^2}{b^2} = \frac{a^2+c^2}{b^2} = \frac{b^2}{b^2} = 1$$

$$\tan \alpha = \frac{\sin \alpha}{\cos \alpha}, \quad \text{for} \quad \frac{\dfrac{a}{b}}{\dfrac{c}{b}} = \frac{a}{c}$$ (2.44)

From this: $\tan(\alpha - 90°) = \dfrac{\sin(\alpha - 90°)}{\cos(\alpha - 90°)} = -\dfrac{\cos\alpha}{\sin\alpha} = -\dfrac{1}{\tan\alpha}$

4. Summary of the calculation rules (without proof):

a. $\sin(\alpha + \beta) = \sin\alpha \cdot \cos\beta + \cos\alpha \cdot \sin\beta$
b. $\sin(\alpha - \beta) = \sin\alpha \cdot \cos\beta - \cos\alpha \cdot \sin\beta$
c. $\cos(\alpha + \beta) = \cos\alpha \cdot \cos\beta - \sin\alpha \cdot \sin\beta$ (2.45)
d. $\cos(\alpha - \beta) = \cos\alpha \cdot \cos\beta + \sin\alpha \cdot \sin\beta$

for the special case that $\beta = \alpha$ it follows:

a. $\sin 2\alpha = 2\sin\alpha \cdot \cos\alpha$
b. $\cos 2\alpha = \cos^2\alpha - \sin^2\alpha$ (2.46)

In combination with rule (2.43) this leads to:

a. $\sin^2\alpha = \dfrac{1}{2}(1 - \cos 2\alpha)$
b. $\cos^2\alpha = \dfrac{1}{2}(1 + \cos 2\alpha)$ (2.47)

By adding or subtracting various versions of (2.45) we get:
a. $\sin(\alpha + \beta) + \sin(\alpha - \beta) = 2\sin\alpha \ \cos\beta$ (2.48)
b. $\sin(\alpha + \beta) - \sin(\alpha - \beta) = 2\cos\alpha \ \sin\beta$
c. $\cos(\alpha + \beta) + \cos(\alpha - \beta) = 2\cos\alpha \ \cos\beta$
d. $\cos(\alpha + \beta) - \cos(\alpha - \beta) = -2\sin\alpha \ \sin\beta$

By replacing $\alpha + \beta$ by p, and $\alpha - \beta$ by q, or $\alpha = \frac{1}{2}(p + q)$ and $\beta = \frac{1}{2}(p - q)$, we get an alternative version of the above four rules:
a. $\sin p + \sin q = 2\sin\frac{1}{2}(p + q)\cos\frac{1}{2}(p - q)$ (2.49)
b. $\sin p - \sin q = 2\cos\frac{1}{2}(p + q)\sin\frac{1}{2}(p - q)$
c. $\cos p + \cos q = 2\cos\frac{1}{2}(p + q)\cos\frac{1}{2}(p - q)$
d. $\cos p - \cos q = -2\sin\frac{1}{2}(p + q)\sin\frac{1}{2}(p - q)$

By means of these rules the existence of sum and difference frequencies in amplitude modulation and nonlinear distortion can be explained. See for example section 5.1.C.

5. Approximation formulae
If α is very small, as in figure 2.6.9., and we write the expressions for
   - the number of radians of α : α(in rad.) $= a/r$
   - the sine of α : $\sin\alpha = y/r$
   - the tangent of α : $\tan\alpha = y/r'$
then we see that $a \approx y$, and $r \approx r'$, in other words it holds for a small α that:
           α (in rad.) $\approx \sin\alpha \approx \tan\alpha$ (2.50)

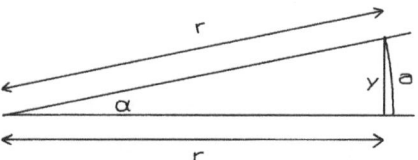

Figure 2.6.9 The sine and tangent of a small angle.

6. Numerical values.

In a calculator or computer the values for sine and cosine are of course not determined by means of the length of lines. They use the following expressions:

$$\sin x = x - \frac{x^3}{3!} + \frac{x^5}{5!} - \frac{x^7}{7!} + \frac{x^9}{9!} - \text{....} \quad (x \text{ in radians}) \qquad (2.51)$$

$$\cos x = 1 - \frac{x^2}{2!} + \frac{x^4}{4!} - \frac{x^6}{6!} + \frac{x^8}{8!} - \text{....} \quad (x \text{ in radians}) \qquad (2.52)$$

Although these series are infinitely long, the terms converge so quickly that after a few terms there is no significant contribution anymore. For example

$$\sin 34.38° = \sin 0.6 = 0.6 - \frac{0.6^3}{6} + \frac{0.6^5}{120} - \frac{0.6^7}{5040} + \text{...}$$

$$= 0.6 - 0.036 + 6.48 \cdot 10^{-4} - 5.55 \cdot 10^{-6} + \text{....} \approx 0.56464$$

7. The derivative of sin en cos

What is the derivative of sin x and cos x? This is easy to find with the help of the above series expansions as with the exponential function e^x. By differentiating the terms one by one we find:

$$(\sin x)' = 1 - \frac{3x^2}{3!} + \frac{5x^4}{5!} - \frac{7x^6}{7!} + \frac{9x^8}{9!} - \text{. . . .}$$

$$= 1 - \frac{x^2}{2!} + \frac{x^4}{4!} - \frac{x^6}{6!} + \frac{x^8}{8!} - \text{. . . . .} = \cos x \qquad (2.53)$$

The derivative of sin x is thus cos x, and in the same way one can show

$$(\cos x)' = - \sin x \qquad (2.54)$$

8. Inverse trigonometric functions

Sometimes the sine (resp. cosine or tangent) of an angle is known, but not the angle itself. An example that looks back to our bicycle-pedal problem: Imagine that instead of the start angle α we know the initial displacement $y(t = 0)$, for example $y(t = 0) = 5$ cm. By substituting $t = 0$ in the general expression we get:

$$y(t = 0) = A \sin(0 + \phi) = 5, \text{ and with } A = 20:$$

$$\sin \phi = \frac{5}{20} = 0.25$$

Again a calculator helps us here. With the so-called 'inverse sine function', usually referred to as $\sin^{-1}$ we can directly calculate that the ϕ we were looking for now has the value 14.5°. An older notation for inverse sine is 'arc sin'. With this we can write: $\phi = \sin^{-1} 0.25 = 14.5°$. In words: ϕ is the angle of which the sine equals 0.25.

Working with inverse cosine and inverse tangent proceeds in the same way. In computer languages a non-exponential style of writing is preferred and here one finds a remainder of the 'arc'-notation. The function $\tan^{-1}$ or arctan is mostly a standard function. A simple trick to introduce the value of π into a computer program is based on the fact that $\tan^{-1} 1 = 45° = \frac{1}{4}\pi$ and thus $\pi = 4 \tan^{-1} 1$.

E. *Sinusoidal signals*

Often (for example when working with isolated signals) the initial phase angle is unimportant. Then ϕ may be given the value 0° (and thus be omitted as in the following derivation of the RMS value). Then it does not matter whether one writes $\sin 2\pi f t$ or $\cos 2\pi f t$. According to rule (2.42) this is just a phase difference of 90°. In those cases where a number of simultaneous sinusoidal signals occur the initial phase may be very important as we will see. A reminder: the value of the initial phase angle can be found by determining the phase angle at $t = 0$; see rule (2.33).

1. The RMS value of a sinusoidal signal
What is the RMS value of a sinusoidal signal? According to rule (2.28):

$$y_{RMS} = \sqrt{\frac{1}{T} \int_0^T y^2(t) \, dt}$$

Because we are dealing with a periodic signal it is sufficient to determine the value of one period, which means that we take for $y(t)$ the following function:

$$y(t) = A \sin 2\pi \frac{t}{T}$$

From this:
$$y^2(t) = A^2\sin^2 2\pi\frac{t}{T} = \frac{1}{2}A^2(1 - \cos 4\pi\frac{t}{T})$$

$$\int_0^T y^2(t)dt = \frac{1}{2}A^2\left(\int_0^T dt - \int_0^T \cos 4\pi\frac{t}{T}dt\right) = \frac{1}{2}A^2 \cdot t\Big|_0^T - \frac{1}{2}A^2\frac{1}{\frac{4\pi}{T}}\sin 4\pi\frac{t}{T}\Big|_0^T$$

$$= \frac{1}{2}A^2 T - \frac{1}{2}A^2\frac{1}{\frac{4\pi}{T}}(\sin 4\pi - \sin 0) = \frac{1}{2}A^2 T$$

and so:
$$y_{RMS} = \sqrt{\frac{1}{T}\cdot\frac{1}{2}A^2 T} = \frac{A}{\sqrt{2}}\ (\approx 0.71A) \qquad (2.55)$$

2. Summation of sinusoidal vibrations

Let us consider an object (e.g. the membrane of a microphone) brought into vibration by two simultaneous sinusoidal vibrations. Let us furthermore assume that the object reacts to each of the two vibrations independent of the other one.

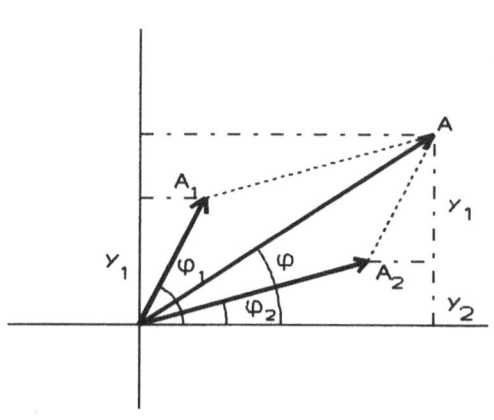

As we shall see in chapter 5 this is a characteristic property of *linear* systems. The resulting movement is the sum of the deviations caused by each of the two driving vibrations. Fig.2.6.10 shows that the resulting deviation can be found by first adding the two rotating arrows ('vectors') with the well-known parallelogram construction, and then projecting the sum vector on the vertical axis. This diagram is a useful tool to find out what happens when sinusoidal vibrations are combined.

Figure 2.6.10
Summing sinusoidal vibrations.

We shall do so, both for vibrations with different and with equal frequencies. The discussion is restricted to combinations of two vibrations:
$$A_1 \sin(2\pi f_1 t + \phi_1) + A_2 \sin(2\pi f_2 t + \phi_2)$$
The addition of more than two vibrations can be performed by first adding two vibrations, then adding the third to this sum and so on.

- equal frequencies, $f_1 = f_2$.

Here both arrows rotate with the same (constant) speed and the same holds for the sum vector. Thus the conclusion is that the sum vibration is sinusoidal as well, with the same frequency:

$$A_1 \sin(2\pi f t + \phi_1) + A_2 \sin(2\pi f t + \phi_2) = A \sin(2\pi f t + \phi) \quad (2.56)$$

We have yet to determine amplitude A and initial phase angle ϕ of the sum vibration. It has advantages (also regarding future applications) first to consider the case that the phase difference between the original vibrations is 90°, i.e. the sum of a sine and a cosine vibration:

$$A_1 \cos 2\pi f t + A_2 \sin 2\pi f t = A \sin(2\pi f t + \phi) \quad (2.57)$$

As $A_1 \cos 2\pi f t + A_2 \sin 2\pi f t = A_1 \sin(2\pi f t + \tfrac{1}{2}\pi) + A_2 \sin 2\pi f t$

the initial phase angle of the first vibration is +90° and that of the second vibrations is 0°. With the vector diagram of fig.2.6.11 we can determine A and ϕ:

$$A = \sqrt{A_1^2 + A_2^2} \quad \text{(Pythagoras' theorem)} \qquad (2.58)$$

$$\tan \phi = \frac{A_1}{A_2} \text{ or } \phi = \tan^{-1} \frac{A_1}{A_2} \qquad (2.59)$$

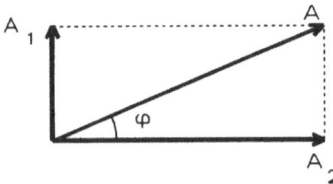

Figure 2.6.11 Addition of a sine and a cosine vibration.

Vice versa A_1 and A_2 can be derived from A and ϕ with the following two formulae, in fact the definition formulae of the sine and the cosine:

$$A_1 = A \sin \phi \qquad (2.60)$$

$$A_2 = A \cos \phi \qquad (2.61)$$

Sometimes it is preferred to write the sum vibration as a cosine function:

$$A_1 \cos 2\pi f t + A_2 \sin 2\pi f t = A \cos(2\pi f t + \Phi) \qquad (2.62)$$

This has no consequences for amplitude A (which can be found with rule (2.58)) but does have consequences for the initial phase angle. We can switch from the sine function in (2.57) to the cosine function in (2.62) in the following way, using 2.42c:

$$A \cdot \sin(2\pi ft + \phi) = A \cdot \sin(2\pi ft + \Phi + \tfrac{1}{2}\pi) \text{ with } \Phi = \phi - \tfrac{1}{2}\pi$$
$$= A \cdot \cos(2\pi ft + \Phi)$$

The value of Φ now follows from rule 2.42b:

$$\tan \Phi = \tan(\phi - 90°) = -\frac{1}{\tan \phi} = -\frac{A_2}{A_1}, \text{ thus } \Phi = \tan^{-1} -\frac{A_2}{A_1} \qquad (2.63)$$

Now we can work out the case of the arbitrary phase angles of the original vibrations:

$A_1\sin(2\pi ft + \phi_1) + A_2\sin(2\pi ft + \phi_2) =$

$= A_1(\sin 2\pi ft \cos \phi_1 + \cos 2\pi ft \sin \phi_1) + A_2(\sin 2\pi ft \cos \phi_2 + \cos 2\pi ft \sin \phi_2)$

$= (A_1\sin\phi_1 + A_2\sin \phi_2)\cos 2\pi ft + (A_1\cos \phi_1 + A_2\cos \phi_2)\sin 2\pi ft$

$= A \sin(2\pi ft + \phi)$

$$A = \sqrt{(A_1\sin \phi_1 + A_2\sin \phi_2)^2 + (A_1\cos \phi_1 + A_2\cos \phi_2)^2}$$
$$= \sqrt{A_1^2 + A_2^2 + 2A_1A_2(\sin \phi_1 \cdot \sin \phi_2 + \cos \phi_1 \cdot \cos \phi_2)} \qquad (2.64)$$
$$= \sqrt{A_1^2 + A_2^2 + 2A_1A_2 \cos(\phi_1 - \phi_2)}$$

The values of A and ϕ follow from rules (2.58) and (2.59):

and $\qquad \tan \phi = \dfrac{A_1\sin \phi_1 + A_2\sin \phi_2}{A_1\cos \phi_1 + A_2\cos \phi_2}$

Amplitude and initial phase angle thus depend upon the amplitudes and initial phase angles of the two original vibrations. Both the vector diagram and the formulae allow the study of special cases like

1. $A_1 = A_2$, $\phi_2 - \phi_1 = 180° \rightarrow A = 0$
   The vibrations cancel each other (destructive interference).
2. $A_1 = A_2$, $\phi_2 - \phi_1 = 0° \rightarrow A = 2A_1$ and $\phi = \phi_1 = \phi_2$.
   The two vibrations amplify each other (constructive interference).

- different frequencies, $f_1 \neq f_2$

In general very little can be said about the sum vibration in these circumstances, but there are some interesting special cases.

A first special situation occurs when the frequencies are multiples of a certain common fundamental frequency f_0. The sum vibration, although not sinusoidal, is then periodic with period $T = 1/f_0$. I will not discuss this subject now as it is dealt with in detail in chapter 4 (the Fourier transform).

Another special situation is the case that the frequencies are very close to each other. Let us proceed from the following addition:

$$\sin 2\pi f_1 t + \sin 2\pi f_2 t = 2 \sin 2\pi \frac{f_1 + f_2}{2} t \cdot \cos 2\pi \frac{f_1 - f_2}{2} t \qquad (2.66)$$

Of course this equivalence holds for all values of f_1 and f_2, but when listening to a pair of sinusoidal vibrations this addition takes place only if the difference between the two frequencies is small, because otherwise the vibrations are separated by the filter action of the inner ear. If the frequencies do indeed lie close to each other we see that the above expression can be interpreted as the product of a 'fast' sine vibration (with a frequency that is the average of f_1 and f_2) and a 'slow' cosine vibration, which acts as a time-dependent amplitude factor, thus as an envelope that has $(f_1 - f_2)$ times per second a maximum. (see fig.2.6.12). The corresponding loudness variation is called *beating*; the beat frequency is $|f_1 - f_2|$.

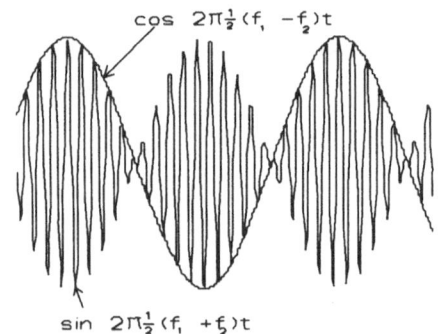

Figure 2.6.12 Beats.

The ear is especially sensitive to such a signal because it belongs in the category of modulated signals that (can) contain low-frequency information. Of the two interpretations (the sum of stationary sine vibrations or an amplitude-modulated sine vibration) the ear 'chooses' the second without hesitation. If the frequency difference increases, the beat frequency increases as well, until the ear is no longer able to follow the ever faster amplitude variation. Then the sound gets a rough or dissonant quality. If the difference in frequency gets larger, all interaction between the two vibrations disappears because they are separated completely by the filter action of the inner ear. The difference tone that is occasionally perceptible is not a continuation of the beating phenomenon, but must be explained by a totally different phenomenon, the non-linear distortion that occurs in the ear.

2.7 Problems

2.1   Determine which vibrations shown in the registrations in §2.1 are periodic (or quasi-periodic) and find the period duration, frequency and corresponding pitch of these vibrations.

2.2   Find the function rule for each of the periodic signal functions shown below.

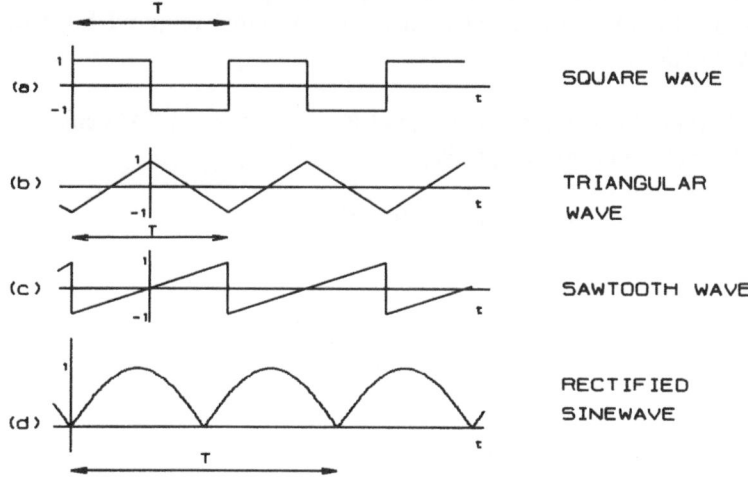

2.3   Determine the R.M.S. values of the signal functions of problem 2.2.

2.4   Determine with the help of a calculator the value of the following exponential and logarithmic expressions:
(a) $7.91^{0.141}$ (b) $145.23^{-3.69}$ (c) $-3^{0.5}$ (d) $^{10}\log 24.966$ (e) $^e\log 24.966$

2.5   Calculate the proportion of the frequencies for the following well-tempered intervals (in brackets the proportion according to natural tuning).
(a) fourth (4/3) (b) sixth (5/3) (c) seventh (15/8)

2.6   The interval width is also measured in cents, a well-tempered semitone interval corresponding to 100 cents. How many cents are contained in one octave and which frequency factor corresponds to one cent?

2.7   What is the level change in dB when two identical vibrations without phase shift are added?

2.8   Calculate the steepness in dB/sec of the exponential envelope $A \cdot e^{-t}$.

2.9   What is the signal level resulting from the addition of two non-coherent signals with levels of 70 and 80 dB SPL respectively?

2.10   As a rule of thumb it can be said that the smallest level change that can be detected under normal circumstances is 1 dB. Assume a noise signal with a level of 60 dB to which a second independent noise signal is added. What should be the level of the second signal required for a detectable level change?

2.11   Calculate the value of the product R · C with $R = 4.7\ k\Omega$ and $C = 0.1\ \mu F$. What is the unit of this product?

2.12   Calculate the extreme values of the function

$$f(x) = \frac{1}{x\sqrt{a^2 + \left(\dfrac{b}{x} - cx\right)^2}} \qquad (x > 0)$$

2.13   Sketch the graph of tan α.

2.14   Sketch the graph of the function $f(x) = (\sin x)/x$. The function value for $x = 0$ can be determined with the series of sin x or with rule (2.50). Where are the zero-crossings of the function? And where are the maxima and minima?

2.15   Calculate the average value over period T of the following products:
a) sin $2\pi nft$ · sin $2\pi mft$ ($f = 1/T$, $n \neq m$)
b) cos $2\pi nft$ · cos $2\pi mft$ (,,   )
c) sin $2\pi nft$ · cos $2\pi mft$ (,,   )

2.16   Determine the amplitude C and the initial phase angle ϕ of the following sum of sinusoidal vibrations:
          1.2cos $2\pi440t$ + 0.7sin $2\pi440t$

2.17   Give the function rule of a time-discrete sinusoidal signal with a frequency of 185 Hz, an amplitude of 2.4 and an initial phase angle of 32° (sample frequency 12 kHz). Is this signal periodic? If so, what is the period?

2.18 Convert the following angle values to radians:
a) $41°$ b) $75°12'$ c) $194°$ d) $-212°$

2.19 Convert the following angle values to degrees:
a) 1.7 r b) -4.2 r c) 5π r

2.20 Calculate the frequency of a sinusoidal signal with a period of
a) 14 ms b) 86 μs c) 1.12 s d) 2 days

2.21 What is the frequency of the signal-function $\sin 1020t$?

2.22 Calculate the following two integrals

$$\int_a^b \cos x \, dx \quad \text{and} \quad \int_a^b \cos^2 x \, dx$$

2.23 Which frequency components occur in the signal
$y(t) = (1+\cos 100t) \cdot \sin 800t$?

2.24 Integration can be shown to be the opposite of differentiation by showing that

$$\frac{d \int_a^x y(t)\, dt}{dx} = y(x)$$

Prove this.

CHAPTER 3

The Harmonic Oscillator

3.1 Undamped vibrations - the time-continuous case

A. *The vibrating string; equation and solution*

We shall now study a few versions of the simple system called 'the harmonic oscillator'. First we will analyse the behaviour of a vibrating string via a simplified model, in which we place a metal ball in the middle of the string. See fig.3.1.1. Thanks to this ball we can ignore the mass of the string itself, which simplifies the calculation considerably. On the ball in both directions a force S is exercised by means of the tension of the string. We now assume now that during the vibration the displacement and therefore the changes in the length of the string are so small that we may consider S as a constant. We indicate the displacement of the ball with $y(t)$. If the string at the place of the ball is pulled to the side the two forces S no longer lie in each others direction and a restoring force F occurs, which is shown in the diagram via the parallelogram construction and can be calculated as follows: $F = 2S' = 2S \sin \alpha \approx 2S\alpha$

If we combine this with $y/\frac{1}{2}L = \tan \alpha \approx \alpha$

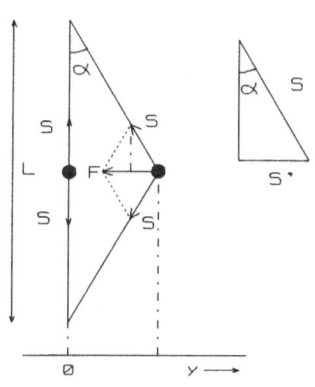

Figure 3.1.1 Vibrating string.

we get: $F = -\dfrac{4yS}{L} = -\dfrac{4S}{L}y = -b \cdot y(t)$

The factor $4S/L$ is abbreviated to b. In this expression both the magnitude and the direction of force F are represented, the latter with the minus sign that shows that the force and the displacement are in opposite directions. We combine this result with Newton's law:

$$\left. \begin{array}{l} F = -b \cdot y \\ F = m \cdot a = m\dfrac{d^2y}{dt^2} \end{array} \right\} \quad \dfrac{d^2y}{dt^2} + \dfrac{b}{m}y(t) = 0 \qquad (3.1)$$

This is a differential equation of the second order, which can be considered as the mathematical model of the system (see also Morse 1948). For the determination of the solution we start with a 'trial' function. This is a function that still contains several unknown coefficients, which we attempt to adjust in such a way that the function provides us with a solution for the equation.

We use the following trial function: $y(t) = A_1 \cos \omega_0 t$

and calculate the first and second derivatives of $y(t)$:

$$\frac{dy}{dt} = -\omega_0 A_1 \sin \omega_0 t, \quad \frac{d^2y}{dt^2} = -\omega_0^2 A_1 \cos \omega_0 t$$

After substitution: $-\omega_0^2 A_1 \cos \omega_0 t + \dfrac{b}{m} A_1 \cos \omega_0 t = 0$

This is true if $\omega_0^2 = b/m$, thus if $\omega_0 = \sqrt{b/m}$.

Thus ω_0 is determined; A_1 is a constant, the value of which is not yet known. In the same way we can show that the function $y(t) = A_2 \sin \omega_0 t$ is also a solution, with the same value of ω_0, thus $\omega_0 = \sqrt{b/m}$.

It is known from the theory of differential equations that the general solution is the sum of the two solutions that have been found:

$$y(t) = A_1 \cos \omega_0 t + A_2 \sin \omega_0 t = A \cos(\omega_0 t + \phi)$$

with

$$\omega_0 = \sqrt{\frac{b}{m}} \text{ or } f_0 = \frac{1}{2\pi}\sqrt{\frac{b}{m}} = \frac{1}{2\pi}\sqrt{\frac{4S}{Lm}}$$

$$A = \sqrt{A_1^2 + A_2^2}, \quad \phi = -\tan^{-1}\frac{A_2}{A_1}$$

(3.2)

Here the sum of the cosine and the sine vibration has been replaced by a single cosine vibration according to rule (2.62). The conclusion is that the string vibrates sinusoidally with amplitude A and initial phase angle ϕ that both depend upon the two original constants (amplitudes) A_1 and A_2. The name of this system, *harmonic oscillator*, is derived from this sinusoidal movement. The frequency depends upon b (the 'stiffness constant') and m, the mass of the ball. When for example $m = 0.001$ kg, $L=1$ m, $S=40$ N (so $b=4S/L=160$ N/m) then (3.2) gives us: $f_0 = 63.7$ Hz.

The two constants A_1 and A_2 can in principle have any value. Only when more information about the system is available, the values of these constants can be determined. Imagine for example that we know that at the start of the experiment ($t=0$) the initial displacement is y_0 and the initial speed is v_0, then it follows that:

$$y(t = 0) = y_0 \quad \text{and} \quad y(t = 0) = A_1, \text{ so } A_1 = Y_0$$

$$v(t) = \frac{dy}{dt} = -A_1\omega_0 \sin \omega_0 t + A_2\omega_0 \cos \omega_0 t \quad \text{and so}$$

$$v(t = 0) = A_2\omega_0 \quad \text{and} \quad v(t = 0) = v_0 \quad \text{thus } A_2 = \frac{v_0}{\omega_0}$$

together: $$y(t) = y_0 \cos \omega_0 t + \frac{v_0}{\omega_0} \sin \omega_0 t$$

We can now substitute the initial conditions y_0 and v_0 in the solution:

$$A = \sqrt{y_0^2 + \frac{v_0^2}{\omega_0^2}}, \quad \phi = -\tan^{-1} \frac{v_0}{\omega_0 y_0}$$

A few special cases:
If the initial speed v_0 is equal to 0 the movement is described by

$$\left.\begin{array}{l} A = y_0 \\ \phi = 0 \end{array}\right\} \quad y(t) = y_0 \cos \omega_0 t \quad \text{(fig.3.1.2)}$$

and if the initial displacement y_0 is equal to 0, by

$$\left.\begin{array}{l} A = \dfrac{v_0}{\omega_0} \\ \phi = -90° = -\dfrac{1}{2}\pi \end{array}\right\} \quad y(t) = \frac{v_0}{\omega_0} \cos (\omega_0 t - \tfrac{1}{2}\pi) = \frac{v_0}{\omega_0} \sin \omega_0 t \quad \text{(fig.3.1.3)}$$

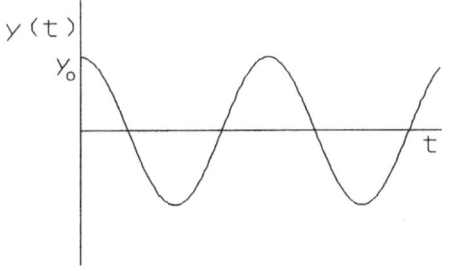

Figure 3.1.2 Vibration 1.

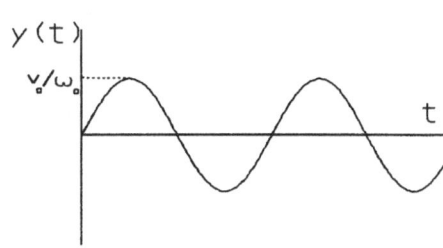

Figure 3.1.3 Vibration 2.

Let us compare the displacement, speed and acceleration:

$$y(t) = A \cos (\omega_0 t + \phi)$$

$$v(t) = - \omega_0 A \sin (\omega_0 t + \phi)$$

$$a(t) = - \omega_0^2 A \cos (\omega_0 t + \phi)$$

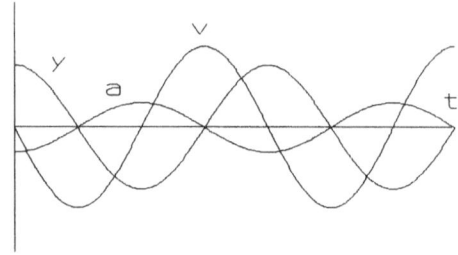

Figure 3.1.4 Displacement, speed and acceleration.

The speed has a phase difference of 90° with the displacement, and is thus maximal when the displacement is 0, that is when the string passes through the rest position. The acceleration (and thus the restoring force) is maximal if the displacement is maximal.

B. *Other harmonic oscillators*

1. Systems with a constant vibration period.

There are more systems in which the 'force proportional to displacement' relation and mass or inertial forces occur so that the outcome is a sinusoidal vibration ('mass-spring systems').

a. The balance in a mechanical watch: the wheel that sways to and fro, and that is pushed back to its rest position by a spring that causes a restoring force, proportional to the rotation angle (instead of the displacement). Here we must use the moment of inertia M instead of mass m. We indicate the amount of rotation with angle Ω and find:

$$\left. \begin{array}{l} F = -b \cdot \Omega \\ F = M \dfrac{d^2 \Omega}{dt^2} \end{array} \right\} \quad \dfrac{d^2 \Omega}{dt^2} + \dfrac{b}{M} \Omega = 0$$

b. The pendulum of a clock. A simple calculation leads here to a differential equation of the known type:

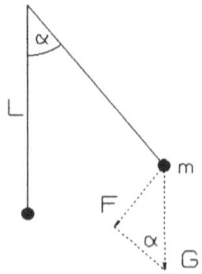

Figure 3.1.5 Pendulum.

$$\left.\begin{array}{l} \sin\alpha = \dfrac{F}{G} \approx \alpha \\[4pt] G = m\cdot g \end{array}\right\} \begin{array}{l} F = m\cdot g\cdot\alpha \\[12pt] \dfrac{y}{l} = \sin\alpha \approx \alpha \end{array} \left.\begin{array}{l} \\[8pt] \end{array}\right\} F = -\dfrac{mg}{l}y = m\dfrac{d^2y}{dt^2}$$

with g = gravity acceleration;   equation: $\dfrac{d^2y}{dt^2} + \dfrac{g}{l}y = 0$

the pendulum moves sinusoidally with frequency $f_0 = \dfrac{1}{2\pi}\sqrt{\dfrac{g}{l}}$

c. The lattice of molecules in a quartz crystal.

In all these three cases purposeful use is made of the fact that the frequency with which a harmonic oscillator vibrates depends upon the system characteristics such as mass, elasticity, etc. and not upon for example the initial conditions. Therefore the vibration period is constant and can be used as a time basis, as a unit for the measurement of time. It is of course possible that the system characteristics depend upon external factors such as the temperature. This holds for example for the length of a pendulum.

2. The Helmholtz resonator.

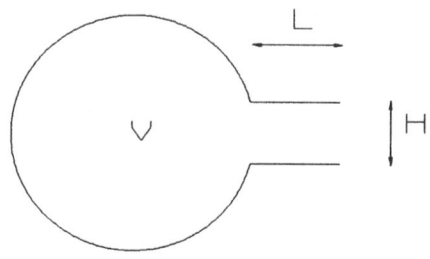

Figure 3.1.6 Helmholtz resonator.

This system, used by Helmholtz as an acoustic filter, played an important role in his investigations. It consists of a glass sphere with one or two tube-like openings. The original drawing is shown in chapter 4, fig.4.3.32. Let us study the system as depicted in fig.3.1.6. The volume of the sphere is V, the length of the tube is L and its cross-section H. We can imagine that we can move the air in the tube to the left, which leads to an increase of the pressure in the sphere by which a force arises which pushes back the air in the tube. Vice versa, moving the air in the tube outwards leads to a lower pressure in the sphere by which the air is sucked backwards. Secondary effects like changes of the air pressure in the tube are neglected.

We have both necessary ingredients for a harmonic oscillator: mass m (of the air in the tube) and elastic force caused by the air cushion in the sphere. If we should know the stiffness constant 'b' (equal to the elastic force if the displacement is one unit of length) we could substitute b and m in our formula. Unfortunately the calculation of b is not easy. Moreover we still have to show that the resulting force is proportional to the displacement. The complications are a result of the fact that with volume changes of gases not only the pressure changes, but also temperature effects occur. From the theory of thermodynamics we know that with rapid changes dP and dV of pressure P and volume V respectively, no heat exchange occurs with the environment. Such a process is called *adiabatic* and the following relationship between the relevant parameters can be proven to exist:

$$\frac{dP}{P} = -\gamma \frac{dV}{V} \quad (\gamma = \frac{c_p}{c_v})$$

(c_p = specific heat with constant pressure, c_v = specific heat with constant volume, the value of γ is 1.40 in air).

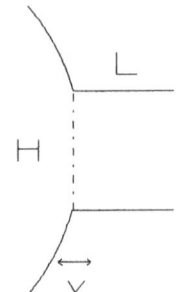

During the vibration the layer that is indicated with a dashed line moves back and forth over a very small distance. This causes a change of volume:

$$dV = y \cdot H \quad \text{thus} \quad dP = -\gamma \frac{P}{V} H \cdot y$$

The force that then occurs follows from:

$$\text{force} = \text{pressure} \cdot \text{area} = dP \cdot H$$

Figure 3.1.7
Helmholtz resonator.

$$= -\frac{\gamma P H^2}{V} y = -b \cdot y \ (b = \frac{\gamma P H^2}{V})$$

The force is thus indeed proportional to the displacement. If we combine this relation with Newton's force law, we get

$$\text{force} = \text{mass} \cdot \text{acceleration} = m \frac{d^2 y}{d t^2} = -b \cdot y$$

Here m is the mass of the air in the neck of the bottle for which we may write

$$\text{mass} = \text{volume} \cdot \text{specific mass} = L \cdot H \cdot \rho_0$$

We arrive again at the well-known differential equation, which of course has the same solution. This means that the air in the neck of the bottle vibrates sinusoidally with frequency f_0: $f_0 = \dfrac{1}{2\pi}\sqrt{\dfrac{b}{m}} = \dfrac{1}{2\pi}\sqrt{\dfrac{\gamma P H}{\rho_0 V L}}$

From acoustics it is known that for the speed of sound v_g the following relation

can be derived: $v_g = \sqrt{\dfrac{\gamma P}{\rho_0}}$

With this, the expression for the frequency becomes $f_0 = \dfrac{v_g}{2\pi}\sqrt{\dfrac{H}{LV}}$ (3.3)

The frequency depends upon v_g and the dimensions of the sphere.

3. The *LC*-circuit

The combination of a capacitor C and a coil L behaves like an electrical harmonic oscillator. To understand this we must know the physical behaviour of these components. We met the capacitor in section 2.3 with the important relations:

$$Q = C \cdot V \ \ (Q\text{: charge, } V\text{: voltage, } C\text{: capacity})$$

and (formula (2.15)): $i = \dfrac{dQ}{dt} = C\dfrac{dV}{dt}$

Now concerning the *coil*.

When one winds a conducting wire, a coil results. A current flowing through a wire generates a magnetic field around that wire. Winding the wire combines the magnetic fields of all windings. The first application of a coil is thus that of an electromagnet (e.g. as it is used as recording and erase head of a tape recorder). Furthermore a coil displays the important feature called 'induction': if the coil is in a changing magnetic field, an electrical voltage is generated in the coil, the induction voltage, which increases as the magnetic field changes more quickly. There are many practical applications of this phenomenon: the dynamo, current generators, the playback head of a recorder. Finally the coil also displays the phenomenon of *self induction*, which is a combination of both phenomena mentioned above. Here the induction voltage is the result of changes of the magnetic field excited by the coil itself. If one connects an electrical voltage to the coil then a current flows through the coil, creating a magnetic field of increasing strength with which an induction voltage is generated which partially cancels the external voltage. Formulated in short: by means of the phenomenon of self induction a coil 'resists' changes in the strength of the current. The induction voltage V is proportional to the speed with which the current i changes (di/dt): $V = L \dfrac{di}{dt}$. The factor of proportionality L is called the (coefficient of)

self induction. The unit is the henry. If we connect a coil with a condenser then we get a '*LC*-circuit' (fig.3.1.8):

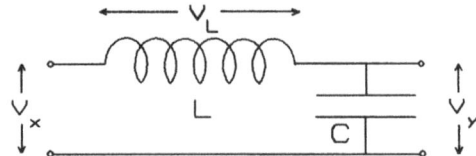

Figure 3.1.8 LC circuit.

A combination of the current/voltage relations

$$V_x = V_L + V_y = L \frac{di}{dt} + V_y \quad \text{and} \quad i = C \frac{dV_y}{dt}$$

gives: $$V_x = LC \frac{d^2 V_y}{dt^2} + V_y$$

If the input voltage V_x is equal to 0 (apart from a very short pulse that is required to cause the vibration) the result will be:

$$\frac{d^2 V_y}{dt^2} + \frac{1}{LC} V_y = 0$$

For the *LC*-circuit the differential equation of the harmonic oscillator holds again and thus the solution is again identical: a sinusoidally fluctuating voltage with a frequency of:

$$f_0 = \frac{1}{2\pi} \sqrt{\frac{1}{LC}} \tag{3.4}$$

We see here that the role of the displacement '*y*' is taken over by the voltage V_y (or the charge Q of the capacitor, because $Q = C \cdot V_y$). We will return to this analogy between mechanical and electrical quantities in section 3.3.

3.2 The undamped vibration - the time-discrete case

In nature macroscopic systems that are described by time-discrete equations and/or functions are quite scarce. This holds a fortiori for time-discrete sine functions that thus require an 'artificial' system (operation, diagram, computer program). If we generate a time-discrete sinusoidal signal (a series of sine values with time distances $1/f_s$, in which f_s is the sampling frequency or clock frequency) by means of a computer, then we have constructed a time-discrete harmonic oscillator. The function rule of this sinusoidal signal is $y(t) = \sin \theta k$ with $\theta = 2f/f_s$ (see section 2.6.B). There are various possibilities for this.

A. *A simple programmed sinewave oscillator*

Most computer languages have the function 'SIN' for calculating sine values. The following FORTRAN-like computer program produces these values based on a clock frequency of 20 kHz and a signal frequency of 440 Hz:

```
        F = 440.0
        FS = 20000.0
        THETA = 6.2831853 * F / FS
        K = 0
3       Y = SIN(THETA * K)
        CALL WAITCL
        OUTPUT Y
        K = K + 1
        GO TO 3
```

Explanation:
The instruction CALL WAITCL is a waiting instruction for the next clock pulse, so that the function values are produced synchronously. With the instruction OUTPUT one can think of just about any form of computer output, such as plotting, display on a screen, conversion in electrical voltage via DA conversion (see chapter 4) etc.

This 'real time' calculation of sine values is relatively time-consuming and could cause timing problems. There is another simple approach that prevents such problems:

B. *The 'look-up table' generator*

Here the sine values are calculated beforehand, placed in a table and then fetched one after the other from this table synchronous with the clock:

```
          DIMENSION Y(10000)
          F = 440.0
          FS = 20000.0
          THETA = 6.2831853 * F / FS
          DO 3 K = 1 , 10000
3         Y(K) = SIN(THETA * K)
          DO 5 K = 1 , 10000
          CALL WAITCL
5         OUTPUT Y(K)
```

This is a simple, fast and much used system, but it is not very flexible, in the sense that for another frequency the table must either be filled anew or the clock frequency must be adapted, or, according to some sort of scheme, samples must be passed over. This fact and the necessary memory space are two disadvantages of the system which, by the way, is not a typical sine generator because the table can be filled in any arbitrary set of values. These disadvantages do not adhere to a third system that is actually a time-discrete sine generator.

C. *A digital sinewave oscillator with feedback*

A block diagram for this generator is shown in fig.3.2.1.

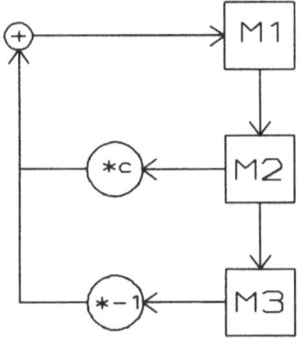

Figure 3.2.1
A digital sinewave oscillator.

M1, M2 and M3 are memory cells. With each clock pulse the contents of M2 shift to M3 and those of M1 to M2. Beforehand the contents of M2 and M3 are made equal to 0 and those of M1 to 1. After the shift, as shown in the diagram, the contents of M1 become that of M2 multiplied by some factor c plus that of M3 multiplied by -1. In M3 we thus find the 'previous' contents of M2 and in M2 those of M1. The scheme can be realized with the computer program shown on the next page, for example.

	k	x(k)	M1=y(k)	M2=y(k-1)	M3=y(k-2)
M1 = 1	0	1	1.00	0.00	0.00
M2 = 0	1	0	1.90	1.00	0.00
M3 = 0	2	0	2.61	1.90	1.00
C = 1.9	3	0	3.06	2.61	1.90
5 M3 = M2	4	0	3.20	3.06	2.61
M2 = M1	5	0	3.02	3.20	3.06
M1 = C * M2 - M3	6	0	2.55	3.02	3.20
CALL WAITCL	7	0	1.82	2.55	3.02
Y = M1	8	0	0.90	1.82	2.55
OUTPUT Y	9	0	-0.11	0.90	1.82
GO TO 5	10	0	-1.11	-0.11	0.90
	11	0	-2.00	-1.11	-0.11
	.	.	.	.	.

It is easy to calculate the subsequent values of M1, M2 and M3. The first 12 are given above. If we graph these values as is done in fig.3.2.2, it appears that we do have a time-discrete sinusoidal signal, but this must of course still be proven. This we shall do by calculating the output signal $y(k)$. Before doing this we draw the circuit of fig.3.2.1 in a slightly different way, with the intention of revealing the delay mechanism (see fig.3.2.3). In connection with later applications we moreover give the system an input to enter the number '1'. Thus we can thus also say that the input signal $x(k)$ (or x_k) looks as follows:

1, 0, 0, 0, 0, 0, 0, 0, ... (see table above)

It is easy to check that in this way the initial value of M1 is indeed '1', and that after this single impulse the input signal plays no role anymore.

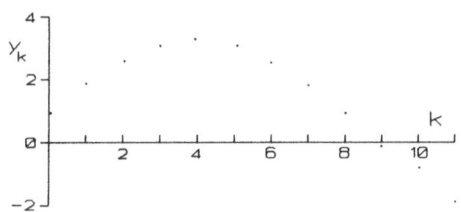

Figure 3.2.2
Output signal digital sinewave oscillator.

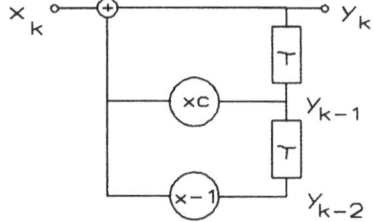

Figure 3.2.3 Alternative circuit for digital sinewave oscillator.

We proceed from M1 = c * M2 - M3 or $y_k = c \cdot y_{k-1} - y_{k-2}$

This is a second-order difference equation. For the determination of the solution we start again with a trial function: $y_k = B_1 \cos \gamma \, k$

Substitution in the equation gives the following result:

$$B_1 \cos \gamma \, k = c \cdot B_1 \cos \gamma \, (k-1) - B_1 \cos \gamma \, (k-2)$$

or $\qquad \cos \gamma \, k + \cos \gamma \, (k-2) = c \cdot \cos \gamma \, (k-1)$

Making use of rule (2.49c) we may write for this:

$$2 \cos \frac{2\gamma k - 2\gamma}{2} \cdot \cos \frac{2\gamma}{2} = c \cdot \cos \gamma (k-1)$$

or $\qquad 2 \cos \gamma (k-1) \cdot \cos \gamma = c \cdot \cos \gamma (k-1)$

This is true when $2\cos \gamma = c$, or $\cos \gamma = \frac{1}{2} c$, or $\gamma = \cos^{-1} \frac{1}{2}c$ $\qquad$ (3.5)

If γ satisfies this condition, then the trial function is indeed a solution. Evidently the absolute value of c may not be greater than 2: $|c| \le 2$

Thus $y_k = B_1 \cos \gamma k$ with $\gamma = \cos^{-1} \frac{1}{2}c$ is a solution of the difference equation . In the same manner it can be shown that $y_k = B_2 \sin \gamma k$ with the same γ is a solution as well. Just as in the time-continuous case the general solution is the sum of these two:

$$y_k = B_1 \cos \gamma \, k + B_2 \sin \gamma \, k \quad (\gamma = \cos^{-1} \frac{1}{2} c) \qquad (3.6)$$

and this expression can be worked into

$$y_k = B \cos (\gamma k + \phi) \text{ with } B = \sqrt{B_1^2 + B_2^2} \text{ and } \phi = -\tan^{-1} \frac{B_2}{B_1}$$

From extra information about the system (the initial conditions), B_1 and B_2 can be derived:

1. $y(k = 0) = x(0) = x_0$
   $\left. \begin{array}{l} \\ y(k = 0) = B_1 \end{array} \right\} B_1 = x_0$

2. $y(k = 1) = x_0 \cdot c = 2x_0 \cos \gamma$
   $\left. \begin{array}{l} \\ y(k = 1) = x_0 \cos \gamma + B_2 \sin \gamma \end{array} \right\} x_0 \cos \gamma + B_2 \sin \gamma = 2x_0 \cos \gamma$

thus $\qquad\qquad\qquad\qquad B_2 = x_0 \dfrac{\cos \gamma}{\sin \gamma}$

From this B and ϕ can be derived (see problem 3.7) and thus the amplitude/phase version of the general solution:

$$y_k = \frac{x_0}{\sin \gamma} \sin (\gamma k + \gamma) \qquad (\gamma = \cos^{-1} \tfrac{1}{2}c) \tag{3.7}$$

The factor $\sin \gamma$ in the denominator may be replaced as follows

$$\sin \gamma = \sqrt{1 - \cos^2 \gamma} = \sqrt{1 - \tfrac{1}{4}c^2}$$

k	y_k
0	1.00
1	1.90
2	2.61
3	3.06
4	3.20
.	.

With this it is shown that this system produces a time-discrete sinusoidal signal. If we calculate the values of y_k using this formula with $x_0 = 1$, $c = 1.90$ or $\gamma = 0.318$ we find the same values as before, as appears in the table on the left. If we compare the expression for y_k with the general formula for a time-discrete sinusoidal signal:

$$B \sin (\gamma k + \phi) \iff \frac{x_0}{\sin \gamma} \sin (\gamma k + \gamma)$$

we see that for this sinusoidal signal
the amplitude is equal to $\qquad x_0/\sin \gamma$
the initial phase is equal to $\qquad \gamma$
the frequency is equal to $\qquad (f_s/2\pi) \cos^{-1} \tfrac{1}{2} c$

There are now two possibilities:
1. c is given. We can then calculate f/f_s. For example $c = 1.90 \rightarrow f/f_s = 0.0505$ From this it is possible to calculate the number of samples per period (f_s/f, here: 19.786). If f_s is given as well, then the frequency f follows; for example

$$f_s = 18000 \text{ Hz} \rightarrow f = f_s/19.786 = 909.74 \text{ Hz}.$$

2. f and f_s are given. Then c can be calculated; for example

$$f = 144 \text{ Hz}, f_s = 40 \text{ kHz} \rightarrow c = 2 \cdot \cos 2\pi f/f_s = 1.9994883.$$

This c-factor must thus be used in our block diagram or computer program to generate a signal frequency of 144 Hz with a clock frequency of 40 kHz.

3.3 Damped vibrations - the time-continuous case

A. *Equation and solution*

The mathematical model that we have derived for the vibrating string (and other harmonic oscillators) is not realistic, because we perceive experimentally that the systems described do not produce infinitely long vibrations, but that the vibrations gradually die out. This is clearly a result of energy loss and we must therefore incorporate this aspect in our model. This can be done by assuming that an energy dissipating frictional force occurs. There are several types of frictional forces: a constant frictional force occurring with sliding movements (Coulomb-type friction), a viscous frictional force important at moderate speeds, and a hydraulic frictional force, occurring at high speeds that is proportional to the square of the speed. With the string vibration only the viscous force is important. For this force we may write: $-R \cdot v$ as it is proportional but opposite to the speed. The total force is now the sum of the elastic force $(-b \cdot y)$ and this frictional force. Because the forces work along the same line we can simply add them. The constant of proportionality R is called the *resistance*.

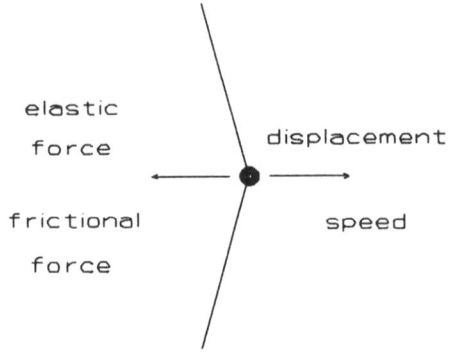

elastic
force

frictional
force

Figure 3.3.1 Forces on the string.

Total force: $-b \cdot y - R \dfrac{dy}{dt}$

This force cause a proportional acceleration:

$$\text{force} = m \frac{d^2 y}{dt^2}$$

This leads to the following differential equation:

$$m \frac{d^2 y}{dt^2} + R \frac{dy}{dt} + b \cdot y = 0$$

or:

$$\frac{d^2 y}{dt^2} + \frac{R}{m} \frac{dy}{dt} + \frac{b}{m} y = 0 \qquad (3.8)$$

For the solution of this equation we proceed from the trial function

$$y(t) = e^{-Pt} A_1 \cos \omega_d t$$

The first and second derivatives of y are:

$$\frac{dy}{dt} = -p A_1 e^{-Pt} \cos \omega_d t - \omega_d e^{-Pt} A_1 \sin \omega_d t$$

and

$$\frac{d^2y}{dt^2} = p^2 A_1 e^{-Pt} \cos \omega_d t + p A_1 \omega_d e^{-Pt} \sin \omega_d t +$$

$$+ A_1 \omega_d p e^{-Pt} \sin \omega_d t - A_1 \omega_d e^{-Pt} \cos \omega_d t$$

After the substitution of these expressions in the equation we arrive at:

$$A_1 \left(p^2 - \frac{R}{m} p + \frac{b}{m} - \omega_d^2 \right) e^{-Pt} \cos \omega_d t - A_1 \left(\frac{R}{m} \omega_d - 2 p \omega_d \right) e^{-Pt} \sin \omega_d t = 0$$

Because $\cos \omega_d t$ and $\sin \omega_d t$ are not (always) 0, the expressions in brackets must be equal to 0. We take first the second expression:

$$\frac{R}{m} \omega_d - 2 \omega_d p = 0 \quad \rightarrow \quad p = \frac{R}{2m} \tag{3.9}$$

and then the first one:

$$\frac{R^2}{4m^2} - \frac{R^2}{2m^2} + \frac{b}{m} - \omega_d^2 = 0; \quad \text{from this:} \quad \omega_d^2 = \frac{b}{m} - \frac{R^2}{4m^2}$$

With

$$\omega_d = 2\pi f_d \quad \text{and} \quad f_0 = \frac{1}{2\pi} \sqrt{\frac{b}{m}}$$

we find:

$$f_d = \frac{1}{2\pi} \sqrt{\frac{b}{m} - \frac{R^2}{4m^2}} = \frac{1}{2\pi} \sqrt{\frac{b}{m}} \cdot \sqrt{1 - \frac{R^2}{4m^2 \frac{b}{m}}}$$

thus:

$$f_d = f_0 \sqrt{1 - \frac{p^2}{4\pi^2 f_0^2}} \tag{3.10}$$

We can further show that $e^{-Pt} A_2 \sin \omega_d t$ is a solution as well, and thus we know the general solution:

$$y(t) = e^{-Pt}(A_1 \cos\omega_d t + A_2 \sin\omega_d t) \quad \text{or} \quad y(t) = A \cdot e^{-Pt} \cos(\omega_d t + \phi)$$

$$A = \sqrt{A_1^2 + A_2^2}, \quad \phi = -\tan^{-1} \frac{A_2}{A_1}$$

A_1 and A_2 follow from the initial conditions.

Interpretation: The figure on the right shows the familiar shape of the exponential function e $^{-pt}$ (in which 'p' determines the 'speed' of the damping). This function must be multiplied by the sinusoidal vibration $A \cos(\omega_d t + \phi)$.

The final result is the exponentially damped sinusoidal vibration. This vibration
- is not sinusoidal
- is not periodic
- has a lower frequency than the undamped vibration
- is identical to the undamped vibration if $R=0$ (no frictional loss).

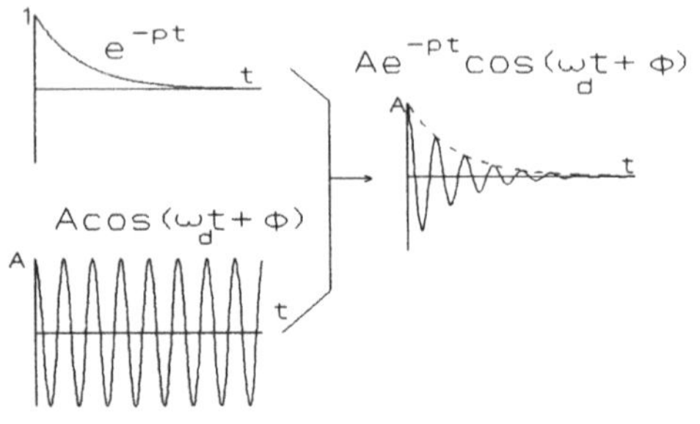

The first three effects are only important if we are dealing with a strong damping. In fig.3.3.3 four damped vibrations are shown with an increasing damping factor p. In the third graph the lengthening of the period (lowering of the frequency) is somewhat visible. The fourth represents a special case that will be dealt with further on.

Figure 3.3.2 Damped sinusoidal vibration.

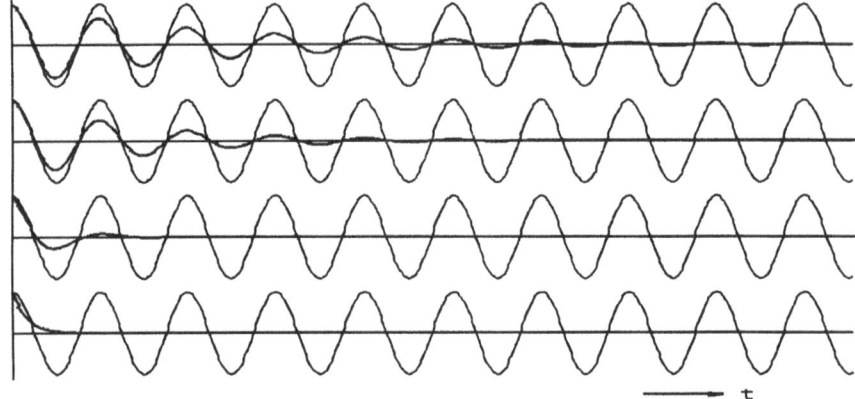

Figure 3.3.3 Damped sinusoidal vibrations with different damping factors.

B. *The damping factor*

All the (time-continuous) harmonic oscillators we have discussed here produce damped vibrations. If one should wish to avoid any decay of the vibration then the energy loss must be compensated for. This is handled in section 3.5. The value of the coefficient p determines the damping rate. Another way of specifying this is by using the decay time t_d. This is *not* the time in which the vibration dies away, because that takes (at least theoretically) an infinitely long time.

Instead one gives the time in which the amplitude decreases to a certain percentage of the initial value. The choice of this percentage is arbitrary. A value used in practice is 36.8%. At first sight this is a strange choice, but it is a result of the fact that for this amplitude decay a time t_d of $1/p$ seconds is required, because in this time the amplitude decay is:

$$e^{-pt_d} = e^{-p \cdot \frac{1}{p}} = e^{-1} = 0.368...$$

The time t_d (also called the 'decay modulus') can thus be calculated from the system constants R and m:

$$t_d = \frac{1}{p} = \frac{2m}{R}$$

The other way around t_d can be derived experimentally by measuring the damping rate of the registration of the signal, from which p follows. The 'vibration' at the bottom of the previous diagram is a special case, which happens when

$$\frac{p^2}{4\pi^2 f_0^2} = 1 \quad (\text{thus} \quad p^2 = 4\pi^2 f_0^2 \quad \text{or} \quad \frac{R^2}{4m^2} = \frac{b}{m})$$

Here the frequency $f_d = 0$ and we speak of 'critical' or 'aperiodic' damping: the system returns to its rest position without passing it, because now $y(t) = A \cdot e^{-pt}$. The movement is exponential. This holds as well when we make the damping still larger, but then the movement is slower. Critical damping is thus the optimal damping for systems (e.g. the needle of a voltmeter or other measuring device) which could, but should not vibrate. With a critically damped system the needle takes the desired position in the shortest possible time.

C. Electrical and mechanical systems

With the LC-circuit we can account for the energy loss by inserting a resistance R in the circuit (which then becomes a LCR-circuit, see fig.3.3.4). According to Ohm's law we have: $V_R = i \cdot R$

and for the coil: $V_L = L \dfrac{di}{dt}$

thus

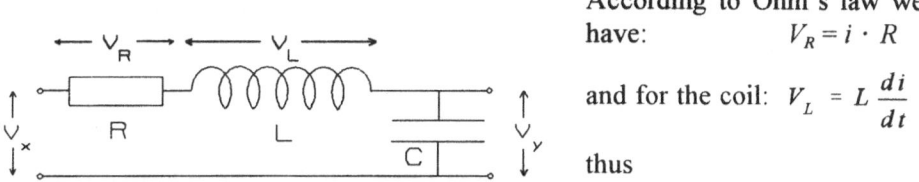

Figure 3.3.4 LCR-circuit.

$$V_x = V_R + V_L + V_y = i \cdot R + L \frac{di}{dt} + V_y$$

With $i = C \dfrac{dV_y}{dt}$:

$$V_x = RC \frac{dV_y}{dt} + LC \frac{d^2V_y}{dt^2} + V_y$$

and if $V_x = 0$ (see section 3.1):

$$\frac{d^2V_y}{dt^2} + \frac{R}{L}\frac{dV_y}{dt} + \frac{1}{LC}V_y = 0$$

With $Q = C \cdot V$ this becomes:

$$\frac{d^2Q}{dt^2} + \frac{R}{L}\frac{dQ}{dt} + \frac{1}{LC}Q = 0$$

The equations are identical to those for the string. We see again here the parallel between mechanical and electrical quantities. Let us put the analogous quantities next to each other:

mechanical	electrical
displacement (y)	charge (Q)
mass (m)	self induction (L)
compliance ($1/b$)	capacity (C)
friction (R)	resistance (R)
force (F)	voltage (V)
speed (v)	current (i)

Thus there is an analogy between mechanical and electrical resistance, between mass (which resists to being moved) and self induction (which resists changes in the current), and between capacity and compliance (b is in fact a measure for the stiffness). The possibility of translating mechanical systems into electrical ones is very useful, because, thanks to the simplification that arises from this and the possibility of describing electrical systems mathematically, the mathematical analysis of mechanical systems is simplified. For the same reason electrical (simulation) models used to be built, but nowadays computer simulations serve the same purpose.

Some examples of situations in which the electrical versions of mechanical systems are used: the study of the vibration of the basilar membrane in the inner ear, the analysis of the acoustic behaviour of the vocal tract and also in electro-acoustics. Fig.3.3.5 shows the electrical model of a loudspeaker, in which the various mechanical quantities are replaced by electrical ones.

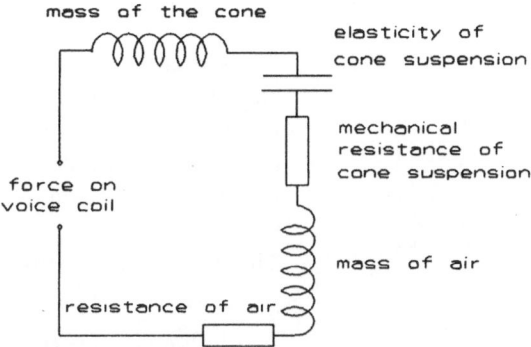

Figure 3.3.5 Electrical model of the mechanical part of a loudspeaker.

With the production of speech the air in the cavities of the vocal tract is brought in vibration by pulse-like air puffs that come from the vocal cords. In the registration below of a speech signal it is clearly seen how each pulse excites a vibration that has some similarity with a damped sinusoidal vibration.

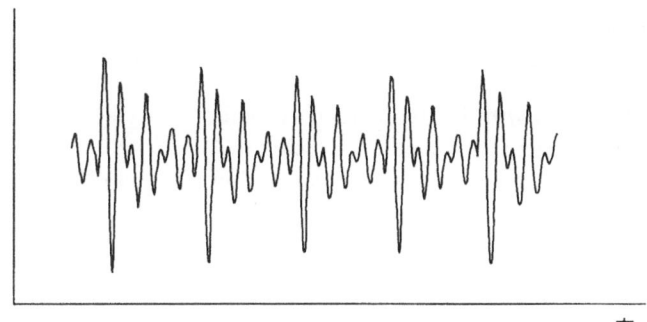

Figure 3.3.6 Registration of a vowel.

3.4 Damped vibrations - the time-discrete case

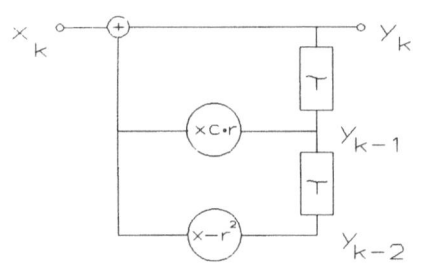

Figure 3.4.1 Circuit for a time-discrete damped sinusoidal vibration.

By setting up a small modification in the circuit for an undamped vibration it is possible to turn it into a damped one. For this purpose the two multiplication factors must be altered as shown in the figure aside.

The difference equation is now

$$y_k = c \cdot r \cdot y_{k-1} - r^2 y_{k-2} \quad (r < 1)$$

A solution is $\ y_k = B_1 \, r^k \cos \gamma \, k$
which means for y_{k-1}:

$$y_{k-1} \ = B_1 \, r^{k-1} \cos \gamma(k-1)$$
$$= B_1 \, r^{-1} \, r^k \cos \gamma(k-1)$$

and for y_{k-2}:

$$y_{k-2} = B_1 \, r^{k-2} \cos \gamma(k-2) \ = \ B_1 \, r^{-2} \, r^k \cos \gamma(k-2)$$

Substitution in the equation gives

$$B_1 \, r^k \cos \gamma \, k \ = \ c \cdot r \cdot B_1 \, r^{-1} \, r^k \cos \gamma(k-1) - r^2 B_1 \, r^{-2} \, r^k \cos \gamma(k-2)$$

or: $\qquad\qquad\qquad r^k \cos \gamma \, k \ = \ c \cdot r^k \cos \gamma(k-1) - r^k \cos \gamma(k-2)$

After division by r^k: $\quad \cos \gamma \, k \ = \ c \cdot \cos \gamma(k-1) - \cos \gamma(k-2)$

This is the same relation as with the undamped vibration, for which it was proven in section 3.2 that this is true if $\gamma = \cos^{-1} \frac{1}{2} c$ (rule (3.6)).

Again it holds here as well that $B_2 r^k \sin \gamma k$ is a solution. The complete solution is therefore: $\quad y_k = r^k (B_1 \cos \gamma \, k + B_2 \sin \gamma k)$

and this can be worked out in the same way as before to

$$y_k \ = r^k \cdot B \cdot \cos(\gamma k + \phi) \text{ with } B \ = \ \sqrt{B_1^{\,2} + B_2^{\,2}}, \quad \phi \ = \ - \tan^{-1} \frac{B_2}{B_1}$$

And finally as in section 3.2 to $\qquad\qquad y_k \ = r^k \dfrac{x_0}{\sin \gamma} \sin(\gamma k + \gamma)$

On the next page a computer program for this circuit is listed. Furthermore a table is shown that contains the first samples y_k of the damped vibration, the first samples $(y_k{}')$ of the undamped vibration and the factor r^k by which the $y_k{}'$-values must be multiplied in order to get the values of the samples of the damped vibration (column 2). The program gives these y-values directly which are easy to check. For r the value 0.99 and for c the value 1.90 is used.

	k	y_k	r^k	y_k'
M1 = 1				
M2 = 0				
M3 = 0	0	1.00	1.000	1.00
C = 1.9	1	1.88	0.990	1.90
R = 0.99	2	2.56	0.980	2.61
CR = C * R	3	2.97	0.970	3.06
R2 = R * R	4	3.07	0.961	3.20
5 M3 = M2	5	2.87	0.951	3.02
M2 = M1	6	2.40	0.941	2.55
M1 = C R * M2 - R2 * M3	7	1.70	0.932	1.82
CALL WAITCL	8	0.83	0.923	0.90
Y = M1	9	-0.10	0.914	-0.11
OUTPUT Y				
GO TO 5				

Comparison between the time-continuous and time-discrete case leads to:

$$y(t) = e^{-pt} A \cos(\omega_d t + \phi) \quad (p = \frac{R}{2m}) \quad \text{and} \quad y_k = r^k A \cos(\gamma k + \phi)$$

It is possible to equate both functions for the time points $t_k = k/f_s$. This will be the case if

1. $A \cos(\omega_d t_k + \phi) = A \cos(\gamma k + \phi)$ thus if $\quad \gamma = \dfrac{\omega_d}{f_s} = \dfrac{2\pi f_d}{f_s}$ (3.11)

just as in the time-continuous case.

2. $e^{-pt_k} = r^k$ If we replace t_k by k/f_s: $e^{-\frac{pk}{f_s}} = r^k$ or $r = e^{-\frac{p}{f_s}}$ (3.12)

and also $\qquad\qquad\qquad \ln r = -\dfrac{p}{f_s}$ thus $p = -f_s \cdot \ln r$ (3.13)

Notice that in contrast with the time-continuous system in the time-discrete system with decreasing r no change in the frequency occurs.

3.5 Forced vibrations - the time-continuous case

A. *The equation and the solution*

Forced vibrations are those that are performed by a system under the influence of a time-varying external force. The difference from free vibrations is that they are caused by a very short, pulse-like force to accomplish the initial displacement, for example. There are various possibilities for exerting this force. A feedback process can be used whereby the driving force is administered so that the energy loss is compensated for and the vibration alters from a damped into an undamped one. While this is in itself very interesting (consider the mechanism for maintaining the action of a watch, clock, *LC*-circuit etc.) we shall delve into what is (for us) more interesting namely the case that the driven system exerts no influence over the driving force and is thus a true 'slave' system. To orient ourselves we can proceed from the familiar string with the ball where the ball is magnetic and is located close to an electromagnet to which we connect an alternating current. (We then get a system that shows some similarity with a loudspeaker.)

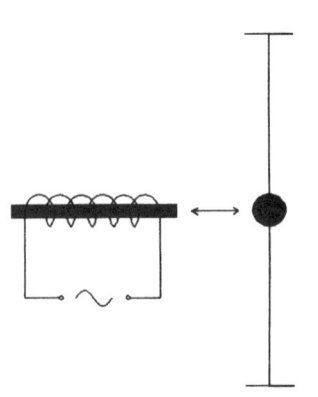

Figure 3.5.1 Forced vibration.

We choose a sinusoidally alternating current in such a way that the external force F_e that works upon the ball, varies sinusoidally with time, with an amplitude F and a frequency ω that we are free to choose, thus: $F_e = F \cos \omega t$

Now the total force is: $F \cos \omega t - b \cdot y - R \dfrac{dy}{dt}$

and as before this is equal to $m \dfrac{d^2 y}{dt^2}$, so that the complete equation is as fol-

lows: $m \dfrac{d^2 y}{dt^2} + R \dfrac{dy}{dt} + b \cdot y = F \cos \omega t$

This is a so-called 'inhomogeneous' differential equation. It is known from the theory of differential equations that the solution is equal to the sum of the solution of the homogeneous equation (the same equation with a zero to the right of the equal sign) plus one solution of the inhomogeneous equation. The homogeneous equation is in fact that of the free vibrating system with damping, and we know already the solution to this (see section 3.3). We must therefore find one solution for the inhomogeneous equation and for this we start with the following trial function: $$y(t) = C \cos(\omega t - \theta)$$

We thus assume that the slave vibration is also sinusoidal, with the same frequency as the driving force, and with a still unknown amplitude C and initial phase angle θ. To check this solution we have to substitute in the equation $y(t)$ and its first and second derivative:

$$\frac{dy}{dt} = -C\omega \sin(\omega t - \theta), \quad \frac{d^2 y}{dt^2} = -C\omega^2 \cos(\omega t - \theta)$$

This gives:

$$-mC\omega^2 \cos(\omega t - \theta) - RC\omega \sin(\omega t - \theta) + bC \cos(\omega t - \theta) = F \cos \omega t$$

or $$(-m\omega^2 + b) \cos(\omega t - \theta) - R\omega \sin(\omega t - \theta) = \frac{F}{C} \cos \omega t$$

On the left side there is again a sum of a cosine and a corresponding sine vibration that can be reduced in the familiar way to

$$B \cos(\omega t - \theta + \phi)$$

$$\text{with } B = \sqrt{R^2\omega^2 + (b - m\omega^2)^2}, \quad \phi = \tan^{-1} \frac{R\omega}{b - m\omega^2}$$

But

$$B \cdot \cos(\omega t - \theta + \phi) = B \cdot \cos \omega t \cdot \cos(\phi - \theta) - B \cdot \sin \omega t \cdot \sin(\phi - \theta)$$
$$= \frac{F}{C} \cos \omega t$$

or $$\{B \cdot \cos(\phi - \theta) - \frac{F}{C}\} \cos \omega t - B \cdot \sin(\phi - \theta) \sin \omega t = 0$$

This is only possible if both $B \cdot \cos(\phi - \theta) - F/C$ and $B \cdot \sin(\phi - \theta)$ are equal to zero. From the second condition we derive $\phi - \theta = 0$ thus $\phi = \theta$. From this:

$$\tan \theta = \tan \phi = \frac{R\omega}{b - m\omega^2} = \frac{R}{\dfrac{b}{\omega} - m\omega}$$

$$\text{thus } \theta = \tan^{-1} \frac{R}{\dfrac{b}{\omega} - m\omega}$$

$$(3.14)$$

With $\phi = \theta$ the first condition leads to

$$B - \frac{F}{C} = 0 \text{ thus } C = \frac{F}{B} = \frac{F}{\sqrt{R^2\omega^2 + (b - m\omega^2)^2}}$$

This can also be worked out further, because $b - m\omega^2 = \omega(b/\omega - m\omega)$ and thus

$$C = \frac{F}{\sqrt{R^2\omega^2 + \omega^2(\frac{b}{\omega} - m\omega)^2}} = \frac{F}{\omega\sqrt{R^2 + (\frac{b}{\omega} - m}} \tag{3.15}$$

With this the solution for the inhomogeneous equation is completely determined, because now we know the amplitude C and also the initial phase angle θ. The general solution is found by adding to this solution the general solution to the homogeneous equation:

$$y(t) = A e^{-Pt}\cos(\omega_d t + \phi) + C\cos(\omega t - \theta)$$

The vibration consists of the sum of a damped sinusoidal, free vibration and the 'slave' vibration (see fig.3.5.2). Initially the vibration is rather irregular, but after some time the free vibration has died away and we do no longer notice anything of the 'transient'. What then is left is a pure sinusoidal slave vibration.

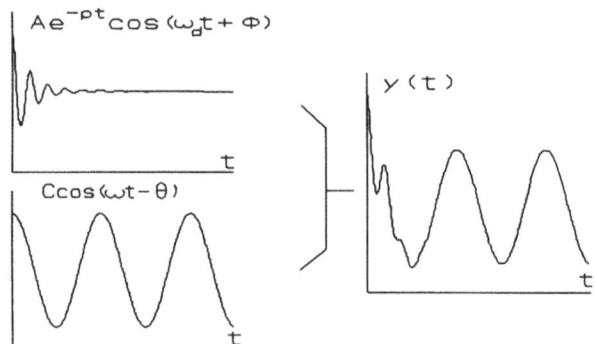

Figure 3.5.2 Forced vibration with transient.

B. *The amplitude of the forced vibration*

We shall first have a look at how the amplitude C of the forced vibration depends upon the drive frequency ω. If in the expression for C we consider the part under the root we see that both terms here are quadratic and thus positive, and that their sum will have a minimum if $b/\omega - m\omega = 0$. Because this expression occurs in the denominator, C will be relatively large, in other words, we expect a maximum of $C(\omega)$ when

$$\frac{b}{\omega} - m\omega = 0 \quad \text{or} \quad \omega^2 = \frac{b}{m} \quad \text{thus} \quad f = \frac{1}{2\pi}\sqrt{\frac{b}{m}}$$

that is when the frequency of the driving force is equal to that of the undamped free vibration. Especially if we calculate several values for $C(\omega)$ and graph them, we see very clearly that if the frequency of the driving force comes near that of the undamped vibration, the system reacts strongly. This phenomenon that we also know from daily practice is called *resonance*. In fig.3.5.3 $C(\omega)$ is plotted for several values of R. We call these curves *resonance curves*. We can also say that such a system is a *filter* because it has a selective behaviour with regard to the drive frequency: only frequencies in a narrow range around ω_0 are transmitted. The graphs of $\theta(\omega)$ shown in fig.3.5.3b will be discussed later.

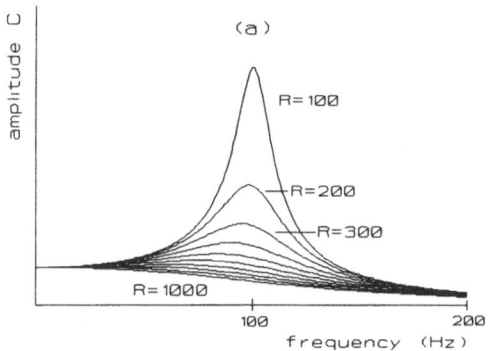
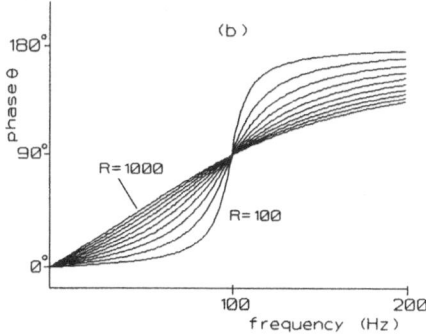

Figure 3.5.3a
Amplitude response of forced vibrations.

Figure 3.5.3b
Phase response of forced vibrations.

Until now we have neglected the role of the factor ω that occurs in the denominator. If we bring this factor into account we see that the place of the resonance peak also depends upon R, something that is also to be seen in fig.3.5.3: with larger R-values (lower resonance peak) the peak shifts towards the left.
The exact location of the maximum can be determined by differentiating $C(\omega)$ to ω and then solving the equation that results from setting this derivative equal to

zero. Using the result from problem 2.12 we find that the maximum is located at

frequency ω_M with $\quad \omega_M = \omega_0 \sqrt{1 - \dfrac{R^2}{2bm}}$

For small values of R holds: $\qquad \omega_M \approx \omega_0$

and for larger values of R: $\qquad \omega_M < \omega_0$

If $R^2 > 2bm$ there is no longer a maximum (see fig.3.5.3).

C. *Analysis of the filter behaviour*

1. The Q-factor

We now concentrate upon the denominator of the expression for C, and in partic-ular upon that part with the root that we abbreviate to W:

$$W = \sqrt{R^2 + (\frac{b}{\omega} - m\omega)^2} = \sqrt{R^2 + (m\omega - \frac{b}{\omega})^2}$$

$$= \sqrt{R^2 + \left(\frac{\omega\sqrt{m}\sqrt{m}\sqrt{b}}{\sqrt{b}} - \frac{1}{\omega}\frac{\sqrt{b}\sqrt{b}\sqrt{m}}{\sqrt{m}} \right)^2}$$

and with $\quad \dfrac{b}{m} = \omega_0^2 \quad$:

$$W = \sqrt{R^2 + \left(\frac{\omega}{\omega_0}\sqrt{bm} - \frac{\omega_0}{\omega}\sqrt{bm} \right)^2} = \sqrt{R^2 + \left(\frac{\omega}{\omega_0} - \frac{\omega_0}{\omega} \right)^2 bm}$$

We now introduce the 'tuning variable' ß:

$$\beta = \frac{\omega}{\omega_0} - \frac{\omega_0}{\omega} \qquad \begin{cases} \beta = 0 & \text{if } \omega = \omega_0 \\ \beta < 0 & \text{if } \omega < \omega_0 \\ \beta > 0 & \text{if } \omega > \omega_0 \end{cases}$$

and find for W:

$$W = \sqrt{R^2 + \beta^2 bm} = \sqrt{R^2\left(1 + \frac{\beta^2 bm}{R^2}\right)} = R\sqrt{1 + \frac{\beta^2 bm}{R^2}}$$

Once again we introduce an abbreviation:

$$\frac{bm}{R^2} = Q^2 \quad \text{or} \quad Q = \sqrt{\frac{bm}{R^2}} = \frac{\omega_0 m}{R}$$ (3.16)

With this, W becomes: $\qquad W = R\sqrt{Q^2\beta^2 + 1}$

Q is called the *quality factor* of the system. It is a dimension-less number that represents important features of the system in a characteristic way, in both the time domain and in the frequency domain.

2. Time-domain interpretation of Q

Let us calculate the amplitude of the free (damped) vibration after the performance of 'Q' periods. The amplitude is then decreased by a factor $e^{-pT'}$, in which T' is the duration of Q periods. We find for this duration:

$$T' = Q \cdot T = Q \frac{1}{f_0} = \sqrt{\frac{bm}{R^2}} \cdot 2\pi \sqrt{\frac{m}{b}}$$

If we multiply this by $-p$ $(= -R/2m)$ to calculate the complete exponent of the amplitude factor we get:

$$-pT' = -\frac{R}{2m}\sqrt{\frac{bm}{R^2}} \cdot 2\pi \sqrt{\frac{m}{b}} = -\pi$$

The amplitude factor is thus $e^{-\pi} \approx 0.0432$. This means that after Q periods the amplitude is decreased to ca. 4% of the start value. This is another way of giving the damping speed rather than the use of the decay modulus $t_d = 1/p$. There the amplitude factor amounted to e^{-1} and here to $e^{-\pi}$; this time duration is thus π times as long. If we have at our disposal a registration of a damped vibration we can make an estimate of the value of Q by counting the number of periods in which the amplitude of the vibration is reduced to about zero.

3. Frequency domain interpretation of Q

We will now determine the width of the resonance peak. Because we can not speak of 'the' width we will define it first as the difference between the two frequencies where the signal is attenuated by 3 dB (we call this the *bandwidth B* of the filter). An amplitude attenuation of 3 dB corresponds to a factor $1/\sqrt{2}$ (because $20^{10}\log 1/\sqrt{2} \approx -3$). If we return to the expression for $C(\omega)$ and substitute in this the derived expression for W, we get: $\quad C(\omega) = \dfrac{F}{\omega R \sqrt{Q^2 \beta^2 + 1}}$

and at the resonance frequency ($\beta = 0$): $\quad C(\omega_0) = \dfrac{F}{\omega_0 R}$

The 3 dB-points ω_A and ω_B (see also fig.3.5.4) are characterized by:

$$C(\omega_A) = C(\omega_B) = \frac{1}{\sqrt{2}} C(\omega_0)$$

If the difference between ω_0, ω_A and ω_B is not too large (which is the case with a reasonably sharp resonance peak) it holds:

$$C(\omega_A) = \frac{F}{\omega_A R \sqrt{Q^2 \beta_A^2 + 1}} \approx \frac{F}{\omega_0 R \sqrt{Q^2 \beta_A^2 + 1}}$$

and this must be equal to $\qquad \dfrac{1}{\sqrt{2}} C(\omega_0) = \dfrac{F}{\omega_0 R \sqrt{2}}$

Thus: $\quad Q^2 \beta_A^2 + 1 = 2$, $\quad Q^2 \beta_A^2 = 1 \quad$ or $\quad Q \beta_A = 1 \quad (Q, \beta_A > 0)$

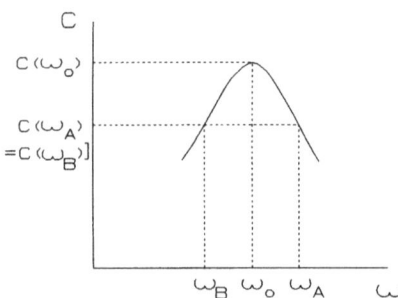

Figure 3.5.4 Bandwidth.

If we now replace β_A by $\dfrac{\omega_A}{\omega} - \dfrac{\omega}{\omega_A}$

and then multiply left and right by $\omega_0 \omega_A$ we get

$$\omega_A \omega_0 Q \left(\frac{\omega_A}{\omega_0} - \frac{\omega_0}{\omega_A} \right) = \omega_A \omega_0$$

or $\qquad Q(\omega_A^2 - \omega_0^2) = \omega_A \omega_0$

In the same way we derive that $Q^2 \beta_B^2 = 1$ thus $Q \beta_B = -1$ $\quad (Q > 0, \beta_B < 0)$ or

$$Q \beta_B = Q \left(\frac{\omega_B}{\omega_0} - \frac{\omega_0}{\omega_B} \right) = -1$$

Multiplying left and right by $\omega_B \omega_0$ now leads to: $\quad Q(\omega_B^2 - \omega_0^2) = -\omega_B \omega_0$

If we subtract both expressions we get:

$$Q(\omega_A^2 - \omega_B^2) = \omega_0(\omega_A + \omega_B) \text{ or } Q(\omega_A + \omega_B)(\omega_A - \omega_B) = \omega_0(\omega_A + \omega_B)$$

$$Q = \frac{\omega_0}{\omega_A - \omega_B} = \frac{f_0}{f_A - f_B} = \frac{f_0}{B} \quad \text{or} \quad B = \frac{f_0}{Q} \tag{3.17}$$

If we replace f_0 by (3.2) and Q by (3.16) we get: $\qquad B = \dfrac{p}{\pi}$ (3.18)

The relationship between time and frequency domain parameters becomes even clearer if we multiply duration T' of the damped vibration by bandwidth B:

$$T' \cdot B = \frac{Q}{f_0} \cdot \frac{f_0}{Q} = 1 \tag{3.19}$$

In chapter 4 we will return to this fundamental relationship.

D. *Global characterization of the resonance curve*

In the resonance curve three ranges can be distinguished: left of the peak, near the peak and right of the peak. We shall see that in the first frequency range the behaviour of the system is mainly determined by b (stiffness-controlled system), in the second by R (resistance-controlled system) and in the third by m (mass-controlled system). To show this, we proceed from

$$C(\omega) = \frac{F}{\omega W} = \frac{F}{\sqrt{R^2\omega^2 + (\omega^2 m - b)^2}} = \frac{F}{\sqrt{R^2\omega^2 + m^2(\omega^2 - \frac{b}{m})^2}}$$

1. Stiffness-controlled system.
   This is the case if $\omega < \omega_0$. With $\omega_0 = \sqrt{b/m}$ and $\omega^2 < \omega_0$ or $\omega^2 < b/m$ we find that the expression in brackets is practically equal to $(b/m)^2$:

$$C \approx \frac{F}{\sqrt{R^2\omega^2 + m^2 \frac{b^2}{m^2}}} \quad \text{and if } \omega^2 \ll \frac{b^2}{R^2} : C \approx \frac{F}{b}$$

2. Resistance-controlled system.

   If $\omega \approx \omega_0$ it holds: $\qquad C = \dfrac{F}{\omega_0 W} \approx \dfrac{F}{\omega_0 R}$

3. Mass-controlled system.

We now assume $\omega > \omega_0$, thus $\omega^2 > b/m$ by which the expression under the root can be worked out to:

$$\sqrt{R^2\omega^2 + m^2\omega^4} = \sqrt{\omega^2(R^2 + m^2\omega^2)}$$

If $\omega \gg R/m$ or $m^2\omega^2 \gg R^2$ it holds: $C = \dfrac{F}{m\,\omega^2}$

With practical systems we can try to give b, R and m values that are optimal dependent on what we want to achieve with that system. For example the resonator used in a vibraphone must have quite a sharp resonance peak, whereas we expect the frequency curve of a loudspeaker to be flat.

E. *The phase behaviour*

As it appears from fig.3.5.3 the input and output signals at frequencies below the resonance frequency (in a stiffness-controlled system) are approximately in phase; at the resonance frequency (in a resistance-controlled system) the phase shift is 90°. For high frequencies (in a mass-controlled system) the phase difference becomes 180°. This is also true when the system is heavily damped. It is therefore often easier to recognize resonances in the phase response than in the amplitude response.

F. *Energy dissipation*

When a driving force brings a system into vibration, also energy will be transferred. It may be expected that this energy will be maximal at the resonance frequency, because the vibration at that point is maximal. This is indeed the case:

Energy (work) $E =$ force $\cdot$ displacement $= F \cdot y$.

If the force is not constant and depends upon the deviation then with a displacement from point a to point b the amount of work is:

$$E = \int_a^b F(y)\,dy$$

If y is a function of time, we can rewrite this expression using the variable t instead of the variable y:

$$y'(t) = \frac{dy}{dt} \quad \text{thus} \quad dy = y'(t)\,dt$$

If we substitute this in the expression for E we get:

$$E = \int_{t_a}^{t_b} F(t)\cdot y'(t)\,dt$$

With a forced vibration the deviation y is equal to:

$$y(t) = C\cos(\omega t - \theta), \quad \text{thus} \quad y'(t) = -\omega C\sin(\omega t - \theta)$$

Together with the time function of the driving force $F\cos\omega t$ this becomes (if we calculate the energy over one period T):

$$E = \int_0^T F\cos\omega t\{-\omega C\sin(\omega t - \theta)\}dt = -\omega FC\int_0^T \cos\omega t\cdot\sin(\omega t - \theta)\,dt$$

$$= -\omega FC\int_0^T \cos\omega t(\sin\omega t\cdot\cos\theta - \cos\omega t\cdot\sin\theta)\,dt$$

$$= -\omega FC\int_0^T (\cos\theta\cdot\tfrac{1}{2}\sin 2\omega t - \tfrac{1}{2}\sin\theta(1 + \cos 2\omega t))\,dt$$

$$= -\omega FC\cdot\tfrac{1}{2}\cos\theta\int_0^T \sin 2\omega t\,dt + \omega FC\cdot\tfrac{1}{2}\sin\theta\int_0^T (1 + \cos 2\omega t)\,dt$$

The first integral is equal to

$$-\frac{1}{2\omega}\cos 2\omega t\,\Big|_0^T = -\frac{1}{2\omega}\cos\frac{4\pi t}{T}\,\Big|_0^T = 0$$

The second integral is equal to

$$t\,\Big|_0^T + \frac{1}{2\omega}\sin 2\omega t\,\Big|_0^T = T$$

Thus we find for E:

$$E = \omega FC\cdot\tfrac{1}{2}\sin\theta\cdot T = \frac{2\pi}{T}FC\cdot\tfrac{1}{2}\sin\theta\cdot T = \pi FC\sin\theta \qquad (3.20)$$

In resonance $\theta = 90°$, thus $\sin \theta = 1$. With other frequencies $\sin \theta$ is less than 1. The energy of the vibrating system is thus indeed maximal at resonance (if the system is not too strongly damped). This energy is taken from the driving force. This is the fact on which the application of the Helmholtz resonator as a sound-absorbing system is based. If the air in this resonator is brought into resonance by a sound coming from outside then it is at the cost of the energy of that sound. In the same way as earlier in fig.3.5.3, fig.3.5.5 shows the relationship between energy dissipation and frequency for a particular value of R, using formula (3.20).

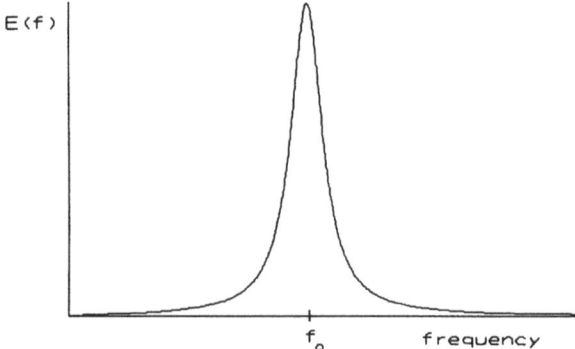

Figure 3.5.5
Energy dissipation as function of frequency.

3.6 Forced vibrations in a time-discrete system

A. *Equation and Solution*

To the input of the time-discrete system that we studied in section 3.4 (see fig.3.4.1) we now connect a time-discrete sinusoidal signal: $x_k = G \cos s k$

The complete difference equation now becomes:

$$y_k - c r y_{k-1} + r^2 y_{k-2} = G \cos sk$$

Once again we have to find one solution for this inhomogeneous equation, and add it to the already known general solution of the homogeneous equation. To determine this special solution we proceed from the trial function:

$$y_k = D \cos (sk + \sigma) \ .$$

As before we assume a sinusoidal output signal with the same frequency as that of the input signal. The general solution of the difference equation is now:

$$y_k = B \cdot r^k \cos (\gamma k + \phi) + D \cdot \cos (sk + \sigma)$$

But we still have to determine the amplitude D and the initial phase angle σ and therefore we substitute the trial function in the equation. After some calculations we find the following expression for D:

$$D = \frac{G}{\sqrt{(r^2 - 1)^2 \sin^2 \gamma + \{(r^2 + 1) \cos \gamma - 2r \cos s\}^2}} \tag{3.21}$$

If we graph D as a function of the frequency s then we get a curve that is very similar to the resonance curve of fig.3.5.3. In fig.3.6.1 this has been done for a few values of r. The peak results from the expression for D that is maximal at frequency s_m with $(r^2 + 1) \cos \gamma = 2r \cos s_m$ or

$$\cos s_m = \frac{r^2 + 1}{2r} \cos \gamma = \frac{r^2 + 1}{4r} c \approx \tfrac{1}{2} c = \cos \gamma \text{ if } r \approx 1$$

It is now possible to choose c and r so that a particular resonance frequency and a particular bandwidth are achieved, because

$$c = 2 \cos \gamma = 2 \cos \frac{2\pi f}{f_s} \tag{3.22}$$

Combining (3.12) and (3.18) we find: $r = e^{-\dfrac{\pi B}{f_s}}$ (3.23)

From this also Q can be calculated: $Q = \dfrac{f_0}{B} = -\dfrac{\gamma}{2 \ln r}$ (3.24)

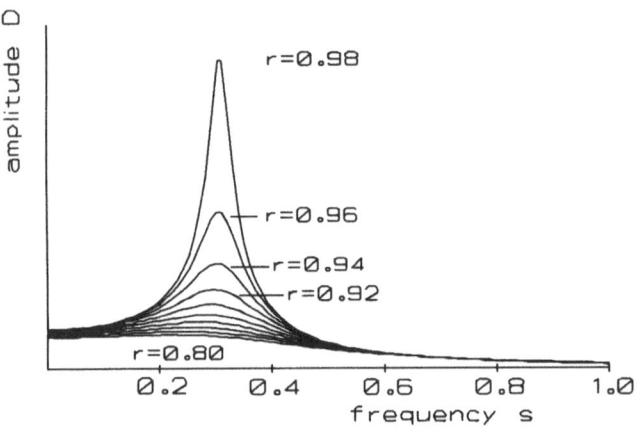

Figure 3.6.1 Resonance curves of a time-discrete harmonic oscillator.

B. *A digital filter*

This time-discrete system is thus a digital filter. One of the applications of this type of filter is the use as a formant filter for synthetic speech production. Let us work out a numerical example. Imagine we want a filter curve with a resonance frequency of 840 Hz and a bandwidth of 200 Hz (the Q-factor of this system is thus $Q = 840/200 = 4.2$). The sampling frequency f_s is 20000 Hz. For the factor c this yields:

$$c = 2 \cos \frac{2\pi \cdot 840}{20000} = 1.93076328 .$$

and for the factor r: $r = e^{-\frac{\pi \cdot 200}{20000}} = 0.96907243$

The computer program for this filter could be the following:

```
        M1 = 0
        M2 = 0
        M3 = 0
        C = 1.93076328
        R = 0.96907243
        CR = C * R
        R2 = R * R
5       INPUT X
        M3 = M2
        M2 = M1
        M1 = CR * M2 - R2 * M3 + X
        CALL WAITCL
        Y = M1
        OUTPUT Y
        GO TO 5
```

The only difference with the program in section 3.4 is that we have to add the input samples x_k to the two feedback values CR*M2 and -R2*M3. The big advantage of this type of filter is its flexibility, the possibility of changing the filter curve by changing the factor c and/or r. With synthetic speech production the adjustment takes place every 20 to 40 ms. With analog filters this is not possible.

3.7 Problems

3.1   A Helmholtz resonator receives as input signal a periodic signal consist-
       ing of short impulses. The filtered output signal is shown below. Using
       this registration give an estimation of
a.   the frequency of the input signal,
b.   the resonance frequency of the system,
c.   the quality factor,
d.   the bandwidth of the filter system.

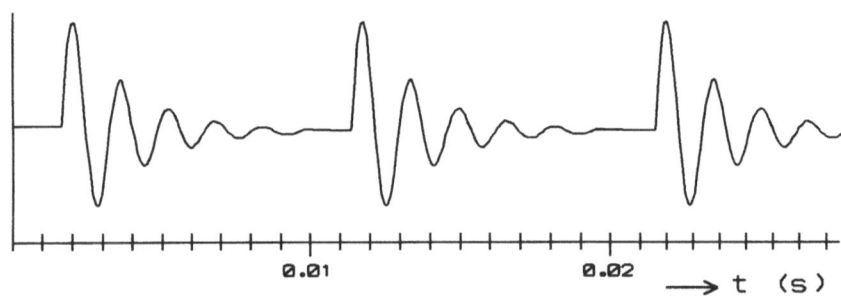

3.2   Give the complete diagram, including the values of the coefficients, of a
       digital system with the same resonance properties as the time-continuous
       system of problem 3.1. The system has a clock frequency of 20 kHz.

3.3
a.   Determine the value of the coefficients A_1 and A_2 that occur in the gen-
       eral solution to the differential equation (3.8) for the harmonic oscillator
       with damping (see section 3.3.A), if the initial displacement $y(t=0)$ is
       equal to y_0 and the initial speed $v(t=0)$ to v_0 (hint: use the same method as
       used for the undamped vibration).
b.   Calculate the values of A and ϕ, the constants of the amplitude/phase
       version of this solution with initial displacement y_0 and initial speed 0.

3.4
a.   For the spectral analysis of a tone recorded on an analogue tape a *LCR*-
       filter can be used, for example. Sketch this filter and calculate the value
       of C required to tune the filter to 800 Hz when self-induction L has the
       value 400 mH and the resistance is 1 Ω.
b.   Derive from the equivalence between mechanical and electrical quanti-

ties an expression for the Q-factor of this filter, and calculate the value of Q.

c.   For a digital analysis we need a digital band filter. Sketch the diagram of such a filter and calculate the coefficients required to tune it to the same frequency (800 Hz) and to give it the same Q-factor. Clock frequency: 25000 Hz.

3.5   Resonating systems usually have a constant Q-factor, but with formants in speech we are dealing with filters with a constant bandwidth. Suppose this bandwidth is 150 Hz:

a.   Calculate the Q-factor at resonance frequencies 750 Hz and 1800 Hz.

b.   Calculate the decay modulus.

3.6

a.   When do we speak of aperiodic damping of a harmonic oscillator?

b.   What is the Q-factor of an aperiodically damped system?

c.   Could a time-discrete harmonic oscillator be aperiodically damped?

3.7   Derive formula (3.7) by calculating amplitude B and initial phase angle ϕ of the time-discrete sine function from the expressions for B_1 and B_2 in section 3.2.C.

3.8   Consider the mouth cavity as a Helmholtz resonator with the following properties: volume: 125 cm^3, length of output tube 2.5 cm and cross section of output opening 2 cm^2. Calculate the resonance frequency (speed of sound: 340 m/s).

3.9   Derive in the same way as in problem 3.4.b the equivalent versions of the formulae (3.14) and (3.15) for the LCR-circuit.

CHAPTER 4

Signal Functions in the Time and Frequency Domains

We have already seen that we are dealing with two types of signal functions: signal functions that consist of an (infinite) collection of pairs of real numbers (the analog, time-continuous signal functions) and signal functions that consist of a (finite) collection of integer numbers (digital, time-discrete signal functions), while often in theoretical analyses something between these is used in which the time values are discrete (and thus integer) and the function values are real.

In the first part of this chapter the differences between these representations will be discussed and the practical consequences of switching from one to the other. For this purpose it is sometimes convenient to consider signals as consisting of a series of infinitely short pulses (adjacent in time-continuous functions, isolated in time-discrete functions). We shall learn more about pulses as elementary signals in chapter 5.

There is still another method to reduce signals to a combination of elementary signals and that is the spectral analysis, where an arbitrary vibration is split into sinusoidal components. This is the subject of the second and most extensive part of this fourth chapter. In the third part several applications of this method are dealt with and in the fourth it is shown that sinusoidal vibrations are not the only elementary vibrations with which such an analysis can be performed.

4.1 The computer; binary number representation and programming

The rise of digital sound technique is the result of the development of computer techniques. If we wish to go into the practical aspects of digital signal representation we must first consider a few fundamental aspects of this technique (see also Hintze, 1966). A computer, as the name says, was in the first place developed for the task of doing calculations. Calculating apparatus has already existed for quite a long time and all the devices, from the most primitive abacus to the most advanced computer, have one thing in common: the numbers with which the calculation should be performed have to be represented somehow 'in' the device, in such a way that they are recognizable and accessible to mathematical operations like addition. In mechanical calculators this occurs for example by means of cogwheels with ten cogs. Such a cogwheel can therefore take ten different positions that can be interpreted as the numbers from 0 to 9.

One of the first applications of this principle was the calculator built by Pascal in 1642 (see fig.4.1.1). Here eight cogwheels were used. The apparatus could only add (with each full rotation of the wheel the one next to it turned one position further) and subtract.

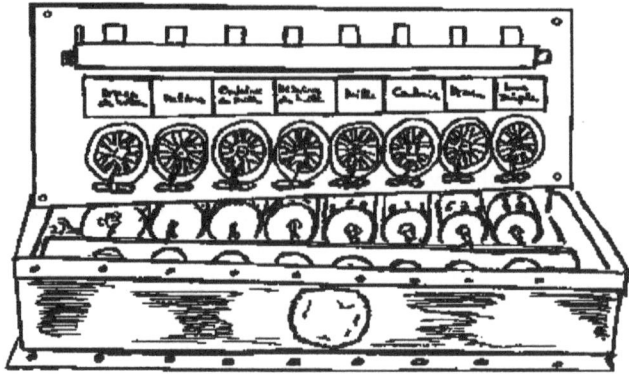

Figure 4.1.1 Pascal's calculator.

Mechanical calculators have many uses as accounting calculators, cash desks, etc. For extensive scientific calculations they were not so suitable, because the calculation speed was limited by the frictional resistance and the mass of the moving parts. A purely electronic calculator would not have these limitations and could thus work much faster. When designing such a calculator, first the problem of number representation must be solved. In principle that is not too difficult: it is possible to take Pascal's example and work with 'wires', on which an electrical voltage is set to represent a number, e.g. 0 volt = '0', 1 volt = '1', etc. To represent the number 729 there would have to be three wires with voltages of 7, 2 and 9 volts respectively. Another possibility is to make use of only one wire and to differentiate between more voltage levels. The value 729 would then be represented for example with a voltage of 7.29 volt. This solution however is not feasible in practice because it is technically nearly impossible to design circuits that higher up in the calculator could distinguish and recognize correctly such minute voltage differences that could occur in the system. Only if a low accuracy (ca. 1%) is sufficient it is possible to build a so-called 'analog' computer in this way. Additions and multiplications of electrical voltages as occur in a mixing desk or a modulator are examples of this application.

As a calculator the analog computer has been replaced by the digital one. This is based on the above described 'multi-wire' system that despite a limited capacity and resolution (depending upon the number of wires) is yet more suitable for accurate calculations because the unmistakable recognition of values by means of the increased tolerance is much easier. The number '5' would be represented by a

voltage of 5 volts, which means in practice a voltage between 4.5 and 5.5 volts.
If this tolerance is so important for the technical realization we can ask ourselves
if the system described above, of 10 voltages (or voltage intervals) for the 10
digits (from 0 to 9) is the optimal solution.

This is not so. In the end the fact that we calculate in a decimal system probably follows from the almost accidental fact that we have 10 fingers at our disposal. Other number systems are possible and in the course of history have in
fact been in use, for example the Babylonian number system with 60 digits
(which we still meet in the subdivision of the degree, the hour and the minute),
while the words 'dozen' and 'gross' (12 dozen) show a grouping of 12 in place
of 10. If it were possible to work in a number system with less than 10 numerical
symbols we would need to differentiate between fewer voltage levels and the
technical realization would be easier. Let us have a look at how we can switch
over from a decimal system to a 'lower' system, for example the octal system,
and what the consequences of this transition are.

First a remark on the notation of numbers. Since antiquity there have been
various systems for representing large numbers by means of a combination of
elementary symbols. The system that is now universally applied is the positional
system, in which the interpretation of a number symbol in a series depends upon
the place in the series. In the decimal system we interpret the number 546 as

$$5 \cdot 100 + 4 \cdot 10 + 6 \quad \text{or} \quad 5 \cdot 10^2 + 4 \cdot 10^1 + 6 \cdot 10^0$$

0	0	This system is very practical for calculations (unlike with, for
1	1	example, the system of Roman numerals). With the positional
2	2	system we can systematically construct all numbers with a
3	3	few rules, starting at the lowest number. First we use all pos-
4	4	sible symbols (in the decimal system the digits from 0 to 9).
5	5	After the '9' the supply is exhausted and we start again, using
6	6	the series 0,1, ... but with the number 1 added to the left (see
7	7	left column in the table to the left).
8	10	Now the octal system. Here we have only eight number sym
9	11	bols at our disposal. Let us for the sake of convenience work
10	12	with the symbols 0, 1, ..., 7. If now, from 0 on, we write all
11	13	the numbers (see right column in the table), the basic supply
12	14	is exhausted when we arrive at the number 7. We apply the
13	15	same method and write directly after that the number '10'
14	16	(which should not be pronounced as 'ten'!) which represents
15	17	the same number as the symbol '8' in the decimal system. It
16	20	means that the number most to the right gives the *units* (as in
17	21	the decimal system), the next one the *octals*, the next one the
. .	. .	*64-als* etc. The (octal) series 746_8 must thus be read:

$$746_8 = 7 \cdot 8^2 + 4 \cdot 8^1 + 6 \ (= 486_{10})$$

(The subscript gives the number system used.)

In general: the series $\quad a_m a_{m-1} ... a_2 a_1 a_0 a_{-1} a_{-2} ...$

must be interpreted as $a_m b^m + a_{m-1} b^{m-1} + .. \ a_2 b^2 + a_1 b^1 + a_0 b^0 + a_{-1} b^{-1} + a_{-2} b^{-2} + ...$

Here 'b' is the *basis* of the number system (10 in the decimal, 8 in the octal system), the a_m each represent one of the b available symbols. Via a few calculations it is possible to switch over from the decimal system (which remains the starting point for us) to any other system, and vice versa. If b < 10 we use a part of the set of the normal number symbols, for b > 10 we must add symbols to the supply. For this the letters A (= 10), B (= 11), C (= 12), etc. are used.

As an example we see in the table to the right the different versions of the decimal number 1779 starting with b = 16 (the hexadecimal system) up to and including b = 2 (the binary system). To move from non-decimal numbers to decimal numbers the above power series must be made use of, for example

$$6F3_{16} = 6 \cdot 16^2 + 15 \cdot 16^1 + 3 \ (= 1779_{10})$$

To switch from decimal tot non-decimal the simplest procedure is to divide the number to be converted repeatedly by the new basis. The remainders that are found by every division form together the number we are looking for. E.g. $1779_{10} = ... {}_9$:

16	6F3
15	7D9
14	911
13	A6B
12	1043
11	1378
10	1779
9	2386
8	3363
7	5121
6	12123
5	24104
4	123303
3	2102220
2	11011110011

Not only is the choice of the basis of a number system arbitrary, but also the elementary mathematical operations are not bound to a particular system although the rules must be adapted to the system in question. The addition 6 + 5 gives in the decimal system the sum of 11, and in the octal system the sum 13_8 (= $1 \cdot 8 + 3 = 11_{10}$). Here are a few operations in the decimal and in the octal system:

$$
\begin{array}{cccc}
756_8 & 494_{10} & 264_8 & 180_{10} \\
124_8 & 84_{10} & 51_8 & 41_{10} \\
+\ \rule{2cm}{0.4pt} & +\ \rule{2cm}{0.4pt} & \times\ \rule{2cm}{0.4pt} & \times\ \rule{2cm}{0.4pt} \\
1102_8 & 578_{10} & 264_8 & 180_{10} \\
& & 16040_8 & 7200_{10} \\
& & \rule{2cm}{0.4pt} & \rule{2cm}{0.4pt} \\
& & 16324_8 & 7380_{10}
\end{array}
$$

Switching from a 'higher' to a 'lower' system (as from decimal to octal) has three consequences:

1. Fewer digit symbols are required.
2. The numbers will be longer.
3. The calculation rules must be adapted.

The reason to look into the possibilities of other number systems was the fact that with the electrical representation of numbers the basis of the number system in use determines the number of digit symbols that must be represented. We have already seen that from a technical viewpoint it is attractive to work with a system that is as 'low' as possible because in this way the number of voltage levels to be distinguished is kept as small as possible. The lowest possible system is the binary system and therefore the choice has fallen on this system. This does not exclude that there will ever be another system in use. There could be technical reasons to choose a three- or four-valued system but at the moment only the two-valued binary system is used. The consequences of this choice are:

1. Only two digit symbols are required. Normally the '0' and '1' are used here. They are called 'binary digits' or simply 'bits'. For electrical representation two voltages are needed, e.g. 0 volt and 1 volt, but any other pair is valid and in practice the type of circuit determines the choice. Recognizing the voltage level is thus in fact a 'threshold decision'.

2. The numbers are on the average a factor 3.322 longer (see the above example: $1779_{10} \rightarrow 11011110011_2$) which means that for the representation of the same number a larger number of wires is needed.

3. The rules for binary calculations must be applied. This is merely an advantage because these rules are exceptionally simple. With the elementary addition of numbers existing of one digit there are only four possibilities:

$$0 + 0 = 0 \qquad 1 + 0 = 1 \qquad 0 + 1 = 1 \qquad 1 + 1 = 10$$

The binary equivalent of the above addition/multiplication example gives:

```
              111101110                        10110100
                1010100                          101001
   +     ─────────────          ×        ─────────────
              1001000010                       10110100
                                              10110100
                                             10110100
                                        ─────────────
                                        1110011010100
```

As can be seen in this example binary multiplication is also quite simple, because the multiplication is done by the value 0 (result: 0) or by the value 1 (result: same value). Multiplication boils down to adding shifted versions of the multiplicand.

There is one problem concerned with the binary system: for a human being the numbers are difficult to read and to interpret. To see what a series of zeros and ones has to tell us a time-consuming calculation is necessary. As a solution to this problem the octal (or the hexadecimal) system can act as a sort of intermediary system. On the one hand the difference between the octal and the decimal system is not that large. The octal system is not difficult to learn to work with; furthermore there are calculators for conversion into and for working within the octal system. On the other hand the following argument shows that there is a simple connection between the octal and the binary system: it can be seen from the table to the left that if one groups the bits of a binary number into threes then each group can take the values 0, 1, 2, . . . 7. This value must be multiplied by (from right to left) 1, 8, 64, . . . and so in this simple manner the conversion from binary to octal is achieved.

```
0  0  0        0
0  0  1        1
0  1  0        2
0  1  1        3
1  0  0        4
1  0  1        5
1  1  0        6
1  1  1        7

|  |  |
4  2  1
```

$$1\,|\,1\;0\;0\,|\,1\;0\;1\,|\,1\;0\;0\,|\,0\;0\;1\,|\,1\;1\;1_2 \quad = \quad 1\;4\;5\;4\;1\;7_8$$
$$1 \qquad 4 \qquad\;\; 5 \qquad\;\; 4 \qquad\;\; 1 \qquad\;\; 7$$

The opposite is just as simple:
$$1\;2\;3\;7\;4_8 \;=\; 0\;0\;1\,|\,0\;1\;0\,|\,0\;1\;1\,|\,1\;1\;1\,|\,1\;0\;0_2$$

In a similar way the hexadecimal (16-valued) system can be used as an intermediary between the binary and the decimal system. The binary number must now be grouped into fours. Because nowadays the number of bits used in most computers is a multiple of eight, this system is very efficient and only slightly more difficult in practice.

For an electrical calculating machine it is not only necessary to have an electrical representation but it is also necessary that operations can be performed with the 'electrical' numbers. We have already seen that calculating with binary numbers is very simple. Because of this the construction of circuits for such calculations is not difficult either. With a few components it is possible to design a circuit for elementary binary additions. This circuit has two inputs, one output for the sum and one for the 'c'-bit (from 'carry').

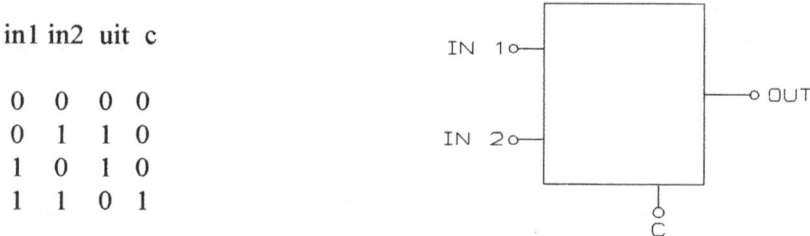

in1	in2	uit	c
0	0	0	0
0	1	1	0
1	0	1	0
1	1	0	1

Figure 4.1.2 Half adder.

Based on this circuit a real 'adder' can be made with two series of input wires and one series of output wires, with a bit pattern equal to the sum of the two input bit patterns (fig.4.1.3). With a similar, but more complicated circuit, designed to shift and add bit patterns, multiplication of two binary numbers can be realized.

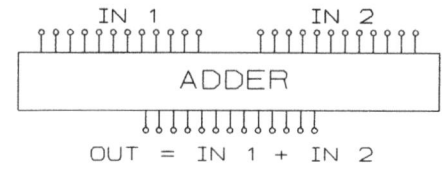

Figure 4.1.3 Complete binary adder.

Also the second condition, the possibility of carrying out calculations, is fulfilled in this way. In principle this is enough to realize an electrical calculator. Although this possibility has been known for some time, the actual development was not begun until after the second world war, a development that has led to a variety of products, from simple calculators for elementary operations to large and incredibly fast programmable calculation machines, the computers.

A computer not only processes binary numbers but can also store them in its memory. A memory consists of a collection of memory elements. Each element is a series of 'cells', simple circuits with the characteristic that they can receive and hold a 0/1-voltage connected to them, so that after the original voltage is taken away, the binary information remains. Such memory elements appear isolated in the processing unit of the computer and are called registers; the actual memory contains a large number of these elements named here words. The purpose of the memory however is not only (and even not in the first place) the storing of numbers during calculations. The crucial moment in the development of the computer came when it was realized that the manipulations necessary for the subsequent steps of a calculation (the transfer of numbers from the memory to the arithmetic registers, and the transfer of the results of the calculation back to the memory etc.) are basically electrical. Therefore it is possible to code these tasks themselves as electrically represented numbers, which can then be stored in the memory as well. The numbers are interpreted by the computer as instructions and the collection of instructions is called the program. If one pushes the button on a calculator with the designation SIN then a program is started which based on the previously entered number calculates:

$$x - \frac{x^3}{3!} + \frac{x^5}{5!} - \frac{x^7}{7!}$$

The number x is thus multiplied a number of times by itself and by a constant factor, and after that the terms are added. The operations are beforehand coded and stored in the memory so that they can be executed very rapidly one after another. With ordinary calculators it is possible to use only such inbuilt programs (for SIN, COS, TAN, LOG, etc.). With a computer it is possible for its users to write their own programs and store these in the computer for execution. This is not the place to delve deeper into the working of computers, therefore a few final remarks.

Clearly, when designing a computer, it is necessary to choose the number of 'wires', the amount of bits to be used. As mentioned earlier the amount of bits that is normally used (the word length) is nowadays a multiple of 8 bits. With a 16-bit computer the largest number that can be represented is a series of 16 'ones':

1 111 111 111 111 111

To see just how large this number is one can convert it into an octal number (177777_8). Here it is easier first to add a 1 to this:

```
  1 111 111 111 111 111
                      1
+ ─────────────────────
  10 000 000 000 000 000
```
$= 1 \cdot 2^{16} = 65536$

The original number was 1 smaller and thus equals to 65535.

Obviously a computer should be capable to represent both smaller and larger numbers. This is, for example, done by using 64 bits with four concatenated numbers of 16 bits. With extremely large and small numbers an exponential representation is used. The exponent is stored separately. To represent negative numbers it is necessary to sacrifice one of the bits in order to indicate the sign. In an 8-bit system +22 is written as 00 010 110 and the number -22 could thus correspondingly be represented by 10 010 110. The bit most to the left (= Most Significant Bit, MSB) serves as sign bit. The negative number itself however is usually represented in another manner following the method known as two's complement. One then reverses not only the sign bit but all other bits as well.

With -22 we then get:

11 101 001 (one's complement)

and then one adds to this number 1:

11 101 010

An advantage of this method of representing negative numbers is that subtraction can be performed via addition of the negative value:

```
  +22        00 010 110
  -22        11 101 010
+ ──       + ──────────
   0       (1)00 000 000
```

We could also use this method in the decimal system. Here the number -22 would in its complementary form (and a word length of 6 places) be written as 999978; the addition of +22 and -22 would look as follows:

```
        000022
        999978
      + ───────
      (1)000000
```

4.2 Time-discrete signal functions

A. *Linear Pulse Code Modulation*

1. Analog to digital conversion.

Our original purpose was to make use of the 'electrical' numbers and their principles for the representation of signal functions. The advantage of this is, as we have seen, the increased invulnerability of the signal because of a greater redundancy, and the possibility of operating on a signal via numerical calculation processes. The price we have to pay for this is that we must work with complex technical systems (computers), which, however, thanks to technical development have become easily available. In the second place it is necessary that we change the signal function from an infinite collection of real number pairs into a finite collection of pairs of integers. We will now deal with the practical and theoretical aspects and their consequences.

In principle we can go on to digitize a signal as follows: we make a pen registration of the function, mark points along the time axis that, for example, lie 25 μs from each other and read the corresponding displacement along the vertical axis for each of the time points. We import the number table that results into a computer via the keyboard. This method is inexact, sloppy and time consuming and is only to be used with a very small number of samples if no other method is available. The same technical development that has lead to the computer has also produced apparatus that can perform this sampling process quickly and accurately. The apparatus that does this is called an *analog-to-digital converter* (ADC, see fig.4.2.1). It has, as shown in the figure, one input to which the (analog) signal voltage is attached. There is one output that consists of a number of parallel wires with which a binary electrical number can be represented. There is a clock input to which a short electrical impulse is connected which serves as a start signal for the converter.

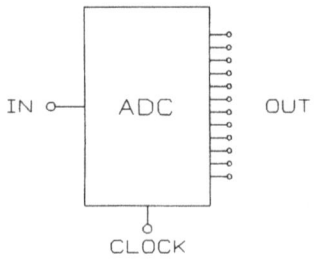

Fig. 4.2.1
Analog-to-Digital converter.

In a very short time the input voltage is measured and a binary number appears at the output representing the result of that measurement. This numerical value does not have to be equal to the voltage value. It is enough that it is proportional to it as will be shown with an example further on.

First this: the conversion process is completed in a matter of a few microseconds. During this time the input voltage must not change. To prevent this from happening a sort of buffer called *Sample and Hold* is used which is usually a component of the AD-circuit. (In chapter 6 I will go deeper into these circuits.)

And another point: the word 'clock' indicates a repeating process. Connected to the clock input is a pulse generator with exact and constant time intervals between the pulses by which also the time between the samples is exactly set. In place of the time interval Δt it is also possible to specify the sample frequency f_s, its reciprocal value. If the time interval is 25 μs, then f_s is equal to 40 kHz.

As an example let us have a look at an ADC that accepts voltage inputs between 0 and 10 volts and gives 12-bit numbers (a '12-bit ADC'). This is an existing type of converter but there are other ones with different word lengths and other voltage ranges (for example from -5 to +5 volts). The two voltage limits mean that with 0 volt at the input the smallest number (a series of 12 zeros) is given, and with 10 volts the largest number, a series of 12 ones, occurs:

$$111\ 111\ 111\ 111_2 = 7777_8 = 4095_{10}.$$

The range between 0 and 10 volts is divided into 4096 equal parts and to each of these parts one of the numbers between 0 and 4095 is assigned. The size of these voltage intervals is indicated with q, the quantization interval.

The value of q is in our case:

$$q = \frac{10}{4096}\ V = 2.4414...\ mV$$

With a certain voltage V and a word length B we have:

$$q = \frac{V}{2^B} \qquad (4.1)$$

With a specific voltage V the number value $N = [V/q]$ corresponds. The brackets indicate that any decimals after the decimal point are omitted. For example:

$$V = 2\ volt,\ N = 819.2 = 819_{10} = 1463_8 = 001100110011_2$$

$$V = 6\ volt,\ N = 2457.6 = 2457_{10} = 4361_8 = 100011110001_2$$

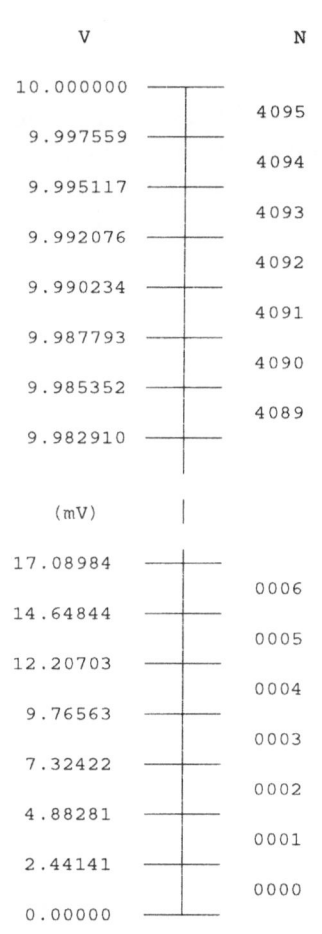

V	N
10.000000	4095
9.997559	4094
9.995117	4093
9.992076	4092
9.990234	4091
9.987793	4090
9.985352	4089
9.982910	
(mV)	
17.08984	0006
14.64844	0005
12.20703	0004
9.76563	0003
7.32422	0002
4.88281	0001
2.44141	0000
0.00000	

Reversely, with the number value N a voltage V with $V = N \cdot q$ corresponds.

Examples: $N = 1000_{10}$, $V_N = 2.4414063$ volt
             $N = 2840_{10}$, $V_N = 6.9335938$ volt

If we start with a voltage of 0 volt, and gradually increase it nothing happens at first. N stays at 0 until V crosses the value 2.4414 mV. At this moment N changes into 1 and stays so until V goes beyond $2q = 4.88281$ mV and so on. In the table the limits of the quantization intervals together with the corresponding numerical values are shown for the lower and upper part of the 0 - 10 volt range.

2. Quantization noise.

If we supply the AD converter with a time-varying voltage it is as if the voltage that is present at the moment of the clock pulse is rounded-off to the lower limit of the corresponding interval. In fig.4.2.2 this process is shown. Important here is that the amount of rounding-off fluctuates at random between 0 and 2.4414 mV.

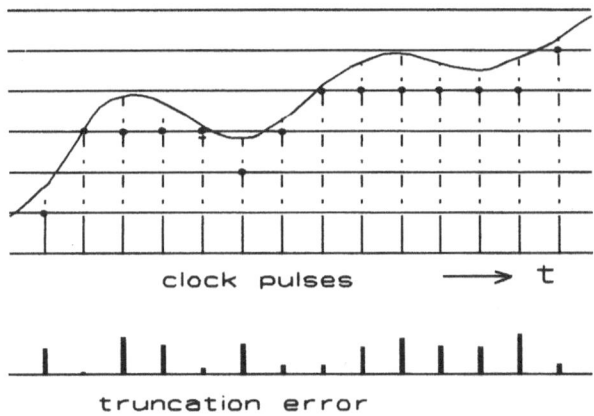

clock pulses   ➡ t

truncation error

Figure 4.2.2 Quantization.

In other words: the effect of the rounding-off can be described as the adding of a random voltage between 0 and -2.4414 mV. Such an uncorrelated signal is of course not periodic and must be considered a noise signal. It is indeed so that due to the rounding-off of the signal a small quantity of noise (quantization noise) is added to the signal that, under certain circumstances, can become audible. The sound level of this noise signal is in general very low, provided that the the word length is not too small.

The noise level can be specified via the signal-to-noise ratio (see section 2.2.D):

$$SNR = 20 \log \frac{y_{RMS}}{r_{RMS}} \qquad (4.2)$$

It is impossible to calculate r_{RMS} in the ordinary way with (2.28) because we do not have a function rule of $r(t)$. Still something is known about $r(t)$: the function values lie between 0 and q and all values between these two limits are equally probable.

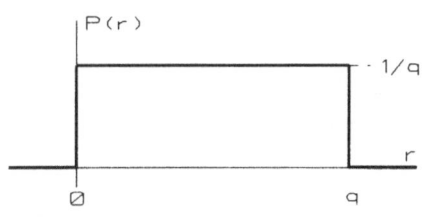

Figure 4.2.3 Probability density.

This is shown in the graph of fig.4.2.3: the probability $P(r)$ of a certain noise signal value is constant for $0 < r < q$ and 0 elsewhere. Furthermore the value of the constant can be determined by the consideration that the area of this graph represents the sum of all probabilities and this sum should be 1. As the horizontal side of the rectangle has length q, the vertical side must have length $1/q$. A graph like this one is called a probability density function. See section 4.4.D.

We now know which values $r(t)$ can have but we do not know in which order they appear. For the calculation of the RMS value the order is however not relevant. We can calculate r_{RMS} with formula (2.32) using the fact that this signal is asymmetrical with a mean value of $\frac{1}{2}q$.

$$r_{RMS} = \sqrt{\overline{r^2} - \frac{1}{4}q^2}$$

$$\overline{r^2} = \int_0^q r^2 P(r)\, dr = \int_0^q r^2 \frac{1}{q}\, dr = \frac{1}{q \cdot 3} r^3 \Big|_0^q = \frac{q^2}{3}$$

$$r_{RMS} = \sqrt{\frac{q^2}{3} - \frac{q^2}{4}} = \frac{q}{\sqrt{12}} \qquad (4.3)$$

Let us furthermore assume that the signal fits exactly within the voltage range of the converter. This means for our example that the signal should be symmetrical around 5 volt (thus in fact an asymmetrical signal) with a peak value $y_p < +5$ volt and > -5 volt. From (4.1) we find

$$q = \frac{2y_p}{2^B}$$

and thus

$$r_{RMS} = \frac{2y_p}{2^B\sqrt{12}} = \frac{y_p}{2^{B-1}\sqrt{12}} = \frac{y_p}{2^B\sqrt{3}}$$

And so the SNR is

$$SNR = 20\log\frac{y_{RMS}}{\dfrac{y_p}{2^B\sqrt{3}}} = 20\log 2^B + 20\log\sqrt{3} - 20\log\frac{y_p}{y_{RMS}}$$

$$= 6B + 4.77 - 20\log\frac{y_p}{y_{RMS}}\ dB \qquad\qquad (4.4)$$

Each increase of the word length with one bit improves the resolution with a factor 2 = 6 dB. Rule (4.4) is often simplified to the well-known rule of thumb: SNR = 6*B* but in this version the role of the proportion between peak value y_p and RMS value y_{RMS} is neglected. With a sinusoidal signal this proportion is $\sqrt{2}$ (see (2.55)) and then the SNR is 6*B*+1.76 dB. With natural signals there is no such fixed proportion. To avoid overmodulation in speech (a very 'peaky' signal) for example, it is necessary to reserve for y_p a range of ca. $4y_{RMS}$, leading to the considerably lower SNR value of 6B - 7.28 dB. The final choice of *B* is determined by its application. For high quality digital music recording 16 bits are used, corresponding to a SNR of ca. 96 dB. For professional applications even higher values ($\geq$ 20 bits) are used. For the analysis of music and speech signals 12 bits are normally sufficient. This word length is also required for the digital transmission of intelligible speech signals.

With very small signal amplitudes the result of the rounding-off has another consequence: with a sinusoidal signal with an amplitude of ca. ½q the converted signal consists of a regular alternation of two sample values, which also occurs with a square wavelike input signal. The change of a sinusoidal signal into a square-like signal indicates a radical distortion of the signal as we shall see. The effect is heard as 'granulation noise'.

The fact that we deal with finite numbers with a limited supply of possible numbers leads to rounding-off which manifests itself as noise. We can also ask what the consequence is of the fact that we also work with a finite number of time points along the time axis. What could possibly go wrong can be seen in fig.4.2.4. The 'bend' in the signal function will not be detected with a low sample frequency (the thick lines) but will be detected with a higher sample frequency (thin lines). Thus there is a relation between the form of the signal function and the sample frequency. For an exact formulation of this relation we need a method by which we can analyse the wave shape. Such a method is spectral analysis that will be discussed in the following section.

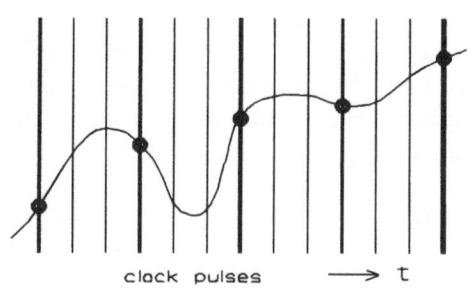

clock pulses ———→ t

Therefore here only a provisional formulation of this relation follows, which will be discussed in more detail further on: the sample frequency must be twice as high as the highest sine frequency that can occur. In this form the statement deals only with sinusoidal signals but in the following section we will see that arbitrary signals can be reduced to sinusoidal ones.

Figure 4.2.4 Sampling.

3. Digital-to-Analog Conversion.

Finally there is the question how we can change a digital signal function into an analog one. Because it is nearly always the intention that the analog signal is in 'real time' (which means the time relation must be the same as with a normal analog signal) this implies we must work here with a circuit that is automatic and such an apparatus indeed exists. It is called a digital-to-analog converter (DAC) and is basically the mirrored version of the AD converter (see fig.4.2.5). The

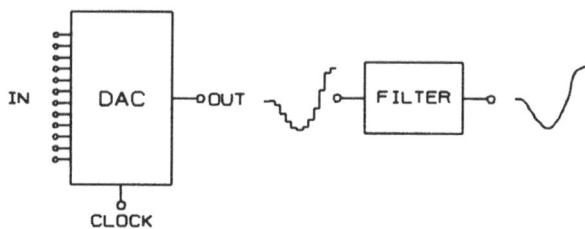

Figure 4.2.5 Digital to Analog conversion.

digital signal is connected to the input. Then with the help of a (clock) pulse the conversion process is started. The result is an electrical voltage that is proportional to the 'electrical number' which is sent to it, and this voltage appears at the output.

In a strict sense this should be an (infinite) narrow impulse, synchronous with the clock pulse, but in practice the voltage level is kept constant until the following clock pulse. The output signal then becomes a staircase signal. With the help of a low pass filter this is changed into a 'smooth' curve. I will return to these items (the staircase and the filter) later in this chapter. With respect to the relation between number value and voltage, the same applies to the AD-converter.

The described method of converting an electrical voltage into a bit pattern is designated with the abbreviation PCM, which stands for Pulse Code Modulation, and because the voltage range is divided into 2^B equal parts (B = word length) one speaks of 'Linear PCM'. This suggests the existence of other conversion systems. These are the subject of the next section.

B. *Other conversion systems* (Blesser, 1978)

For an optimal sound reproduction 16-bit linear PCM is made use of. This does indeed give excellent results but it is a rather 'expensive' process because of the number of bits per second that has to be processed. With a sample frequency of 50 kHz this number is $2 \cdot 50000 \cdot 16 = 1600000$ bits per second. The high quality is not always needed; for the digital registration of speech, for example, lower standards can be kept to. In such cases it would be possible to decrease the word length to say 12 bits but there are specialized conversion systems with a smaller number of bits that come very close to the 16 bit linear system in quality, on condition however that there is a priori knowledge about the signal to be processed, for example the fact that it is a speech signal. If information about the signal source is used for an improved coding scheme we speak of *source coding*. A system like PCM that does not ask for information about the signal is indicated with the name *waveform coding*. There are also techniques that combine both approaches. Source coding is most efficient, but imposes restrictions to the application possibilities. What kind of a priori knowledge is needed? Useful information is the fact that in speech and music signals the dynamic variations are relatively slow, or that they sometimes exhibit regular waveform patterns. In slightly more official terms: the 'short time' RMS value fluctuates considerably and the signals display a certain degree of correlation.

In the first case some form of signal compression/expansion can be used, in the second case predictive coding may be applied. It is also possible to take into account certain properties of the hearing organ. Signal components that are 'masked' by other components can for example be omitted. This technique will be discussed in the final chapter of the book.

1. Compression/expansion.

With linear PCM the noise level is fixed. It is determined by q as the RMS value is $q/\sqrt{12}$. We should make q so small that even with a weak signal the SNR is large enough. With strong signals q is then unnecessarily small. We can actually work more efficient by not giving q a fixed value but to let it depend on the amplitude of the signal. The same effect can be achieved by using compression/expansion ('companding'). Here weak parts of the signal are extra amplified before conversion. This reduces the proportion between peak and RMS value y_p/y_{RMS}. According to (4.4) this gain may be used to reduce B. The simplest way would be to apply an analog companding system, like those used in analog cassette players for noise reduction. Here a continuously variable gain factor (controlled by the signal level) with a certain delay time accomplishes an increase of the level of weak signal fragments only when a certain duration is exceeded.

Another possibility is to use an amplifier with a small number of discrete gain factors. Which factor is selected depends again upon the signal level, and the selected factor is coded as a binary number coupled to the sample value. A specific sample value is thus represented by two numbers: the number produced by the AD-converter, and the number that stands for the gain factor. This resembles the exponential coding of numbers involving a value ('mantissa') and an exponent and is called 'floating point' representation. This conversion method is correspondingly named 'floating point conversion'. With a 10-bit mantissa and a 3-bit exponent to distinguish between 8 gain factors (for example 0 dB, 6 dB, 12 dB etc.) a SNR of 102 dB can be achieved. With linear PCM 17 bits are required for this.

The consequence of a varying gain (equivalent with a variable q value) is that the noise level is no longer constant but depends upon the signal level. Normally this is inaudible due to the masking effect of the sound signals. With some signals that have large amplitude fluctuations but a small masking effect (for example strong low tones) this so-called *modulation noise* can be audible and annoying.

An attractive alternative is the purely digital version of this principle, *block coding*. A signal is sampled with for example a 16-bit ADC. Then a group of subsequent samples is examined. If the bits most to the left ('most significant bits' or MSB) bits of all samples are equal to zero, all bit patterns are shifted one position to the left, and this is repeated until at least one MSB is unequal to zero. Then the word length is reduced by omitting, for example, the first six bits on the right (the least significant bits). The remaining ten bits words are transmitted together with the size of the left shift.

Still another method is to use an amplifier that amplifies less when the signal amplitude increases ('instantaneous companding'). Thus, no switching between fixed gain factors but a continuously variable gain without any delay time. This causes a serious nonlinear distortion of the signal that is cancelled by applying exactly the opposite distortion at the end of the chain. The shape of the transfer function should of course be exactly defined. In Europe this relation is fixed in the so-called 'A-law' and in America in the 'μ-law'. See fig.6.4.5 and 6.4.6.

2. Predictive coding.

If the signal function has a more or less regular structure it is possible to predict its course to a certain extent. The principle of predictive coding is shown in fig.4.2.6, left: the signal *s* is sent through a circuit which, based on the current and/or past signal values, can predict the next value. The new predicted signal value *p* is compared with the actual new value by subtracting the latter from the former.

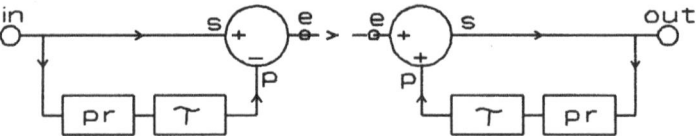

Figure 4.2.6 Predictive Coding.

The difference between the two is called the 'error' *e* (*e = s - p*). For transmission we use the *e*-signal instead of the original one, which we can reconstruct on the side of the receiver by means of the predictor. This generates *p*. To this we add *e* and get *s* again: *s = e + p*. All of this, though, only makes sense if the error signal *e* has a smaller dynamic range and can thus be represented with fewer bits than required for the original signal *s*. With speech signals a particular version of predictive coding, LPC or Linear Predictive Coding has proven to be very efficient. See section 7.3.

Differential coding.

The simplest prediction is that the 'new' signal value is equal to the previous one. Then block 'pr' disappears from the diagram and *e* is equal to the difference between two consecutive signal values. In normal speech and music signals the low frequencies contain more energy than the higher ones. In those cases this way of coding is indeed more efficient than linear PCM.

Delta modulation.

The dynamic range of error signal *e* can be further reduced by increasing the sampling frequency because then the difference between consecutive samples will decrease. Using very high sampling frequencies allows one bit coding (+1: the signal value increases, -1: the signal value decreases) and this form of differential coding is called *delta modulation*. The attractive aspect of delta modulation is that the technical realization is very simple (fig.4.2.7).

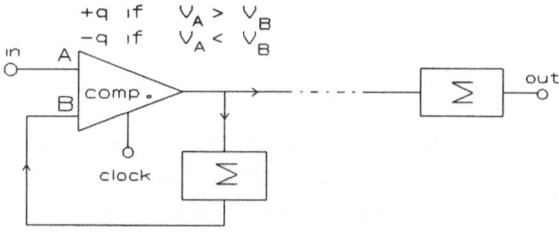

Figure 4.2.7 Delta modulation.

The modulator consists of just two components, a comparator and an adder (indicated with Σ). Fig.4.2.8 shows that V_B is a staircase-like approximation of V_A. The decoding circuit consists of a second adder that generates V_{out}, which is thus identical to V_B as it is derived from the same bitstream.

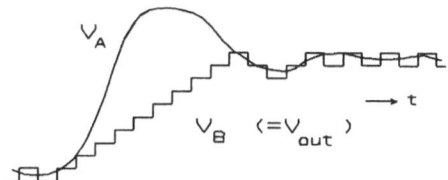

Figure 4.2.8 Input and output signal with delta modulation.

The two problems that appear with delta modulation: steep flanks cannot be followed and the substitution of a constant voltage by a square wave (with granulation noise as a result) can be partially solved by making the step size variable (larger with steep flanks and smaller with slow fluctuations). This is called adaptive delta modulation.

4.3 The Fourier Transform

A. The relation between arbitrary and sinusoidal functions.

At the beginning of the 19th century the French mathematician Fourier discovered the possibility of describing an arbitrary function as a sum of sine functions. (See also Hsu, 1970) The principle of this is depicted in fig.4.3.1:

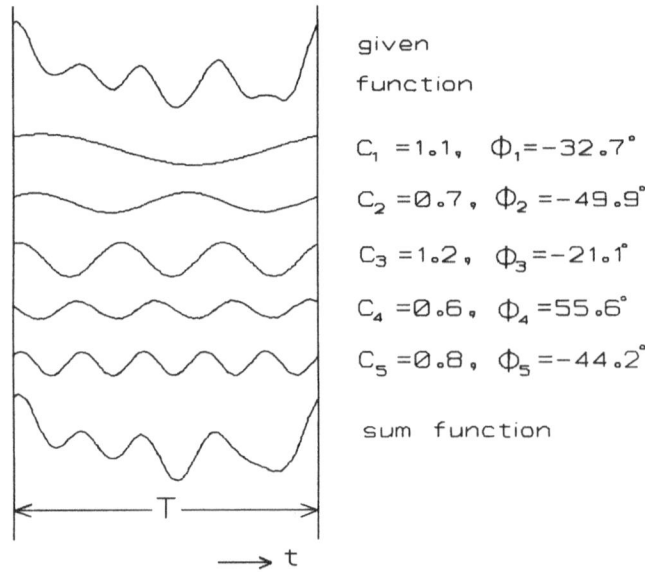

given function

$C_1 = 1.1$, $\Phi_1 = -32.7°$

$C_2 = 0.7$, $\Phi_2 = -49.9°$

$C_3 = 1.2$, $\Phi_3 = -21.1°$

$C_4 = 0.6$, $\Phi_4 = 55.6°$

$C_5 = 0.8$, $\Phi_5 = -44.2°$

sum function

Figure 4.3.1 Fourier analysis.

The topmost curve is the given function that has an arbitrary shape; below this function five sine functions are shown, consisting of 1, 2, 3, 4 and 5 complete periods respectively. A similar systematic statement cannot be made about the five amplitudes and initial phase angles; these are independent of each other. The sum function that is the result of adding the 5 sine functions is shown below and looks very much like the given function.

The theory of Fourier says that the dissimilarity between the given function and the sum function can be reduced by bringing more sine functions into play. Let us write down the function rules of the five sine functions, using the fact that the total duration of the given function is equal to T. As usual the frequency f corresponds with T via $f = 1/T$.

A sine function with period duration T, amplitude C_1 and initial phase angle ϕ_1 is described by:

$$C_1 \cos\left(2\pi \frac{t}{T} + \phi_1\right) = C_1 \cos\left(2\pi ft + \phi_1\right)$$

This is the function rule of the first of the five sine functions. For the second it is:

$$C_2 \cos\left(2\pi \frac{t}{\frac{1}{2}T} + \phi_2\right) = C_2 \cos\left(2\pi 2ft + \phi_2\right)$$

In the same way we find for the following three functions:

$$C_3\cos(2\pi 3ft + \phi_3)$$

$$C_4\cos(2\pi 4ft + \phi_4)$$

$$C_5\cos(2\pi 5ft + \phi_5)$$

Thus the sum function is: $\quad y'(t) = \sum_{n=1}^{5} C_n \cos\left(2\pi nft + \phi_n\right)$

Such a series is called a *Fourier series*. If we for the moment forget the difference between the function $y(t)$ and the sum function $y'(t)$ then we can say that in this way we have found a function rule for the given, arbitrary function. Using an appropriate number of sine functions we ourselves can determine how 'good' the approximation will be. Therefore we write the sum function simply as $y(t)$ without the prime.

In fig.4.3.2 on the left again a summation of the same type is to be seen, again with five sine functions, and the sum function shown at the top. According to rule (2.56) the sum of a cosine and a sine function each with its own amplitude can be written as one sine function with an amplitude and initial phase angle that depend on the original amplitudes.

Vice versa, a sine function with a certain amplitude (C) and initial phase angle (ϕ) can naturally also be split into a cosine and a sine function with amplitudes respectively equal to a and b, with $a = C \cos \phi$ and $b = - C \sin \phi$.

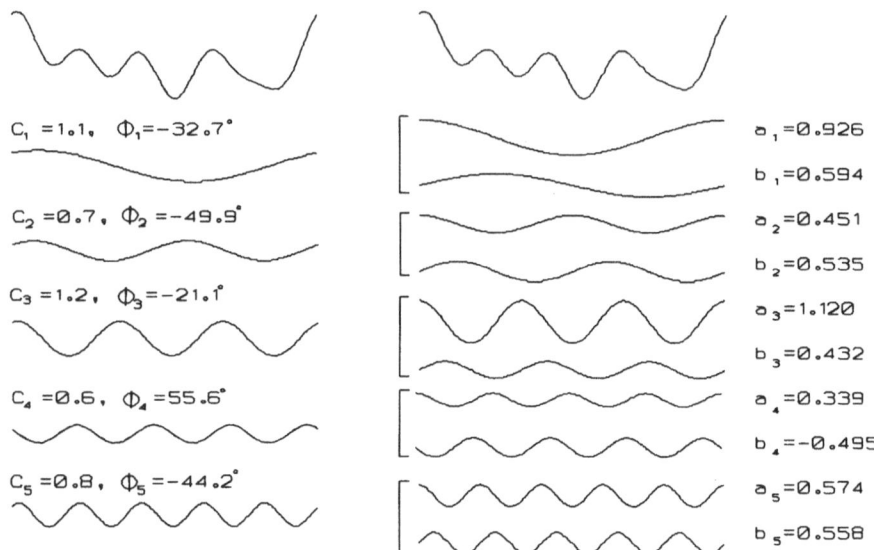

$C_1 = 1.1, \quad \Phi_1 = -32.7°$

$C_2 = 0.7, \quad \Phi_2 = -49.9°$

$C_3 = 1.2, \quad \Phi_3 = -21.1°$

$C_4 = 0.6, \quad \Phi_4 = 55.6°$

$C_5 = 0.8, \quad \Phi_5 = -44.2°$

$a_1 = 0.926$

$b_1 = 0.594$

$a_2 = 0.451$

$b_2 = 0.535$

$a_3 = 1.120$

$b_3 = 0.432$

$a_4 = 0.339$

$b_4 = -0.495$

$a_5 = 0.574$

$b_5 = 0.558$

Figure 4.3.2 The two equivalent Fourier series.

This is exactly what happened with each of the 5 functions in the above figure. Corresponding to each of these functions on the left-hand side, a sine and a cosine function are drawn on the right-hand side, which, when added, yield the given function:

$$C_n \cos(2\pi nft + \phi_n) = a_n \cos 2\pi nft + b_n \sin 2\pi nft$$

$$a_n = C_n \cos \phi_n, \quad b_n = -C_n \sin \phi_n$$

$$C_n = \sqrt{a_n^2 + b_n^2}, \quad \tan \phi_n = -\frac{b_n}{a_n}$$

If we apply this to every term of the Fourier series we find the alternative version:

$$y(t) = \sum_{n=1}^{5} (a_n \cos 2\pi nft + b_n \sin 2\pi nft)$$

B. *The constant term*

The summation of sine functions always leads to a sum function that is *symmetrical* regarding the zero line (where, in other words, the average value is zero). This seems to limit the applicability of the Fourier series to symmetrical functions but there exists also a Fourier representation for asymmetrical functions because any asymmetrical function can be written as the sum of a symmetrical function S and a constant term D (see example in fig.4.3.3).

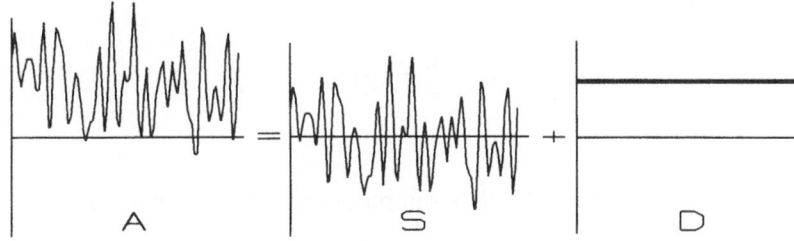

Figure 4.3.3 The constant term.

For an arbitrary symmetrical function a Fourier series exists. For an arbitrary asymmetrical function we need only to add a constant (which, for practical reasons to be discussed later, we shall designate as $\frac{1}{2}a_0$) to find the general form of the Fourier series. By setting the upper limit of the summation to infinity and thus adding an infinite number of sine functions, an exact agreement can be achieved between the given function $y(t)$ and the sum function:

$$y(t) = \frac{1}{2}a_0 + \sum_{n=1}^{\infty} (a_n \cos 2\pi nft + b_n \sin 2\pi nft) \qquad (4.5)$$

or

$$y(t) = \frac{1}{2}a_0 + \sum_{n=1}^{\infty} C_n \cos (2\pi nft + \phi_n) \qquad (4.6)$$

C. *Determination of the Fourier coefficients*

For practical applications of the Fourier series of an arbitrary function it is not enough just to know that we can split a function into sine functions, it is also necessary to know how to determine the amplitude and initial phase angles of the participating sine functions.

It is not possible to directly determine the amplitude and phase coefficients. It is possible though to determine the a_n- and b_n-coefficients and to derive from these the amplitudes and initial phase angles. This is actually why we need the (a_n, b_n)-version of the Fourier series.

Let us begin by determining the constant term a_0. This is in fact easy to find. As we know that $\frac{1}{2}a_0$ is the difference between the average value of the function and 0 we only have to calculate the average value. According to rule (2.56) the average value of the time-continuous function $y(t)$ over the interval 0 - T is equal to:

$$\text{average value} = \frac{1}{T}\int_0^T y(t)\,dt = \frac{1}{2}a_0$$

For a time-discrete function $y(k)$ (with N samples in this time interval) this is:

$$\text{average value} = \frac{1}{N}\sum_{k=0}^{N-1} y(k) = \frac{1}{2}a_0$$

For a_0 we thus have the following two expressions for the time-discrete and the time-continous case respectively:

$$a_0 = \frac{2}{N}\sum_{k=0}^{N-1} y(k) \qquad\qquad a_0 = \frac{2}{T}\int_0^T y(t)\,dt$$

The calculation of the other coefficients is somewhat more complicated but still comparable with that of a_0. The principle is as follows: for the determination of a_n (resp.b_n) the function is multiplied by cos $2\pi nft$ (resp. sin $2\pi nft$), after which the average value of this product is determined.

I shall explain the whole procedure with an example. First I will introduce some abbreviations. Let us replace

the given function $y(t)$		by	Y
the sinusoidal components $C_n\cos(2\pi nft + \phi_n)$		by	H_n
the cosine term $a_n\cos 2\pi nft$ of each component		by	A_n
the sine term $b_n\sin 2\pi nft$ of each component		by	B_n
the function cos $2\pi 2ft$ by which we multiply		by	K_2

Let us suppose that there are 5 components and that we want to calculate the coefficient a_2. To achieve this, we have to multiply Y by K_2:

$$Y \cdot K_2 = (H_1 + H_2 + H_3 + H_4 + H_5) \cdot K_2$$

$$= H_1 K_2 + H_2 K_2 + H_3 K_2 + H_4 K_2 + H_5 K_2$$

$$= (A_1 + B_1)K_2 + (A_2 + B_2)K_2 + (A_3 + B_3)K_2 + (A_4 + B_4)K_2 + (A_5 + B_5)K_2$$

$$= A_1 K_2 + B_1 K_2 + A_2 K_2 + B_2 K_2 + A_3 K_2 + B_3 K_2 + A_4 K_2 + B_4 K_2 + A_5 K_2 + B_5 K_2$$

This procedure is shown graphically in fig.4.3.4. Topleft, the function Y is shown and below Y the 5 cosine and sine terms A_1 to B_5. The function K_2 can be seen in the middle, topright is the product $Y \cdot K_2$ and below all products $A_1 K_2, \ldots, B_5 K_2$. According to the above derivation the topmost product function is thus equal to the sum of the 10 product functions shown below.

The next step is that of determining the average value of the product $Y \cdot K_2$. As the average value of a sum is equal to the sum of the average values of all terms we find:

$$\overline{Y \cdot K_2} = \overline{A_1 K_2} + \overline{B_1 K_2} + \overline{A_2 K_2} + \overline{B_2 K_2} +$$
$$+ \overline{A_3 K_2} + \overline{B_3 K_2} + \overline{A_4 K_2} + \overline{B_4 K_2} + \overline{A_5 K_2} + \overline{B_5 K_2}$$

This result can be drastically simplified because it can be shown that the average value of products of sinusoidal functions of this kind is zero. This is proven in section 2.7, problem 2.15; it is demonstrated in fig.4.3.4: all product functions but one shown to the right are symmetrical around the zero line. The only exception is the average value of the product $A_2 K_2$:

$$\overline{A_2 K_2} = \frac{1}{T} \int_0^T a_2 \cos 2\pi 2ft \cdot \cos 2\pi 2ft \, dt = \frac{a_2}{T} \int_0^T \cos^2 2\pi 2ft \, dt$$

On page 53 we have seen that this integral is equal to $\frac{1}{2}T$, and thus

$$\overline{A_2 K_2} = \frac{a_2}{T} \frac{T}{2} = \frac{1}{2} a_2$$

In the above sum of average values all terms are zero except the third, so:

$$\overline{Y \cdot K_2} = \overline{A_2 \cdot K_2} = \frac{1}{2} a_2 \quad \text{or} \quad a_2 = 2 \overline{Y \cdot K_2}$$

In the same way it is possible to calculate b_2 by means of the average value of the product of the function Y with the function $S_2 = \sin 2\pi 2ft$: $b_2 = 2\overline{Y \cdot S_2}$

As a_0 can be calculated from the average value of Y so it is possible to calculate a_n and b_n from the average value of $Y \cdot K_n$ and $Y \cdot S_n$:

$$a_n = 2\overline{Y \cdot K_n} \qquad b_n = 2\overline{Y \cdot S_n}$$

We can work this out because we know how to calculate the average value of these product. If time-continuous: by means of integrating the product and dividing by T, if time-discrete: by means of adding the samples of the product function and dividing by the number of samples. In this last case K_n is equal to:

$$K_n = \cos 2\pi nft_k = \cos 2\pi \frac{n}{T}k\Delta t$$

If we assume again that there are N samples in the time interval from 0 up to and including T it holds that $T/\Delta t = N$ and K_n becomes: $\quad K_n = \cos 2\pi \frac{n}{N}k$

and in the same way $\qquad\qquad\qquad\qquad\qquad\qquad S_n = \sin 2\pi \frac{n}{N}k$

Now we can formulate the general expressions for a_n and b_n. The complete list of the rules to calculate the Fourier coefficients is shown below:

time–discrete	time–continuous	
$a_0 = \dfrac{2}{N}\displaystyle\sum_{k=0}^{N-1} y(k)$	$a_0 = \dfrac{2}{T}\displaystyle\int_0^T y(t)\,dt$	(4.7)
$a_n = \dfrac{2}{N}\displaystyle\sum_{k=0}^{N-1} y(k)\cos 2\pi \dfrac{n}{N}k$	$a_n = \dfrac{2}{T}\displaystyle\int_0^T y(t)\cos 2\pi nft\,dt$	(4.8)
$b_n = \dfrac{2}{N}\displaystyle\sum_{k=0}^{N-1} y(k)\sin 2\pi \dfrac{n}{N}k$	$b_n = \dfrac{2}{T} y(t)\sin 2\pi nft\,dt$	(4.9)

The time-discrete calculation is called the DFT (Discrete Fourier Transform).

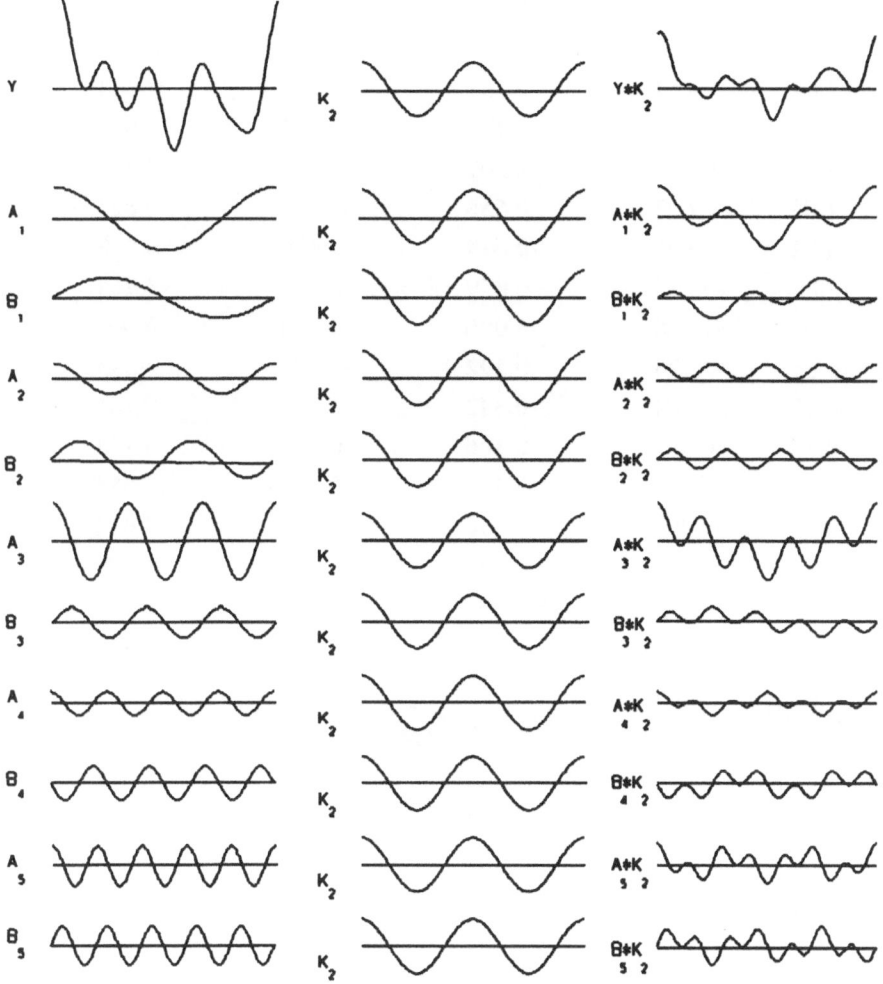

Figure 4.3.4 Determination of Fourier coefficients.

Now follows an example of the calculation of Fourier coefficients with these formulae. Imagine that we wish to derive from a fragment of a time-discrete signal function specified as a list of 14 samples the amplitude and initial phase angle of the third component, thus $N = 14$ and $n = 3$. In the second column a list of samples is shown. Try to find out how the products, sums and average values are calculated.

k	$y(k)$	$\cos 2\pi \dfrac{3}{14}k$	$y(k)\cdot\cos 2\pi \dfrac{3}{14}k$	$\sin 2\pi \dfrac{3}{14}k$	$y(k)\cdot\sin 2\pi \dfrac{3}{14}k$
0	1.42	1.000	1.420	0.000	0.000
1	2.63	0.223	0.585	0.975	2.564
2	4.19	-0.901	-3.775	0.434	1.818
3	3.54	-0.623	-2.207	-0.782	-2.768
4	1.05	0.623	0.655	-0.782	-0.821
5	-0.72	0.901	-0.649	0.434	-0.312
6	-3.86	-0.223	0.859	0.975	-3.763
7	-5.09	-1.000	5.090	0.000	0.000
8	-2.66	-0.223	0.592	-0.975	2.593
9	-0.38	0.901	-0.342	-0.434	0.165
10	0.17	0.623	0.106	0.782	0.133
11	0.71	-0.623	-0.443	0.782	0.555
12	0.95	-0.901	-0.856	-0.434	-0.412
13	1.23	0.223	0.274	-0.975	-1.199
	$+\underline{\quad}$		$+\underline{\quad}$		$+\underline{\quad}$
	3.18		1.309		-1.447

With formula (4.7): $\qquad\qquad\qquad a_0 = \dfrac{2}{14}\cdot 3.18 = 0.454$

With formulae (4.8) and (4.9):

$$a_3 = \frac{2}{14}\cdot 1.309 = 0.187, \quad b_3 = \frac{2}{14}\cdot -1.447 = -0.207$$

From this the amplitude and initial phase angle can be calculated:

$$C_3 = \sqrt{0.187^2 + 0.207^2} = 0.279 \qquad \phi_3 = \tan^{-1}\frac{0.207}{0.187} = 0.836\,r = 47.9°$$

As an example of the calculation of the Fourier coefficients for a time-continuous function we take the signal function that is shown in fig.4.3.5, a (co-)sine function *with a non integer number of cycles* (in this case 4½ period). The period duration of this cosine function is $T/4½$, thus the frequency is $9/2T$ and the function rule is:

$$y(t) = \cos 2\pi \frac{9}{2T}t = \cos 9\pi \frac{t}{T}$$

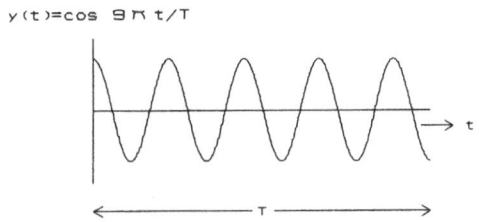

$y(t)=\cos 9\pi t/T$

Figure 4.3.5 A truncated sine function.

When the number of periods would have been integer (e.g. 5) the signal function would have been identical with the fifth Fourier component, thus $C_5 = 1$, $\phi_5 = 0$ and $C_n = \phi_n = 0$ for n ≠ 5. Now, however, our task is to construct this truncated sine wave with sine functions that all have an integer number of periods. Let us calculate the Fourier coefficients.

$$a_0 = \frac{2}{T} \int_0^T \cos 9\pi \frac{t}{T} \, dt = \frac{2}{T} \frac{T}{9\pi} \sin 9\pi \frac{t}{T} \bigg|_0^T = \frac{2}{9\pi} (\sin 9\pi - 0) = 0$$

$$a_n = \frac{2}{T} \int_0^T \cos 9\pi \frac{t}{T} \cdot \cos 2\pi \frac{n}{T} t \, dt =$$

$$= \frac{2}{T} \int_0^T \{ \frac{1}{2} \cos (9+2n) \frac{\pi}{T} t + \frac{1}{2} \cos (9-2n) \frac{\pi}{T} t \} \, dt =$$

$$= \frac{1}{T} \frac{T}{(9+2n)\pi} \sin (9+2n) \frac{\pi}{T} t \bigg|_0^T + \frac{1}{T} \frac{T}{(9-2n)\pi} \sin (9-2n) \frac{\pi}{T} t \bigg|_0^T =$$

$$= \frac{1}{(9+2n)\pi} \{ \sin (9+2n)\pi - 0 \} - \frac{1}{(9-2n)\pi} \{ \sin (9-2n)\pi - 0 \} = 0$$

$$b_n = \frac{2}{T} \int_0^T \cos 9\pi \frac{t}{T} \cdot \sin 2\pi \frac{n}{T} t \, dt =$$

$$= \frac{2}{T} \int_0^T \{ \frac{1}{2} \sin (9+2n) \frac{\pi}{T} t - \frac{1}{2} \sin (9-2n) \frac{\pi}{T} t \} \, dt =$$

$$= -\frac{1}{T} \frac{T}{(9+2n)\pi} \cos (9+2n) \frac{\pi}{T} t \bigg|_0^T + \frac{1}{T} \frac{T}{(9-2n)\pi} \cos (9-2n) \frac{\pi}{T} t \bigg|_0^T =$$

$$= -\frac{1}{(9+2n)\pi} \{ \cos (9+2n)\pi - 1 \} + \frac{1}{(9-2n)\pi} \{ \cos (9-2n)\pi - 1 \} =$$

$$= \frac{2}{(9+2n)\pi} - \frac{2}{(9-2n)\pi} = -\frac{8n}{(81-4n^2)\pi}$$

From $a_n = 0$ follows $C_n = |b_n|$ and $\phi_n = -90°$.

Let us calculate the first 10 C_n values:

n	1	2	3	4	5	6
C_n	0.0331	0.0784	0.1698	0.5992	0.6701	0.2425

n	7	8	9	10
C_n	0.1550	0.1164	0.0943	0.0798

D. *The importance of the Fourier Transform*

Now that the principles of the Fourier series are known, we shall have a look at the practical applications of it. Let us begin with a short glance at its importance. First a remark about the terminology:
- The splitting of an arbitrary function into sinusoidal components is called *Fourier analysis*, or more general the *Fourier transform*.
- The reverse, the addition of sinusoidal components into the 'complex' function is called *Fourier synthesis* or the *inverse Fourier transform*.

The word 'complex' is used here to designate any non-sinusoidal function that thus consists of more than one component.

1. By means of the Fourier transform we can find a function rule for an arbitrary given function which generally contains a finite set of coefficients, making the given function more manageable.
2. Because arbitrary signal functions can be reduced in this way to sinusoidal vibrations it is often sufficient to know what a particular system 'does' with these elementary signals. This 'sine behaviour' determines the way in which other signal functions are processed. I will come back to this in chapter 5.
3. Fourier analysis is a physical reality in the sense that we can split an arbitrary signal with the help of sharp filters (like harmonic oscillators with large Q-factors) into sinusoidal components. We encounter a natural Fourier analyser in the ear. The inner ear contains a filter system that splits a sound signal into sinusoidal components ('overtones').
4. Fourier synthesis is one of the methods for generating synthetical sounds.

E. *Practical applications*

Fourier analysis theory does not say how large the 'analysis window' T must be. Although we are free in principle to choose any T, not every choice makes sense. If for example we arbitrarily segment a signal (see fig.4.3.6), it is possible to apply Fourier analysis to each of these segments but the analysis results are in fact meaningless.

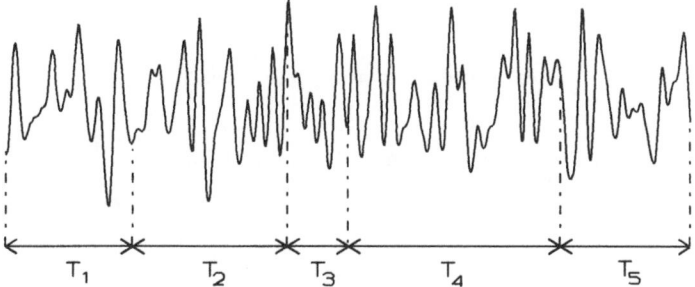

Figure 4.3.6 Arbitrarily chosen analysis windows.

As we shall see the choice of T depends on what we know about the given function (and in particular about its periodicity):
- If the signal is truly periodic T should be made equal to the duration of the period ('pitch synchronous analysis').
- If the signal is not periodic T should be (very or infinite) long.
- The important group of quasi-periodic signals forms a bridge between these two other cases. Both possibilities for T are useful. Which of these is the best choice for T cannot be decided a priori but the second is more often used because the period duration does not have to be known.
- Fourier analysis is also possible with physical methods (thus without calculations). Here also the (possible) period duration does not need to be known. We shall study this case, working with spectrum analysers, separately.

We can now set up the following scheme:

$$Fourier\ analysis \begin{cases} calculation\ method \begin{cases} T = period\ duration \\ T = 'long' \end{cases} \\ measuring\ method \end{cases}$$

The question how to choose the window duration, also plays a role with the measuring method. Usually several time constants can be chosen on the analyser.

F. *Fourier analysis of periodic signals*

1. Amplitude and phase spectrum.

If a signal function is purely periodic (with period duration T) then it is sufficient to analyse one single period. As for the wave shape the next one is identical with the previous one and so the analysis result will be the same as well. As moreover each of the participating sine functions starts and ends at the same point of the cycle (with the same phase angle) these sine functions can be continued endlessly in both directions. Periodic repetition not only holds for the function itself but also for the sinusoidal components (see fig.4.3.7).

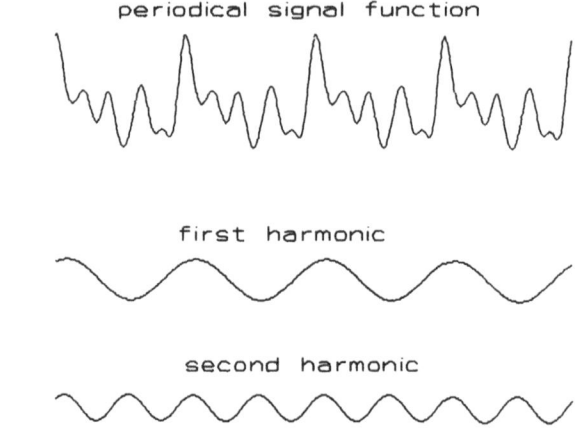

Figure 4.3.7 Fourier analysis of a periodical function.

The sinusoidal components have frequencies that are multiples of the fundamental frequency $f_o = 1/T$ and are called the *harmonics* of the vibration. The Fourier series is now valid for the entire time axis; t may now take any positive or negative value ($-\infty \le t \le \infty$) in the formulas (4.5) and (4.6). The description of the given function thus consists of a list of amplitudes and of initial phase angles, respectively, the *amplitude spectrum* and the *phase spectrum* of the vibration.

A graphical representation of these data, with the frequency as independent variable along the horizontal axis, is illustrative. Because the amplitude and phase values are only found at the frequencies of the harmonics the graphs are bar diagrams or *line spectra*. For the vibration of fig.4.3.1. these spectra look as follows (fig.4.3.8):

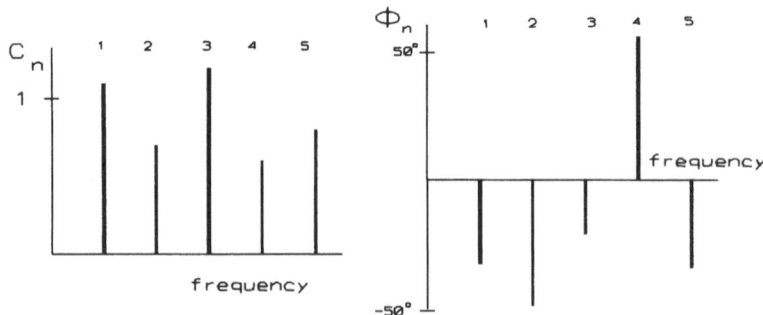

Figure 4.3.8 Amplitude and phase spectrum of the vibration of fig.4.3.1.

2. The power spectrum.

Often the squares of the amplitude coefficients C_n are plotted instead of these coefficients themselves. This leads to what is called the *power spectrum*. The reason for this is that the amount of energy in a signal function can not only be derived from the time function by means of the formula (2.3) but also from the amplitude spectrum based on the *Theorem of Parseval*:

$$\overline{E} = \overline{y^2} = \frac{1}{T}\int_0^T y^2(t)\,dt = \frac{1}{2}\sum_{n=1}^{\infty} C_n^2 \tag{4.10}$$

For the proof of this theorem we replace $y(t)$ in this expression by the corresponding Fourier series:

$$y(t) = \sum_{n=1}^{\infty}(a_n\cos 2\pi nft + b_n\sin 2\pi nft) = \sum_{n=1}^{\infty}(A_n + B_n)$$

in which we use the same short notation as in section 4.3.C.

$$\overline{E} = \overline{\{\sum_n (A_n + B_n)\}^2}$$

$$= \overline{(A_1 + B_1 + A_2 + B_2 + A_3 + B_3 + ...)^2}$$

$$= \overline{A_1^2} + \overline{2A_1B_1} + \overline{2A_1A_2} + \overline{2A_1B_2} + ... + \overline{B_1^2} + \overline{2B_1B_2} + \overline{2B_1A_2} + ...$$

We saw there and in problem 2.15 that all the terms of this series are zero with the exception of the quadratic ones, and as

$$\overline{A_n^2} = \overline{a_n^2\cos^2 2\pi nft} = \frac{1}{2}a_n^2 \; ; \quad \overline{B_n^2} = \overline{b_n^2\sin^2 2\pi nft} = \frac{1}{2}b_n^2$$

we find as the final result:

$$\overline{E} = \frac{1}{2}\sum_{n=1}^{\infty}(a_n^2 + b_n^2) = \frac{1}{2}\sum_{n=1}^{\infty} C_n^2$$

The mean energy of a signal is thus determined by the squared amplitude coefficients. The phase angles are unimportant. Because the power spectrum is related to the energy content, the vertical axis is normally calibrated in dB's.

3. Determination of the amplitude and phase spectrum of periodic signals.

We will now calculate the Fourier coefficients of some periodical signal functions for which we know the function rule. Then the calculation can be performed in the same way as with the example in section 4.3.C, the truncated sine function shown in fig.4.3.5.

- a square wave signal (fig.4.3.9).

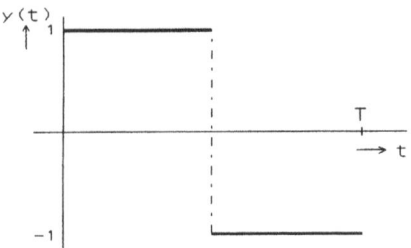

Figure 4.3.9 Square wave.

The function rule is very simple indeed here: $y(t) = +1$ for the t-values between 0 and $\frac{1}{2}T$, and $y(t) = -1$ for the t-values between $\frac{1}{2}T$ and T. The calculation is demonstrated in problem 4.10 and the corresponding solution in the Appendix. The result of the calculation reads as follows:

$$a_0 = 0, \quad a_n = 0, \quad b_n = \frac{1}{\pi n}(2 - 2\cos\pi n)$$

This means:
$b_n = 0$ if n is an even number and $b_n = 4/\pi n$ if n is an odd number. Only the odd terms with b_1, b_3, etc. remain from the Fourier series. We can write this series as follows:

$$y(t) = \sum_{n=0}^{\infty} \frac{4}{\pi(2n+1)} \sin 2\pi(2n+1)ft \qquad (4.11)$$

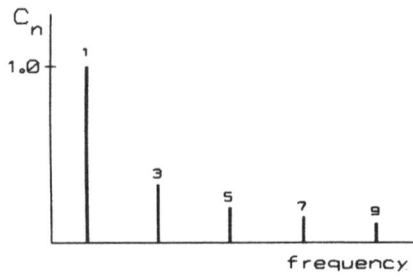

*Figure 4.3.10
Amplitude spectrum square wave.*

Because the a_n-coefficients are equal to zero, C_n is equal to $|b_n|$ and all phases are equal to $-90°$. The spectrum of a square wave signal thus contains only odd harmonics (this explains the rather nasal timbre of this signal) of which the amplitudes are inversely proportional to their number (see fig.4.3.10).
In the same way one can calculate the spectrum of a

- triangular wave signal (fig.4.3.11):

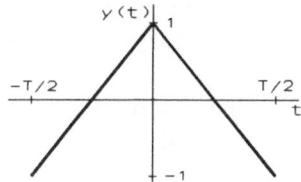
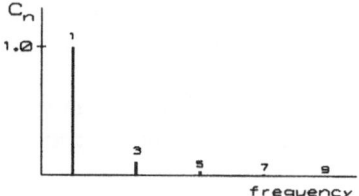

Figure 4.3.11 Time function and amplitude spectrum of a triangular vibration

Function rule: $y(t) = \dfrac{4t}{T} + 1$ if $-\dfrac{1}{2}T < t \leq 0$, $y(t) = -\dfrac{4t}{T} + 1$ if $0 \leq t < \dfrac{1}{2}T$

Fourier series: $y(t) = \displaystyle\sum_{n=0}^{\infty} \dfrac{8}{\pi^2 (2n+1)^2} \cos 2\pi(2n+1)ft$ (4.12)

This signal also consists only of odd harmonics, of which the amplitudes are now inversely proportional to the square of the rank number.

- sawtoothwave signal (fig.4.3.12):

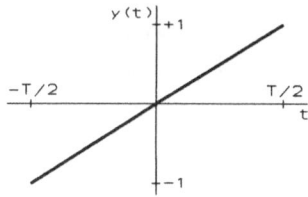
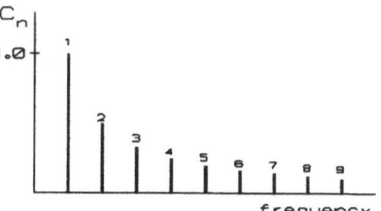

Figure 4.3.12 Time function and amplitude spectrum of a sawtooth vibration

Function rule: $y(t) = \dfrac{2t}{T}$

Fourier series: $y(t) = \displaystyle\sum_{n=1}^{\infty} (-1)^{n+1} \dfrac{2}{\pi n} \sin 2\pi nft$ (4.13)

This spectrum contains 'all' harmonics. As with the square wave signal, the amplitudes are inversely proportional to the rank number. The timbre is rather pleasing and full. The sawtooth vibration was much used in the production of electronic sounds, also because the production of it is so simple. With the help of filters one can derive other timbres from this signal. This technique is called *subtractive synthesis.*

The $1/n$-amplitude factor of the square wave and sawtooth signal implies a 6 dB/octave slope of the spectrum (factor ½ for double frequency), the factor $1/n^2$ a -12 dB/octave slope.

In general it holds that the spectral slope is 6 dB/octave when a signal function has jump discontinuities, and -12 dB/octave when it has bends.

- pulse signal (fig.4.3.13)

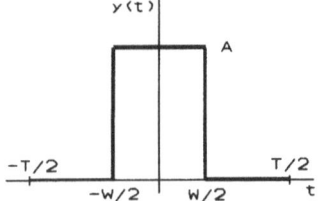
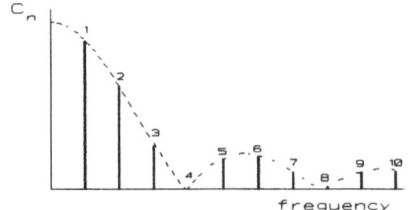

Figure 4.3.13 Time function and amplitude spec-trum of a pulse signal

The function rule is now:

$$y(t) = A \text{ if } -\tfrac{1}{2}W < t < \tfrac{1}{2}W, \quad y(t) = 0 \text{ for other t-values}$$

The impulse signal is characterized by the fact that the pulsewidth W is constant and does not change when T is altered. The calculation of a_0, a_n and b_n gives (see problem 4.18):

$$a_0 = \frac{2AW}{T}$$

$$a_n = \frac{2A}{\pi n} \sin \pi n \frac{W}{T}, \quad b_n = 0$$

or

$$C_n = |a_n|, \quad \phi_n = 0$$

Thus the Fourier series is:

$$y(t) = \frac{AW}{T} + \sum_{n=1}^{\infty} \frac{2A}{\pi n} \sin \pi n \frac{W}{T} \cdot \cos 2\pi nft \qquad (4.14)$$

Like the squarewave signal, the pulse signal has jump discontinuities, but the spectrum is more complex because the expression for a_n not only contains the expected factor $1/n$ but also the factor $\sin \pi nW/T$. Due to this factor the a_n-value can be negative. According to rule (2.42) the minus sign represents a phase shift of 180° and may thus be neglected in a discussion about the amplitude spectrum. A further analysis of the amplitude spectrum is simplified by rewriting the expression for the amplitude coefficient in a slightly more complicated way:

$$ C_n = |\frac{2A}{\pi n} \sin \pi n \frac{W}{T}| = |\frac{2AW}{T} \frac{\sin \pi n \frac{W}{T}}{\pi n \frac{W}{T}}| $$

This amplitude coefficient thus contains the factor sin x/x with $x = n\pi W/T$. The graphic representation of this function looks somewhat like a damped sine wave (see fig.4.3.14), but that impression is wrong. The envelope follows the rule $1/x$ instead of e^{-x}. Plotting the C_n-values leads to a line spectrum, of which the function sin x/x forms the envelope (see fig.4.3.13, the dashed line).

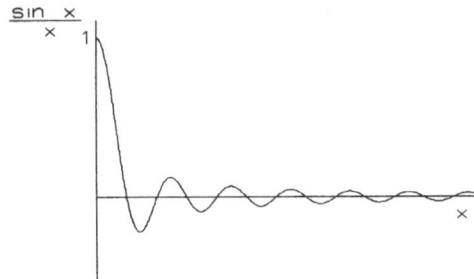

Figure 4.3.14 The function sin x/x.

Another way to construct the amplitude spectrum is to draw the (continuous) spectral envelope first and then fit the spectral lines at the appropriate frequencies.

The function rule for the spectral envelope $E_s(f)$ is: $E_s(f) = \frac{2AW}{T} \frac{\sin \pi fW}{\pi fW}$

The C_n-coefficients can be calculated from this by substituting multiples of $1/T$ for the frequency f: $C_n = |E_s(\frac{n}{T})|$

The function rule for $E_s(f)$ is the product of an (unimportant) scale factor $2AW/T$ and a factor only containing W. This means that with a constant pulsewidth W the spectral envelope is fixed except for the scale factor. This is especially clear when we check the position of the zeropoints of $E_s(f)$. These are located at those frequencies f_z where $\pi f_z W$ is a multiple of 180° or π radians:

$$\pi f_z W = k\pi \quad \text{thus} \quad f_z = \frac{k}{W} \quad (k = 1, 2, 3, \ldots)$$

A change in T leads to a shift of all the spectral lines below the spectral envelope. It could then happen that a harmonic coincides exactly with a f_z making it disappear. This is the case when

$$\frac{n}{T} = \frac{k}{W} \quad \text{or} \quad \frac{W}{T} = \frac{k}{n} = \text{a rational number}$$

The number of the suppressed harmonic is then: $\quad n = k\dfrac{T}{W}$

Examples:

1. $\quad W = \dfrac{T}{5} \rightarrow n = k\dfrac{T}{W} = k\cdot 5$

2. $\quad W = \dfrac{2}{5}T \rightarrow \dfrac{T}{W} = \dfrac{5}{2} \rightarrow n = k\dfrac{T}{W} = 5, 10, 15, \ldots \quad (k = 2, 4, 6, \ldots$

In both cases all harmonics with numbers that are multiples of 5 are suppressed.

3. $\quad W = \dfrac{T}{2} \rightarrow n = 2k = 2, 4, 6, \ldots \quad (k = 1, 2, 3, \ldots)$

All even harmonics are suppressed (see square wave spectrum).

We know that speech sounds (vowels) too have the property that the spectral envelope is independent of the repetition frequency. It is indicated by the term *formant structure*; formants are maxima in the spectral envelope. Each vowel has its characteristic pattern of maxima.

We have seen that lengthening the period of the pulse signal of fig.4.3.13 by moving the $+\frac{1}{2}T$ border to the right and the $-\frac{1}{2}T$ border to the left has no consequences for the spectral envelope; the harmonics shift under this curve to the left. When we are primarily interested in the spectral envelope it is allowed to lengthen the signal by adding zero segments. It can be shown that this is true for any signal: lengthening the signal by adding zero segments changes the position of the spectral components but does not affect the envelope. Later this fact will be used several times. A first example is the VOSIM-signal to be discussed now.

- VOSIM signal (Tempelaars, 1977)

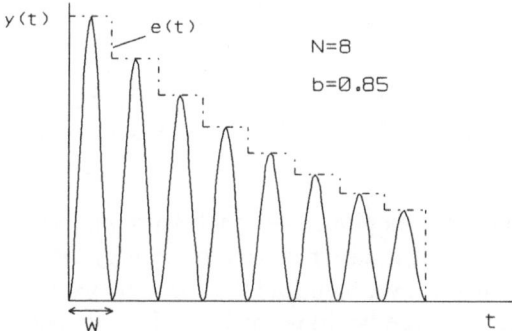

Figure 4.3.15 The VOSIM signal function.

The VOSIM-sound synthesis system (VOice SIMulation) is based on the time function shown above. A period of this signal contains a number of periods of a sine²-function with a staircase-shaped envelope. The proportion between two consecutive levels of the envelope is constant ('b'), making the envelope very similar to an exponentially decaying one. If we indicate the number of sin²-pulses with N, and the duration of each pulse with W, we get the following function rule for the envelope $e(t)$:

$$e(t) = u(t) - b^{N-1} \cdot u(t - NW) + (b-1) \sum_{n=1}^{N-1} b^{n-1} \cdot u(t - nW)$$

Here $u(t)$ is the so-called 'step function':

$$u(t) = 1 \text{ if } t \geq 0 \text{ and } u(t) = 0 \text{ if } t < 0$$

The envelope $e(t)$ must be multiplied by the sin²-pulses. If we square a sine function we get a 'raised cosine' function, because (see rule 2.47):

$$\sin^2\alpha = \tfrac{1}{2}(1 - \cos 2\alpha) = -\tfrac{1}{2}\cos 2\alpha + \tfrac{1}{2}$$

In this new version we have a sinusoidal function with half the amplitude, double the frequency, an offset of ½, mirrored and its phase shifted over ½π. If we designate the new frequency with F it holds that $F = 1/W$ and the function rule for the sin²-pulses is:

$$p(t) = \frac{1}{2}(1 - \cos 2\pi F t)$$

The complete function rule $y(t)$ is now the product of $e(t)$ and $p(t)$. In the same way as with the pulse signal the proportion between the pulsewidth W and the period duration T plays here too, an important role. Let us abbreviate this proportion to γ. With this function rule the spectrum can be calculated. For the amplitude coefficients C_n we find the following result:

$$C_n = \left| \frac{\sin \pi n \gamma}{\pi n (n^2 \gamma^2 - 1)} \right| \sqrt{\frac{1 - 2b^N \cos 2\pi N \gamma + b^{2N}}{1 - 2b \cos 2\pi n \gamma + b^2}}$$

This expression gives the amplitude for each harmonic with the number n. The interpretation of this result is easier if we replace n by the continuous variable x. Just as in the previous case we then find the spectral envelope (see figure below). The actual line spectrum can be drawn as a bar diagram below this contour. This is the preferred method, because here too the important features of the spectral envelope, the minima and maxima, and in particular the pronounced maximum at frequency F, are independent of the repetition frequency. Thus, this spectrum has also a formant structure with one dominating maximum.

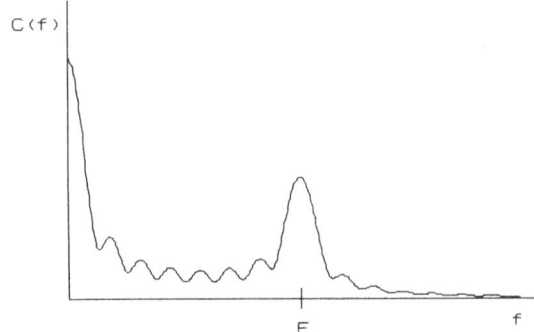

Figure 4.3.16 Amplitude spectrum of the VOSIM signal.

Other characteristics: the spectrum displays minima of which the depth is dependent on 'b'. If $b = 1$ (which means that all $\sin^2$-pulses have the same amplitude) the minima extend up to the zero line. So eventually harmonics with frequencies corresponding to these zero points are totally suppressed. What is more, the spectrum is less wide than the pulse spectrum. It extends from 0 to roughly $2F$ Hz. Higher frequency components are weak and can be ignored.

G. *Fourier analysis independent of (possible) periodicity*

This method is used if the signal is non-periodic, or when nothing is known about the periodicity of the signal.

1. Fourier analysis of time-continuous, non-periodic signals.

How to proceed in order to apply Fourier analysis to non-periodical signals follows from the considerations about the impulse spectrum in the previous section. If we increase the period duration of the pulse signal ever more, the distance between the harmonics (which is equal to $1/T$) will decrease and the frequency of the first harmonic (which is also equal to $1/T$) will approach 0. The spectral envelope however does not change. In the limit situation when T becomes infinite, a continuum of frequencies below this envelope results. Individual components can no longer be distinguished. With our ears we cannot hear any difference either between a periodical impulse with a period duration of e.g. 1 hour and a non-periodical impulse. Mathematically the non-periodical impulse can be considered as a periodical impulse with an infinite long period duration. In the formulae for the Fourier series the sum is therefore to be replaced by an integral.

$$y(t) = \int_0^\infty (a(f) \cos 2\pi ft + b(f) \sin 2\pi ft) df \qquad (4.15)$$

The discrete Fourier coefficients a_n and b_n are replaced by the continuous frequency functions $a(f)$ and $b(f)$. The same holds for the amplitude/phase version:

$$y(t) = \int_0^\infty C(f) \cos (2\pi ft + \phi(f)) df \qquad (4.16)$$

The graphic representation of the amplitude and the phase spectrum shows also continuous curves that must be interpreted as spectral envelopes, instead of bar diagrams. We shall shortly see examples of this. The relation between $C(f)$ and $\phi(f)$ on the one side, and $a(f)$ and $b(f)$ on the other is as before:

$$C(f) = \sqrt{a^2(f) + b^2(f)} , \quad \tan \phi(f) = -\frac{b(f)}{a(f)} \qquad (4.17)$$

while $a(f)$ and $b(f)$ can be calculated with:

$$a(f) = 2 \int_{-\infty}^{\infty} y(t) \cos 2\pi ft \, dt, \quad b(f) = 2 \int_{-\infty}^{\infty} y(t) \sin \tag{4.18}$$

Examples:

a) the spectrum of a single pulse (fig.4.3.17)

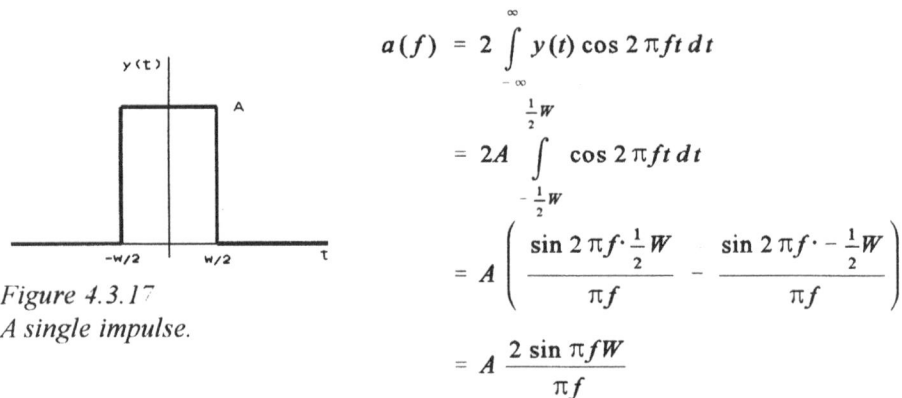

$$a(f) = 2 \int_{-\infty}^{\infty} y(t) \cos 2\pi ft \, dt$$

$$= 2A \int_{-\frac{1}{2}W}^{\frac{1}{2}W} \cos 2\pi ft \, dt$$

$$= A \left(\frac{\sin 2\pi f \cdot \frac{1}{2}W}{\pi f} - \frac{\sin 2\pi f \cdot -\frac{1}{2}W}{\pi f} \right)$$

$$= A \frac{2 \sin \pi fW}{\pi f}$$

Figure 4.3.17
A single impulse.

It can be shown in the same way that $b(f) = 0$, and thus $C(f) = |a(f)|$. If we use the same trick with $C(f)$ as at the time with C_n we can write for $C(f)$:

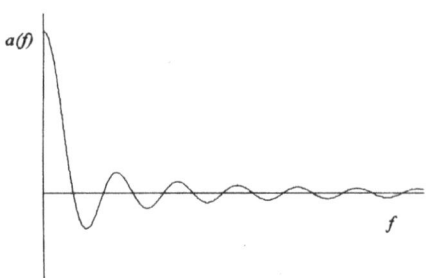

$$C(f) = \left| 2A W \frac{\sin \pi fW}{\pi fW} \right|$$

As expected, the continuous amplitude spectrum of the single impulse (fig.4.3.18) is equal to the envelope of the line spectrum of the periodically repeated impulse.

Figure 4.3.18 Spectrum of a single impulse.

b. the spectrum of an exponentially damped sinusoidal signal.

We encountered this vibration in section 3.3. Then it was noticed that this signal is not periodic. A continuous spectrum is thus the consequence. Fig.4.3.19 shows two damped sinusoidal signals. The first (fig.4.3.19a) is described by the function rule $y(t) = e^{-Pt} \sin 2\pi f t$ and begins at $t = 0$ thus with the value 0. The second signal has the function rule $y(t) = e^{-Pt} \cos 2\pi f t$ and begins with a jump from 0 to 1.

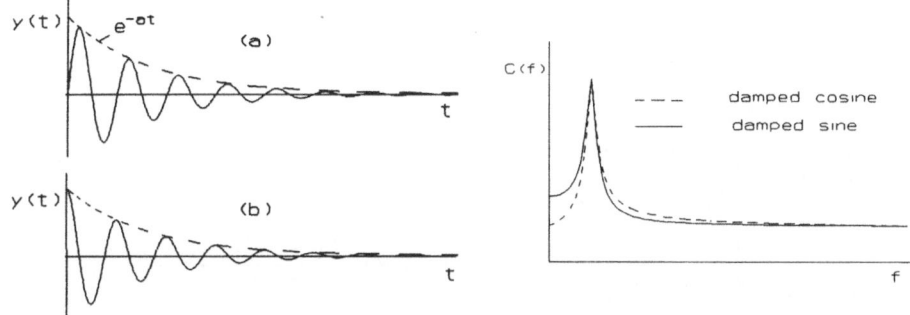

Figure 4.3.19 Exponentially damped sine and cosine vibrations.   *Figure 4.3.20 Amplitude spectrum of damped sinusoidal signals.*

The continuous spectrum of both signals is shown in fig.4.3.20. Both spectra have a sharp peak that coincides with the frequency of the (undamped) vibration and that determines the pitch impression when listening to these signals. The fact that the cosine spectrum has a little bias to the higher frequencies is audible as a clearly pronounced click at the beginning of the sound. The other signal starts with not so sharp a click.

Statements about the relation between the continuous spectrum and the audible impression of the corresponding vibration are also possible with respect to noise signals where due to the non-periodicity we are dealing with continuous spectra as well (see section 4.4). In general it is advisable here to proceed from the signal *energy*. With a periodic signal we take for that reason the power spectrum. With a continuous spectrum

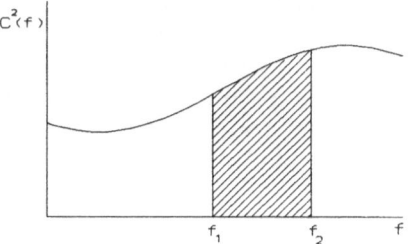

Figure 4.3.21 Power density spectrum.

we can plot $C^2(f)$. Here we speak of the power density spectrum, as we can no longer consider the energy of a particular frequency component, but only the amount of energy between two frequencies, which can be derived from this spectrum by determining, as shown in fig.4.3.21, the area under the curve between f_1 and f_2.

c. The spectrum of a finite sinusoidal signal.

Fig.4.3.22 shows a finite sinusoidal signal consisting of a non-integer number of periods. We already studied a signal like this. See fig.4.3.5. When we calculate the spectrum of this signal for $t = -\infty$ to $t = +\infty$ we get a continuous spectrum shown in fig.4.3.23 with the smooth curve.

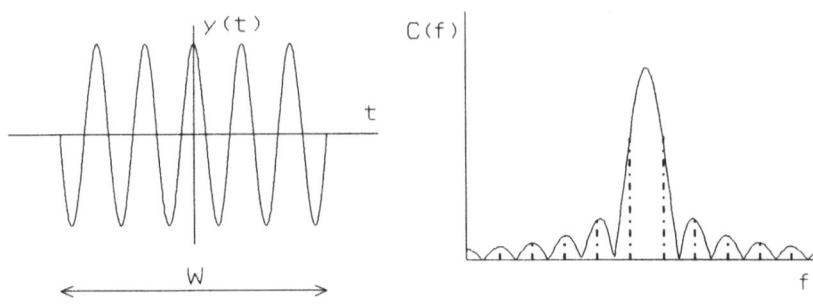

Figure 4.3.22
Finite sinusoidal signal.

Figure 4.3.23 Amplitude spectrum of
the finite sinusoidal signal.

This spectrum exhibits a central peak that would become narrower and sharper with an increasing value of *W*, and would change into a single spectral line when *W* is infinite large. The top of this peak thus corresponds in amplitude and frequency with the sine vibration.

We could also calculate the spectrum for $t = -\tfrac{1}{2}W$ to $t = +\tfrac{1}{2}W$. We then find discrete frequency components, just as in the already mentioned example. They are indicated in fig.4.3.23 with dashed lines. Here we see, as with the pulse spectrum, that the continuous spectrum of the infinite long signal functions as spectral envelope of the discrete spectrum of the finite signal. Let us look again at the pulse spectrum. Because $b(f)$ is equal to 0 we have next to each other:

$$y(t) = \int_{-\infty}^{\infty} a(f)\cos 2\pi ft\, df \quad \text{and} \quad a(f) = \int_{-\infty}^{\infty} y(t)\cos 2\pi ft\, dt$$

The symmetry between the two expressions is obvious. This holds as well for the general expression for the time function and the spectrum function when use is made of complex functions (which in this book are not dealt with). We call the time function $y(t)$ and the spectrum function $a(f)$ a *Fourier transform pair*, and we can symbolically indicate the integral operations used above with the operator F and F^{-1}, thus:

$$a(f) = F\,|y(t)| \quad \text{and} \quad y(t) = F^{-1}|a(f)|$$

This is the most general formulation of the Fourier transform. The Fourier series turns out to be a special case: it is the Fourier transform of a periodic function.

It is surprising that an infinite number of infinitely long sine vibrations must be added to each other to arrive at something as simple as a single pulse. Obviously the 'infinite sum' (integral) gives the value 0 except during the duration of the pulse. This is indeed the case. Let us go back to the integral formula (4.16) and proceed from a particular time point t_0:

$$y(t_0) = \int_0^\infty C(f)\cos\left(2\pi ft_0 + \phi(f)\right) df$$

If we investigate what happens when we vary the frequency f over the integration range we see that $C(f)$ in general changes only slowly with f, while at the same time $\cos(2\pi ft_0 + \phi(f))$ fluctuates quickly between $+1$ and -1.

The product of $C(f)$ with the cosine function fluctuates approximately symmetrically between positive and negative values, and the average value and thus also the integral will indeed be 0 (fig.4.3.24).

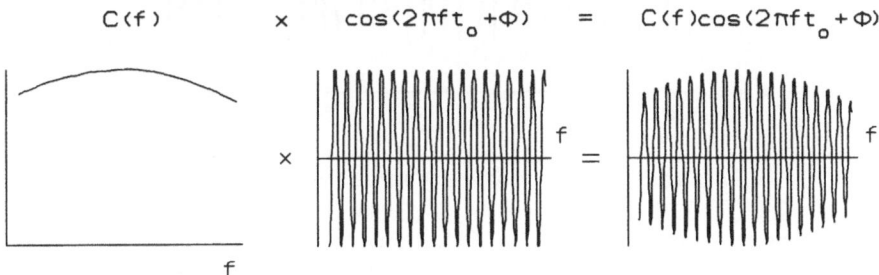

Figure 4.3.24 The principle of stationary phase.

This is not the case for the t-values, where the time function is not equal to 0; thus with the pulse if $-\frac{1}{2}W < t < \frac{1}{2}W$. A possible explanation for this is that $\phi(f)$ depends upon f in such a way that the effect of $2\pi ft$ is compensated for. The whole expression $2\pi ft + \phi(f)$ is then almost constant and the rapid fluctuations fail to occur. This is called the principle of stationary phase (Goldman, 1948). It implies that the derivative of this function to f is equal to 0, and this gives us a method to calculate for which t_p-values this is the case:

$$\frac{d}{df}(2\pi ft_p + \phi(f)) = 0 \rightarrow 2\pi t_p + \frac{d\phi}{df} = 0 \rightarrow t_p = -\frac{1}{2\pi}\frac{d\phi}{df} \qquad (4.19)$$

In this way we can derive from $\phi(f)$ where the 'centre of gravity' of the signal function occurs.

2. Fourier analysis of time-discrete signals with 'hidden' periodicities.

It is again not possible to use the period duration as analysis window T. On the other hand neither is it possible to use a computer calculation to make T infinitely large. In such cases the usual technique is to make T (relatively) large, so that it may be assumed that the chosen signal fragment will contain a (large) number of periods of the unknown signal. Here is an example to clarify this:

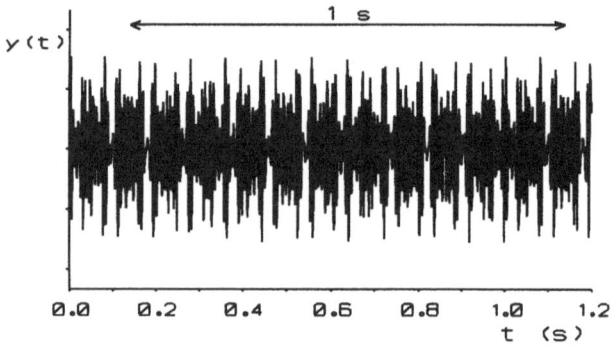

Figure 4.3.25 A mixture of three sine vibrations.

Imagine that the signal is a mixture (addition) of three sinusoidal vibrations of 112 Hz, 148 Hz and 173 Hz. These are not harmonically related in the usual meaning of the word. Let us choose a fragment of 1 second for the Fourier analysis (fig.4.3.25). This means that we assume that there is a fundamental frequency of 1 Hz. If we calculate the amplitude spectrum with the help of the DFT and, for example, go up to the 200-th harmonic, we find the result shown in fig.4.3.26. The three sinusoidal vibrations are encountered as resp. the 112th, 148th and 173th harmonic in this fundamental. In this manner frequencies of sinusoidal components can be traced, and this technique can even be used if the frequency of such a component is not exactly a multiple of $1/T$ Hz (in this case thus 1 Hz).

Imagine that we add to the three sine vibrations a fourth with a frequency of 102.3 Hz. Then no longer does an integer number of periods fit within the duration of 1 second. This causes a dis-

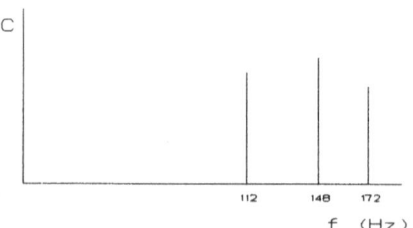

Figure 4.3.26 The amplitude spectrum of the signal of fig.4.3.25.

continuity between the start and end of the signal comparable with the situation of the signal of fig.4.3.5. In that example we have seen that when calculating the spectrum of a truncated sine wave, we find a large number of spectral components. The two with the largest amplitudes are those on both sides of the truncated sine wave. See the C_n-values of that example.

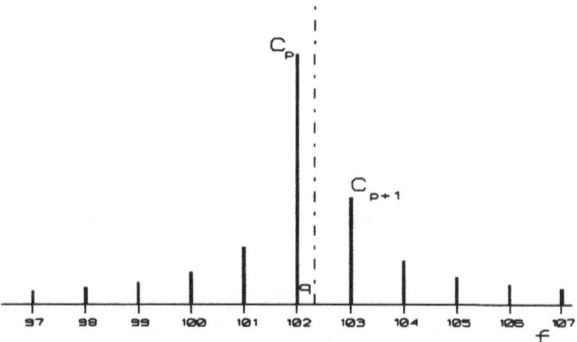

Figure 4.3.27 Leakage.

Here we have a comparable situation; instead of a single spectral line we find a large number of components. In our case that spectrum is shown in fig.4.3.27. This phenomenon is called 'leakage'. It makes the interpretation of the spectrum more difficult. The first problem is how the amplitude and frequency of the actual sinusoidal component can be estimated. For this approximation formulae exist based on the observation made above on fig.4.3.23: the location of the 'true' frequency component corresponds with the peak of the spectral envelope. If we could reconstruct that peak from the discrete frequency components we would have an estimate of the frequency and the amplitude. In fig.4.3.27 the location of the peak is indicated with a dashed line. There are several estimation algorithms of which one developed by Burgess (1975) will be discussed. It uses the proportion of the amplitudes of the two main components, C_p and C_{p+1}. If these have about the same value then the actual frequency value lies directly half way between them. The general formula for determining the place q of this component is:

$$q = \frac{1}{1 + \dfrac{C_p}{C_{p+1}}} \qquad (4.20)$$

In the example $p = 102$, $C_p = C_{102} = 879$ and $C_{p+1} = C_{103} = 377$, thus

$$q = \frac{1}{1 + 2.33} = 0.30016$$

If we know q, we can determine the frequency: $(p + q) \cdot 1\text{Hz} = 102.3 \text{ Hz.}$

The general formula is: $$f = \frac{p + q}{T} \qquad (4.21)$$

The amplitude is estimated with:
$$A = \frac{C_p \pi q}{\sin \pi q} \quad \text{if } q < 0.5 \qquad (4.22)$$

or
$$A = \frac{C_{p+1} \pi (1 - q)}{\sin \pi q} \quad \text{if } q > 0.5 \qquad (4.23)$$

In the example we find: $A = \dfrac{879 \cdot \pi \cdot 0.3}{\sin \pi \cdot 0.3} = 1024.2$

Another example:

Let us assume a signal fragment consisting of $N = 512$ samples with a clock frequency $f_s = 8000$ Hz. The duration of the fragment is thus $T = 512 \cdot (1/8000) = 0.064$ s, and the fundamental frequency is $1/T = 15.625$ Hz. Imagine that we find for the two principal peaks: $C_{71} = 418$, $C_{72} = 632$, then q, f and A become:

$$q = \frac{1}{1 + \dfrac{418}{632}} = 0.602$$

$$f = (71 + 0.602) \cdot 15.625 = 1118.8 \text{ Hz}$$

$$A = \frac{632 \cdot \pi \cdot (1 - 0.602)}{\sin \pi \cdot 0.602} = 832.7$$

In the second place we can ask ourselves if something can be done about the 'leakage'. The source of this was as we have seen, the discontinuity between the start and the end of the signal. One way to proceed is to provide the signal fragment with an envelope that is zero at both the beginning and the end.

If we give the signal $y(t)$:

the envelope $e(t)$:

then after multiplication we arrive at:

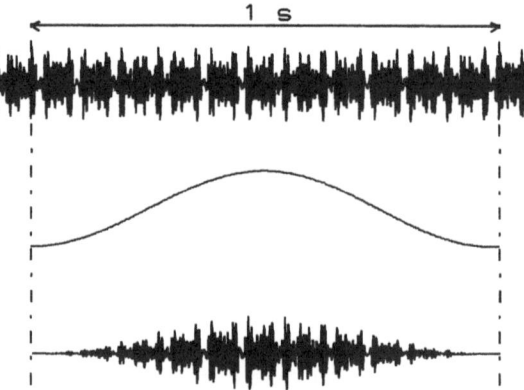

Figure 4.3.27 Hanning window.

This signal naturally differs from the original one and this gives a positive and also a negative effect to the spectrum:
- the advantage is that the amplitude coefficients between the spectral peaks are practically equal to 0.
- the disadvantage is that the peaks have become broader.

Both effects are visible in fig.4.3.29.

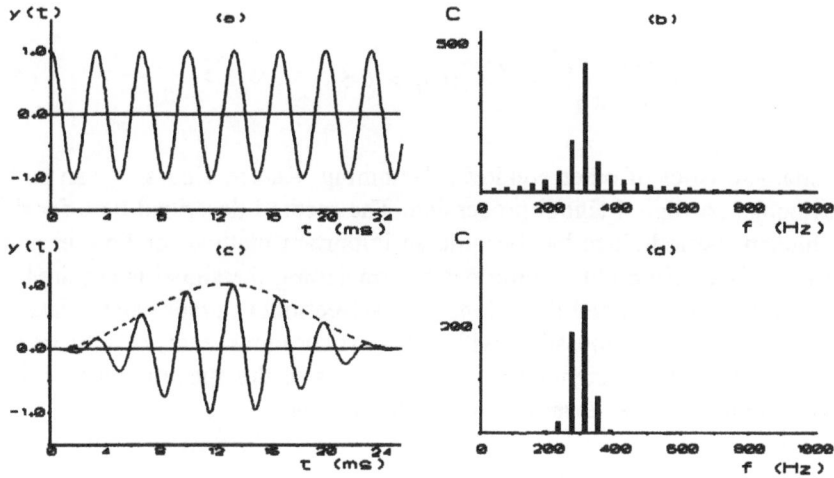

Figure 4.3.29 The effect of windowing.

For the envelope one uses the name 'window'. There are many possible forms and each window forms another compromise between effective suppression of the unwanted components and broadening of the peak. A window that is often used (also in the above figure) is the sin²- or 'raised cosine'- function:

$$e(t) = \frac{1}{2}(1 - \cos\frac{2\pi t}{T})$$

This is called the *Hanning window*. Because the form of the peak is altered through application of the window, the formulas for estimating f and A must be adapted:

$$q = \frac{2 - \dfrac{C_p}{C_{p+1}}}{1 + \dfrac{C_p}{C_{p+1}}} \qquad (4.24)$$

$$f = \frac{p + q}{T} \tag{4.25}$$

$$A = \frac{C_p \pi q (1 - q^2)}{\sin \pi q} \quad \text{if } q < 0.5 \tag{4.26}$$

$$A = \frac{C_{p+1} \pi q (1 - q)(2 - q)}{\sin \pi q} \quad \text{if } q > 0.5 \tag{4.27}$$

The characteristics of other windows (Hamming, Kaiser, Gauss,...) can be found in textbooks on digital signal processing. The method described here for detecting 'hidden' periodicities has become an important method for Fourier analysis because only a minimum of information concerning the signal is required. Non-harmonic components and deviations in quasi-periodic signals can be detected in this way, which is impossible with pitch-synchronous analysis. By means of analysing overlapping signal fragments changes in the spectra can be followed closely. Fig.4.3.30 is an example of such a spectrum.

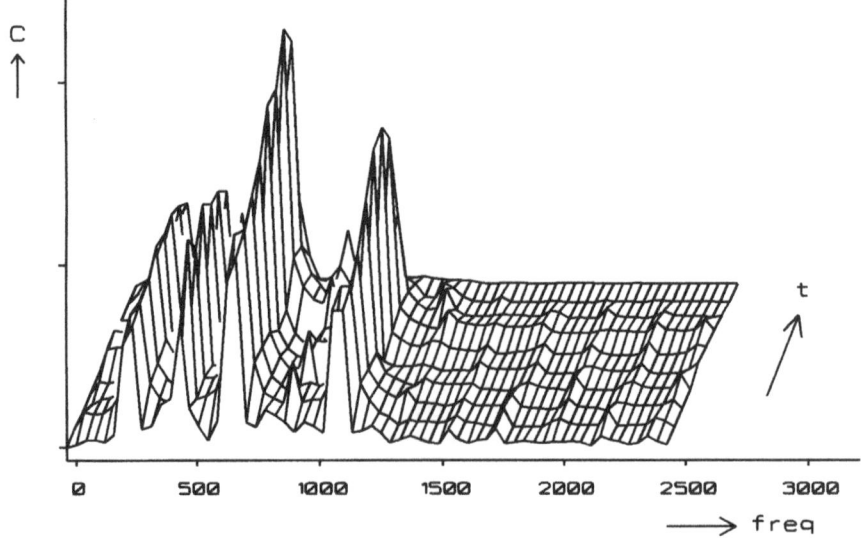

Figure 4.3.30 Time-variant spectrum of a vowel.

The above example shows that with this method it is necessary to calculate hundreds of harmonics. Even with the use of a computer this can become a time-consuming process.

Luckily there is a possibility of performing the necessary calculations within an acceptable time period with the help of the *Fast Fourier Transform* (FFT) which is dealt with in the next section.

First a few tips for the choice of the appropriate values of N, T, etc. It is necessary to know at the start that with N samples the maximum number of harmonics that can be calculated is $\frac{1}{2}N$. This is a consequence of the sampling theorem that we shall discuss in detail in the following section and that says that the highest allowed signal frequency is equal to half the sample frequency. With N samples and a clock frequency f_s the duration of the signal fragment is:

$$N \cdot \frac{1}{f_s} = \frac{N}{f_s}$$

The corresponding fundamental frequency f_0 is equal to: $\quad f_0 = \dfrac{f_s}{N}$

All spectral components have frequencies that are multiples of this value. Conversely the number of a harmonic can be found by dividing its frequency by the fundamental frequency. The highest possible frequency is $\frac{1}{2} f_s$. The number of this component is therefore:

$$\frac{\frac{1}{2}f_s}{f_0} = \frac{\dfrac{f_s}{2}}{\dfrac{f_s}{N}} = \frac{1}{2}N$$

If we have a look at the previous example $(N = 512, f_s = 8 \text{ kHz}, T = 0.064 \text{ s}, f_0 = 15.625 \text{ Hz})$, then due to this rule we can calculate 256 harmonics, and indeed the frequency of the 256th harmonic is:

$$256 \cdot 15.625 = 4000 \text{ Hz } (= \tfrac{1}{2}f_s)$$

We get thus a frequency 'grid' that will look as follows:

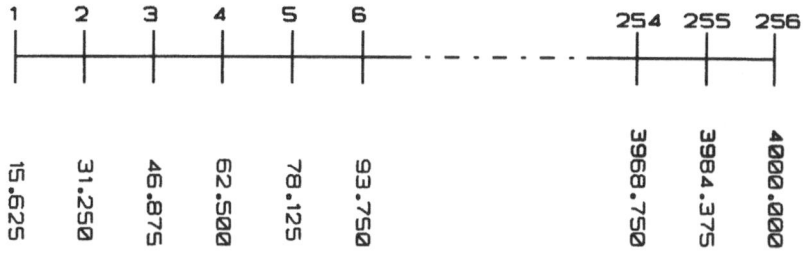

It appears from this:
1. f_s determines the upper limit of the frequency range to be analyzed. Procedure: determine from the given signal which frequencies the analysis should encompass. Remove eventual higher frequencies by means of a lowpass filter and set f_s at minimum the double value of the filter cutoff frequency.
2. N determines the frequency resolution and with this the accuracy of the analysis. One is often not free in the choice of N. For FFT-analysis N must for example be a power of 2. Then $N = 512$ or 1024 is usually a good choice. For accurate results N must be large. With a given f_s also $T (= N \cdot 1/f_s)$ is small and this means that if the signal is not stationary it can change so essentially within the time duration T that the analysis can not be used effectively.

We are here confronted with the fundamental fact that time and frequency resolutions cannot simultaneously be very high. Improving the time resolution makes the frequency resolution worse and vice versa. When we characterize the time resolution with duration T, and the frequency resolution Δf with the frequency spacing of the grid, for which it holds that $\Delta f = 1/T$ it follows that

$$\Delta f \cdot T \geq 1$$

We have already encountered this relationship in section 3.5.C, rule (3.19) and will discuss it again in section 4.3.1.

Considering that the time scale of the fluctuations in speech signals and musical signals is about some tens of milliseconds, we see that with a f_s between 8 and 30 kHz $N = 512$ is usually an appropriate choice. With 8 kHz we find $T = 64$ ms, with 30 kHz $T = 17$ ms.

H. *The Fast Fourier Transform* (Cooley et al. 1965)

This is not a 'different' Fourier transform, only a faster version of the normal DFT We have seen that the determination of the Fourier coefficients takes place via multiplication of the signal function by sin $2\pi nft$ or cos $2\pi nft$, and by subsequently determining the average value of these products. In fig.4.3.4 this procedure is demonstrated. We have also seen that with time-discrete functions this average value is determined by adding the product of signal samples and samples from the sine- or cosine-function and dividing the sum by the number of samples (see rules (4.8) and (4.9)). Below this procedure is shown once again, in this case for a time-discrete function $x(k)$ with 8 samples. These samples are multiplied by the samples from sin $2\pi nk/8 (= S_n)$ and from cos $2\pi nk/8 (= C_n)$ with n = 0, 1,..,7 (fig.4.3.31).

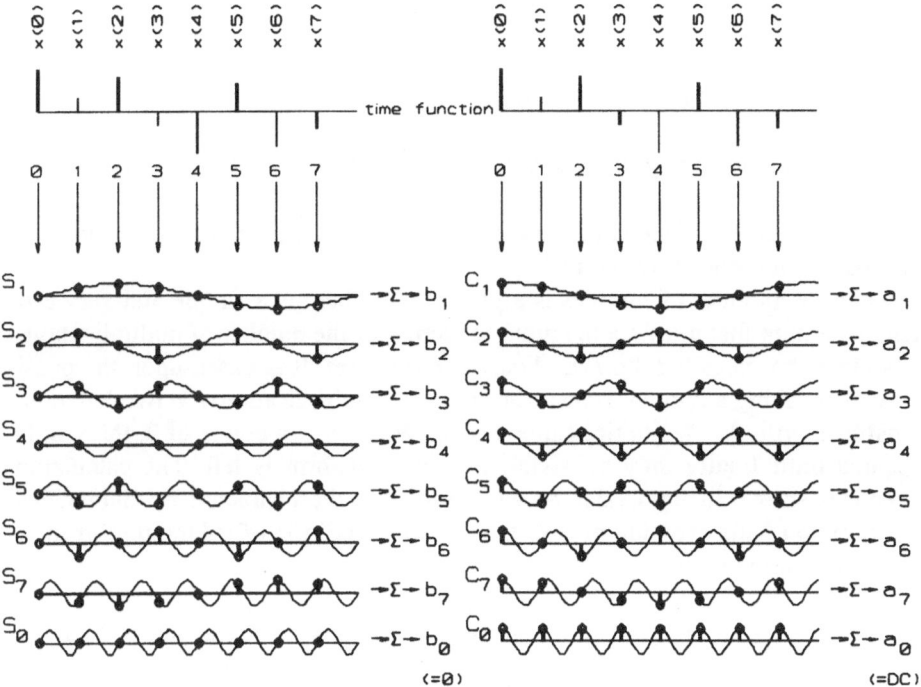

Figure 4.3.31   The DFT.

At first view it seems that an error has been made. Eight harmonics are shown and we have just seen that with 8 samples only four harmonics can be calculated. Indeed, inspection of fig.4.3.31 shows that the calculation of S_5 to S_8 and of C_5 to C_8 is useless, because

$S_5 = -S_3$, $S_6 = -S_7$, $S_7 = -S_1$, $S_8 = 0$, $C_5 = C_3$, $C_6 = C_2$, $C_7 = C_1$, $C_8 =$ average value. The theory of the Fourier transform however also allows the transformation of complex functions, consisting of a real and an imaginary part and then this symmetry no longer exists. Although this case is not important for us (signal functions are always real) is it better to consider the most general case and use all the coefficients indicated.

We see that for every coefficient 8 multiplications are required. The calculation of all 16 components thus asks for 128 multiplications, or in general with N samples $2N^2$ multiplications. The fact that most multiplications are by 0 or 1 (and a few with the factor 0.707, which in fig.4.3.32 is indicated with the letter k) is misleading as it is caused by the small number of samples. With a larger value of N this happens less often. It is therefore not possible to base an eventual reduction of the number of calculations on this fact. The numerical example in section 4.3.C shows that a reduction of the number of multiplications can be achieved by a more efficient organization of the operations. For the calculation of a_3 for example in the fourth column four multiplications by the factor (+ or -) 0.623 occur, but

$$-0.623 \cdot 3.54 + 0.623 \cdot 1.05 + 0.623 \cdot 0.17 - 0.623 \cdot 0.71 =$$

$$0.623(-3.54 + 1.05 + 0.15 - 0.71) = 0.623 \cdot -3.03 = -1.89$$

By first adding (or subtracting) and then multiplying the number of multiplications has been reduced from 4 to 1.

When the number of samples is a power of 2, there is a simple and very efficient algorithm that allows a maximal reduction of the number of multiplications. This algorithm is called the *Fast Fourier Transform*. It is based upon the possibility of reducing a N-point transform (with $2N^2$ multiplications) to two $\frac{1}{2}N$-point transform, with $2 \cdot \frac{1}{4}N^2$ multiplications each. When N is a power of 2, this can be repeated until finally only a (trivial) 2-point transform is left. The calculation scheme is shown in fig.4.3.32. At each transition the characteristic 'butterflies' occur, oblique and crossing arrows that indicate addition of subtraction of number in the following way:

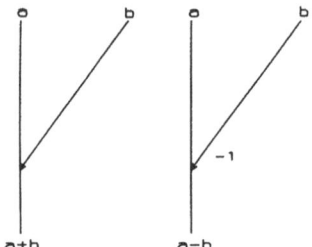

It can easily be checked that in this way the same results are found, by following the calculation of a specific component, for example a_4 in both schemes.

In the scheme of fig.4.3.31 we find:

$$a_4 = x(0) - x(1) + x(2) - x(3) + x(4) - x(5) + x(6) - x(7)$$

In fig.4.3.32 we start eight times at the bottom with a_4 and follow the eight possible trajectories upwards to each of the eight x-samples. Allowing for the indicated multiplication factors, we now find:

$$a_4 = x(0) + x(2) + x(4) + x(6) - x(1) - x(3) - x(5) - x(7)$$

which is the same result. It can be checked that this is also the case with all other coefficients. See for example problem 4.16.

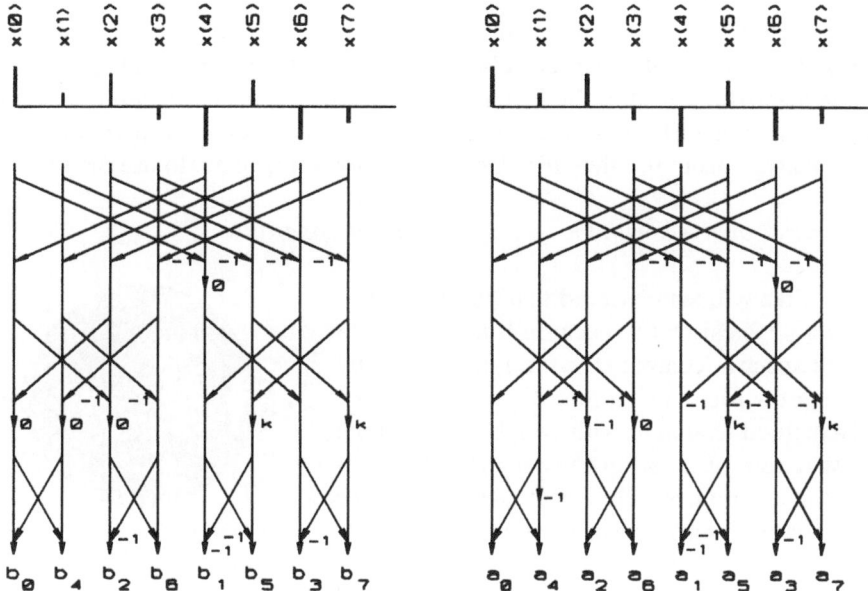

Figure 4.3.32 The FFT.

The scheme shows that now $2 \cdot 8 \cdot 3$ multiplications are required, and in the general case (N samples) the number of multiplications M using the FFT algorithm is:

$$M = 2N \cdot {}^2\log N \qquad (4.28)$$

The reduction of the number of multiplications is small when $N = 8$. With larger values of N this difference and thus the attractiveness of the FFT method grows as can be seen in the following table:

	DFT	FFT	proportion
N	$2N^2$	$2N \cdot {}^2\log N$	
8	128	48	2.67
16	512	128	4.00
32	2048	320	6.40
64	8192	768	10.67
128	32768	1792	18.29
256	131072	4096	32.00
512	524288	9216	56.89
1024	2097152	20480	102.40
2048	8388608	45056	186.18

With $N = 1024$ the reduction factor is ca. 100. This not only means a corresponding reduction in the computing time, the result is also more accurate than with the DFT because with fewer calculations there are also fewer rounding-off errors. A characteristic aspect of the FFT is that all Fourier coefficients are found simultaneously. In the case that one is only interested in one or a few harmonics it can be advantageous to use the DFT. Usually however it is better to use the FFT.

I. *Fourier analysis with the help of measuring apparatus*

It has already been observed that Fourier analysis is a physical reality because it is possible to isolate spectral components from each other by means of a filter. For a long time this was the only method available with which to determine the spectrum of an arbitrary vibration. The registration techniques were insufficient to produce an accurate description of the time function required for calculating the spectrum. Helmholtz was the first to use the filter method. He took several

Figure 4.3.33
Helmholtz resonator.

resonators named after him, which he tuned to the spectral components of the sound he wanted to analyse, and then by placing his ear at the (second) opening of the resonator he attempted to detect the presence and to estimate the strength of the relevant component (see fig.4.3.33).

When it became possible to convert acoustic signals into electrical ones and to filter and measure them, it was possible to build exact and relatively simple spectrum analysers. The principle is shown in fig.4.3.34:

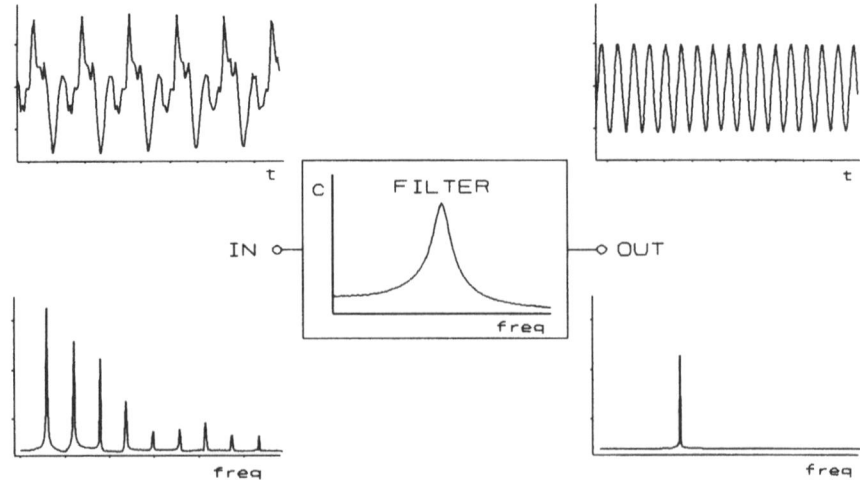

Figure 4.3.34 Spectral analysis.

The filter is for example a harmonic oscillator with a high Q-factor. The choice of Q must be considered well. Increasing the sharpness of the filter improves the frequency resolution but reduces the time resolution because with a sharper filter the transient is longer. Measurement of the amplitude of the forced vibration can only be performed when this transient has sufficiently decayed. If we set the duration of the transient to T' (see section 3.5.C.2) and use this value for the time resolution Δt, then the relationship between time resolution Δt and frequency resolution Δf is given by the 'uncertainty' relation (3.19): $\Delta t \cdot \Delta f \geq 1$

With strongly modulated signals like those in speech and music, sharp filtering is not allowed otherwise the temporal structure that contains the modulation information would be affected. In this respect the filter bandwidth of our auditory system (the so-called 'critical' bandwidth) is a good compromise between frequency resolution, required to separate signals and signal components, and time resolution, required to detect the modulation.

Before the introduction of digital analysis methods, spectrum analysis was performed with analog analysers. Not many of these have survived the competition with their digital successors. To get only a swift and global impression of the spectral composition of a signal the (analog) *real time analyser,* is still a useful device. This analyser consists of several fixed band filters with adjacent passbands. See fig.4.3.35.

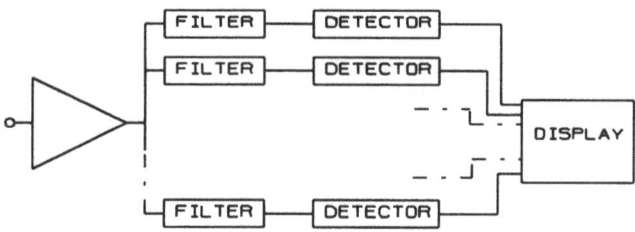

Figure 4.3.35 Real-time spectrum analyzer.

All the outputs of all the filter sections are (practically) simultaneously scanned and the results of the measurements are displayed (LED-display, oscilloscope, see for example fig.4.3.36).

The *spectrograph* has been especially developed for the analysis of speech sounds (in speech research the necessity of paying attention to the time-dependence of the vibration parameters and thus to the modulation aspects was understood much earlier than in music research) by which the time dependence of frequency (via the frequency axis) as well as of amplitude (via the blacking of the image) are reproduced. See fig.4.3.37.

The filters used for spectral analysis usually introduce large phase shifts (see the phase response of the harmonic oscillator, fig.3.5.3b), which means that it is difficult to measure the amplitude and the phase angle simultaneously. This is possible, though, with special measuring devices. For the same reason one can assume that the phase information given by the Fourier analysis in the inner ear is not very useful, which actually means that we are 'phase deaf'. This is indeed one of Helmholtz's hypotheses, certainly in relation to the perception of timbre.

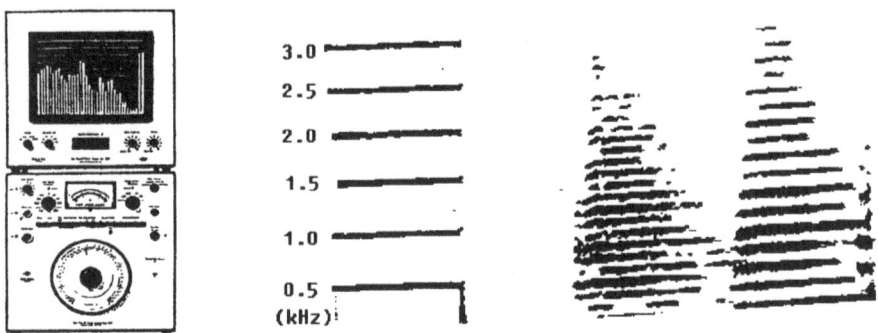

Figure 4.3.36   *Figure 4.3.37*
Brüel & Kjær Real Time *Spectrograms of two trombone tones.*
Spectrumanalyser.

Some quotations from Helmholtz's book:
"Wir können demnach das wichtige Gesetz aufstellen, dass die Unterschiede der musikalischen Klangfarbe nur abhängen vor der Anwesenheit und Stärke der Partialtöne, nicht von ihren Phasenunterschieden."
("We can therefore formulate the important law that the differences in musical timbre depend only upon the presence and strength of the partials, not upon their phase differences.")
This hypothesis was generally accepted, with the convenient consequence that one need not be bothered with the annoying problem of measuring the phase. Therefore often only the amplitude spectrum is used, which is even termed 'the' spectrum. Based on this hypothesis signal functions with equal amplitude spectra and different phase spectra will have the same timbre and thus sound identical, although the wave shape will be different. We have learned in the mean time that this is not entirely true. If changes in the phase spectrum lead to a change of the signal's envelope, then the difference is audible. This problem also crops up in Fourier synthesis as a method for generating complex sounds. In chapter 7 this method will be dealt with, but it is already clear here that one must pay attention to the relevance of the phase angles of the component sine vibrations. Although not everything concerning phase perception is known, the realization that one

must be careful in this respect is growing. In a space with many reflections one can indeed assume that the phase spectrum for the stationary part of the ear signal is determined by chance, because it is determined by the sum of all contributions of all reflections. As the reverberation component of the received signal is reduced, this shall be less the case, and when listening with headphones it does not happen. At such a moment the effect of filters must be critically looked into because the effect of an equalizer for example (see section 5.4) upon the phase spectrum can be disastrous. As we shall see these problems can sometimes be avoided with digital filters because their phase behaviour is much better controlled.

4.4 Time and frequency domain aspects of some signal-theoretical subjects

A. *The sampling theorem*

The formulae (4.7), (4.8) and (4.9) of section 4.3.C allow the calculation of spectral coefficients for both time-continuous and time-discrete signal functions. In practice a time-discrete signal function is derived from a time-continuous one via sampling, and the DFT serves as an approximation of the time-continuous formulas. For that reason we used the a' and b' notation.
We shall now ask ourselves what the difference between both spectra actually is. To that end the following argumentation:

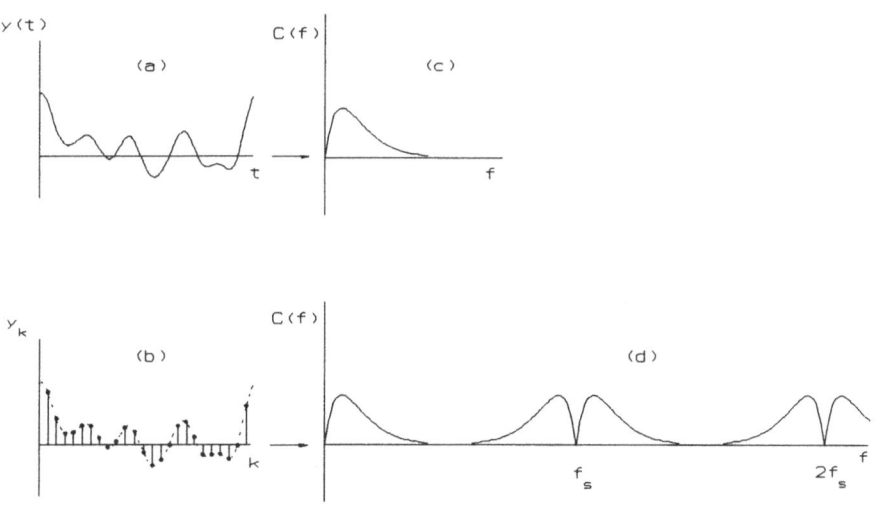

Figure 4.4.1 Spectra of a time-continuous and the derived time-discrete signal.

In figure 4.4.1 the (continuous) spectrum shown at (c) corresponds to the signal function shown at (a). Next we derive from the time-continuous signal function (a) a time-discrete signal function (b). In the same way as with the analog signal we can analyse (b) by considering it as a series of infinitely narrow pulses. If we calculate the spectrum of this then it appears that (as is to be expected with infinitely narrow pulses) it stretches infinitely over the whole frequency axis and consists of periodically repeated (and mirrored) versions ('aliases') of the spectrum of the time-continuous signal. See fig.4.4.1d. The repetitions occur around the points of symmetry f_s, $2f_s$, $3f_s$.. The spectrum is thus periodic with f_s as period. (Pay attention to the correspondence between periodic signal vs. line spectrum and 'line' signal vs. periodic spectrum.) Considering the figure it appears further that we can reconstruct the original spectrum from the 'time-discrete' one by removing all aliases with a filter. This can be done only if there is no overlap between the original spectrum and the first mirrored version, stretching from from f_s downward. As can be seen overlapping would take place if the original spectrum would extend beyond ½f_s. This situation is shown in fig.4.4.2.

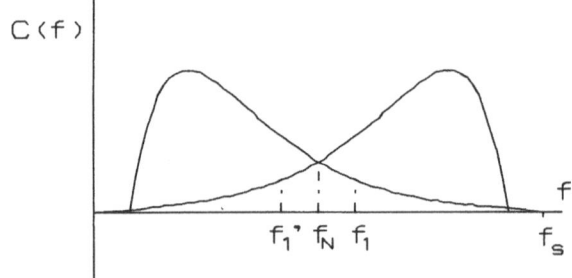

Figure 4.4.2 The alias effect in the frequency domain.

The consequence of overlapping is that frequency components of the original spectrum become mixed with mirrored components. Afterwards it can no more be seen what was the origin of a particular component and thus it is not possible to retrieve the original signal. This is thus the foundation of the sampling theorem:

In principle the original time-continuous signal can be reconstructed from the time-discrete signal if the sampling frequency is more than twice the highest signal frequency.

A good example of the consequences of a sampling frequency that is too low is the well-known stroboscope effect in films, for example, in rotating wheels. In fig.4.4.3 can be seen what happens if a sinusoidal signal is sampled with a frequency that is too low. The samples that are found correspond to a sinusoidal vibration with a low frequency (= the mirrored component).

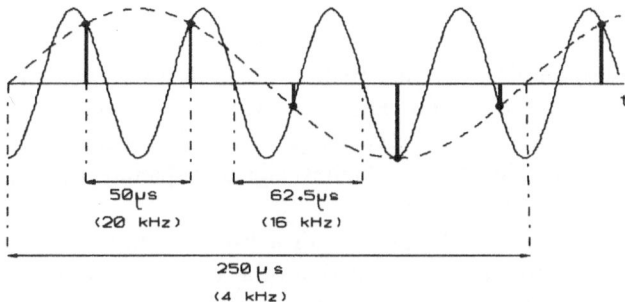

Figure 4.4.3 The alias effect in the time domain.

B. *Digital-to-analog conversion*

How does the reconstruction work? In principle rather simply: with an ideal lowpass filter with a cutoff frequency f_c that is equal to $\frac{1}{2}f_s$ (this frequency is also called the 'Nyquist-frequency' f_N) all copies of the original spectrum are removed. With this, the spectrum is again identical with the original and this of course holds as well for the signal function. The effect of the filter can also be described in the time domain: the impulse response (see chapter 5) of the ideal lowpass filter has the well-known shape of the (sin x)/x function. Every infinitely narrow pulse of the time-discrete signal is spread out over the time axis and gets as a time function:

$$\frac{\sin \dfrac{\pi(t - nT)}{T}}{\dfrac{\pi(t - nT)}{T}} \qquad (T = \frac{1}{f_s}, \ nT = \text{time coordinate n-th impulse})$$

If all the (sin x)/x curves of all pulses are added, the original time function is found back (this is the so-called 'Whittaker reconstruction', see fig.4.4.4).

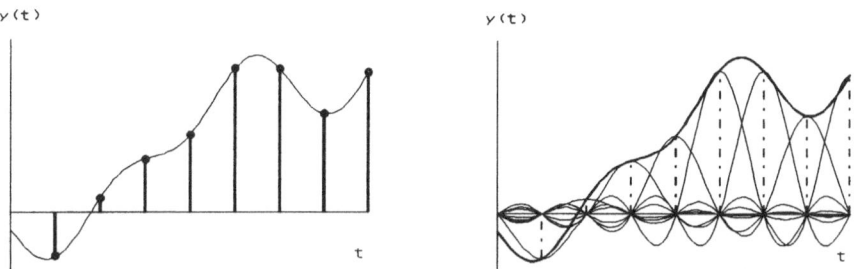

Figure 4.4.4 Sampling and reconstructing the signal.

In practice unfortunately it is not so simple. Exact reconstruction is not possible because a) every finite signal has an infinite bandwidth (there is thus no 'highest' frequency), and b) an ideal lowpass filter (with an infinitely steep flank) does not exist. Such a filter would be non-causal. See section 5.1. The second point has as a practical consequence that a larger margin is kept between the clock frequency and what may be considered as the highest signal frequency, the cutoff frequency f_c of the filter, e.g. $f_s = 2.2 \cdot f_c$.

Most DA-converters are supplied with a filter because a practical DA-converter can be considered as a combination of a 'true' converter (which produces very short voltage peaks, pulses, which are proportional to the input number) and a so-called zero-order hold filter that holds the voltage value until the following clock pulse, by which the well-known staircase output signal originates. This circuit has the following impulse response from which the amplitude characteristic can be derived using techniques that will be discussed in chapter 5. It is called a 'zero order hold filter'.

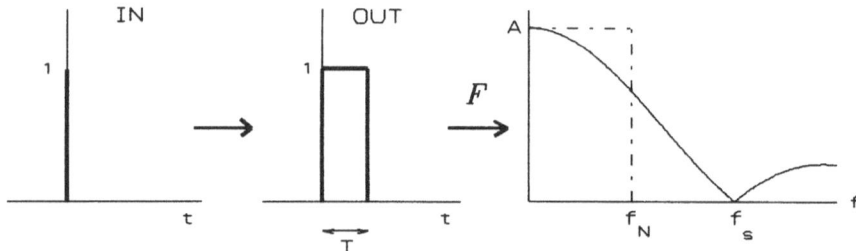

Figure 4.4.5 Zero order hold filter.

This filter characteristic is a very bad approximation of the ideal lowpass characteristic (dashed line). Moreover in the frequency band to be transmitted an unwanted attenuation takes place. Therefore on the one hand extra filtering e.g. with the help of an analog lowpass filter is required, and on the other hand the loss in the transmission band is sometimes corrected ('aperture correction').

C. *Changing the sampling frequency* (Rabiner, 1982).

In certain circumstances it can be desirable to alter the sampling frequency of a given digital signal. This is for example the case when devices working with different clock frequencies are connected, or for the realization of the digital equivalent of a variable tape speed with the corresponding changes in pitch and duration. This effect can be attained by changing the number of samples per unit of time, while still keeping the original clock frequency.

A well-known procedure for changing f_s consists of 2 steps: first f_s is raised by an integer factor M and then lowered by another integer factor L. The total change is:

$$f_s' = \frac{M}{L}f_s \qquad (4.29)$$

Increasing f_s ('interpolation') occurs by inserting M-1 zero samples (samples with the value 0) between two samples of the given signal. The spectrum is not altered by this because when calculating the a- and b-coefficients with the formulae (4.8) and (4.9) the contribution of the additional samples is zero. This is the case because

$$\text{if } y(k) = 0, \text{ then } y(k) \cdot \cos 2\pi \frac{nk}{N} = y(k) \cdot \sin 2\pi \frac{nk}{N} = 0$$

Still the number of samples per time unit has become M times as large. That means that also the Nyquist-frequency that gives the 'usable' part of the spectrum has become M times as large. As it appears in fig.4.4.6a the spectrum now contains unwanted frequency bands (the original aliases) which must be removed with a filter (fig.4.4.6b).

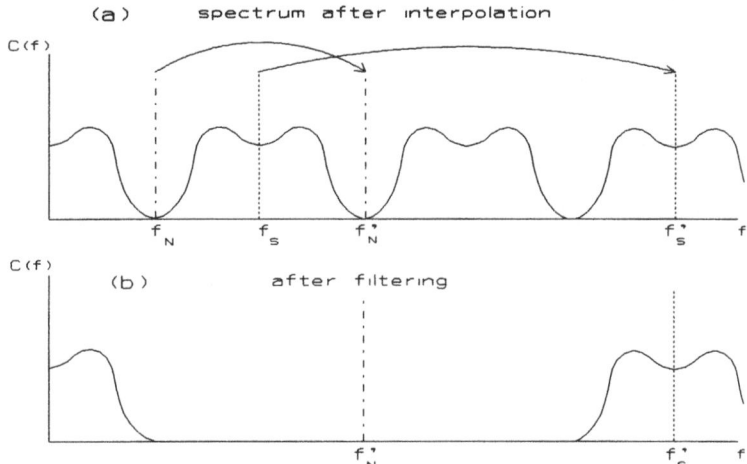

Fig.4.4.6 (a) Effect of interpolation on the spectrum.
(b) Removal of the unwanted frequency bands.

The reduction of the sampling frequency ('decimation') is if possible still simpler. It boils down to removing $L-1$ samples from each group of L samples. Before this can take place the signal must be filtered so that no frequency components occur above the new, lower Nyquist frequency (f_N / L). In fig.4.4.7 both operations are shown schematically.

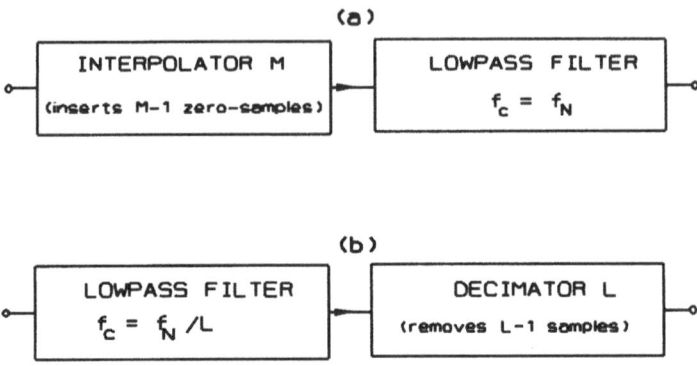

Figure 4.4.7 Interpolation and decimation.

D. *Noise*

All signal functions of which the course of values is either totally or to a large degree determined by chance, form together the set of noise signals. These signals are naturally not periodic and have thus a continuous spectrum. The power density spectrum is often flat, by which the characteristic hiss sound is audible which often marks noise signals. An exact specification of the signal function is only possible by means of a precise description of the course of values. For obvious reasons this is not attractive and normally the description is restricted to a statistical one in either the time or the frequency domain.

1. Time domain description.

For this the so-called *amplitude density distribution* is used. This statistical signal description (of which the applications are not restricted to noise signals) refers to the question as to the probability of the occurrence of a particular signal function value. If we are dealing with real-valued signal functions then the number of possible values is infinitely large and the probability of a particular value infinitely small. Then the question should be reformulated and should refer to the probability that a signal function value will lie between two given values. One can estimate this probability by measuring for which percentage of the total duration of the signal the signal value lies between the two given limits. With a 'natural' signal this percentage will rise if the limits are chosen close to the zero line. When in this case the signal function is determined by a large number of random factors (e.g. collisions between elementary particles) one can prove that the amplitude density function is a so-called Gauss or 'normal' distribution. This curve can be seen in fig.4.4.8. For amplitude density distribution one also uses the term 'probability density function' (pdf). In the case that the pdf is a Gauss curve one speaks of 'random noise'.

To indicate the amplitude of such a signal use is made of the standard deviation of the normal distribution. This corresponds to the RMS value that we are already familiar with. Random noise is just one possibility. In fig.4.4.8 another signal function is shown which alternates at arbitrary moments between two set levels. To this belongs an amplitude density distribution with two (equally large, namely corresponding to 50%) peaks at the two possible amplitude values. This is also shown in fig.4.4.9. We speak here of 'binary' noise.

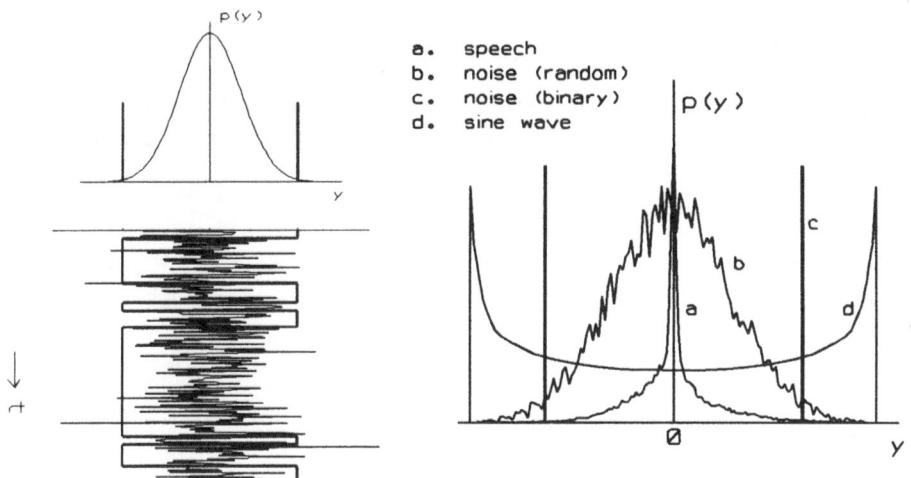

Figure 4.4.8
Probability density function.

Fig.4.4.9
Probability density functions of some signals.

A third possible distribution occurs in noise signals which result from random values generated by a computer that are sent to a DA-converter. For this a computer program is needed. There are programs that produce numbers that have a normal distribution but with most programs the random numbers are equally distributed between two boundaries. This corresponds with a flat pdf as we saw already when we dealt with quantization noise (section 4.2.A.2, fig.4.2.3). Amplitude density distributions can also be determined for other signal functions than noise signals, to arrive at a statistical description of that particular function. In fig.4.4.9 such distributions are shown for several signals (under which we find random noise and binary noise). As it already appears from out these examples, the distributions say nothing about the sound characteristics of the signals. Signals that have totally different distributions are sometimes not distinguishable by ear. In this respect the frequency domain specification, to be discussed now, is more relevant than the time domain specification.

2. Frequency domain description.

In the frequency domain a noise signal is described by a continuous amplitude spectrum or a continuous power spectrum (the so-called power density spectrum or pds). Based on this spectrum one distinguishes several different types of noise. We shall concentrate on the pds and will place two graphic representations beside each other: one in which the energy is measured with a filter with a constant bandwidth (e.g. 100 Hz) and one in which a filter with a constant frequency interval (e.g. a third octave) is used.

a. white noise.

Here the continuous spectrum consists of a horizontal line (fig.4.4.10a) with the first filter, and a rising line of 3 dB per octave (fig.4.4.10b) with the second.

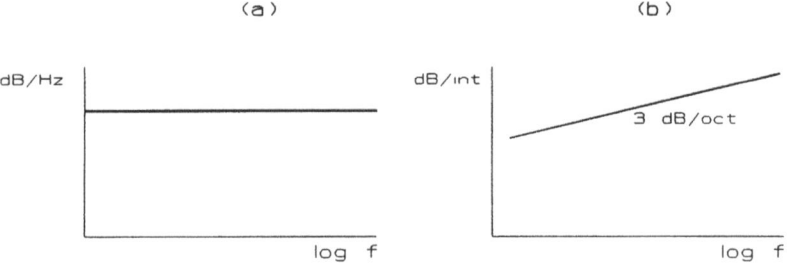

Figure 4.4.10 White noise.

The qualification 'white' has its origins in a comparison with light where a mixture of all frequency components (colours) gives white light. The rising of the spectrum in the second case is explained as follows: with white noise there is a constant energy density which means that the quantity of energy 'per Hz' is constant.

   If we compare the amount of energy in the octave 100 - 200 Hz with that in the octave 200 - 400 Hz than the second octave contains twice as much energy, and this holds for each consecutive octave. Because pitch perception is linked to the logarithm of the frequency relationship it holds that the same interval (the octave) has double the amount of energy. In our perception of white noise the high frequency energy thus dominates, which gives the noise a sharp hissing sound.

   The factor 2 in the energy relationship per octave corresponds to 3 dB, and the rising characteristic has thus a steepness of 3 dB/octave. The constant-interval filter thus simulates the working of the hearing organ. Natural noise signals have often a 'white' spectrum and that holds e.g. also for the noise produced with uniformly distributed random numbers (see section 6.3.B.5).

b. pink noise.

For our ears a much more balanced noise signal occurs if we send white noise through a lowpass filter with a falling 3 dB/octave characteristic. The spectra of fig.4.4.11 are the result. Such noise is called 'pink' noise (to keep the analogy with light) because the low frequencies (=red light) are now more strongly represented. Instead of speaking of a decay with 3 dB/octave we can also say that the energy is proportional to $1/f$. These two statements are equivalent, because doubling f leads to a factor $0.5 = -3$ dB in energy.

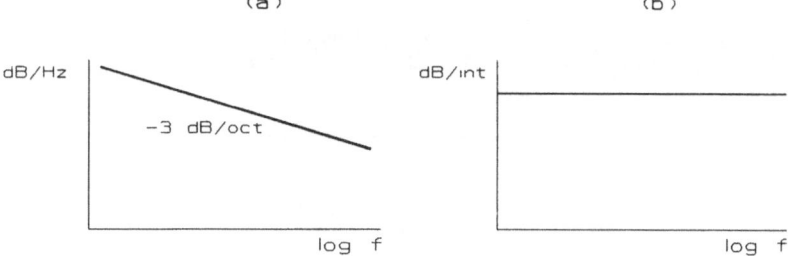

Figure 4.4.11 Pink noise.

Pink noise also occurs in nature. With such very different phenomena such as the activity of sunspots, fluctuations of the electrical current in semiconductors, nerve membrane potentials, the water level of the river Nile and also with certain acoustic parameters in music occur statistical fluctuations that have a characteristic of approximately $1/f$.

Another statistical phenomenon that is well-known in physics is the so-called Brownian movement of microscopically small particles in liquid, that has a p.d.s. that is proportional to $1/f^2$ (decays with 6 dB/octave). For obvious reasons, this noise is called 'brown' noise. Pink and brown noise are examples of noise that do not have a flat density spectrum. Another example of this is arrived at if we send white noise through a band filter. In this way we get band noise where the energy is concentrated in a small part of the frequency range. We can also use a number of adjustable filters (an 'equalizer') and adjust the spectrum to our wishes ('shaped noise').

With the determination of the spectrum of a given noise signal we have the choice of two possibilities: calculation via the DFT (or FFT) or measurement with an analyser. In both cases it holds that the spectrum of a short signal fragment shows large statistical fluctuations. To get more insight into the global shape of the spectrum we must average it out. With the DFT we do this literally by determining the average of a (large) number of spectra. With the analog method the built-in (and sometimes adjustable) response time (the so-called 'time constant') of the system takes care of the averaging. After averaging detail information is lost. It is then not possible anymore with the DFT to transform back to the original time function. Below in fig.4.4.12 registrations of white, pink and brown noise can be seen.

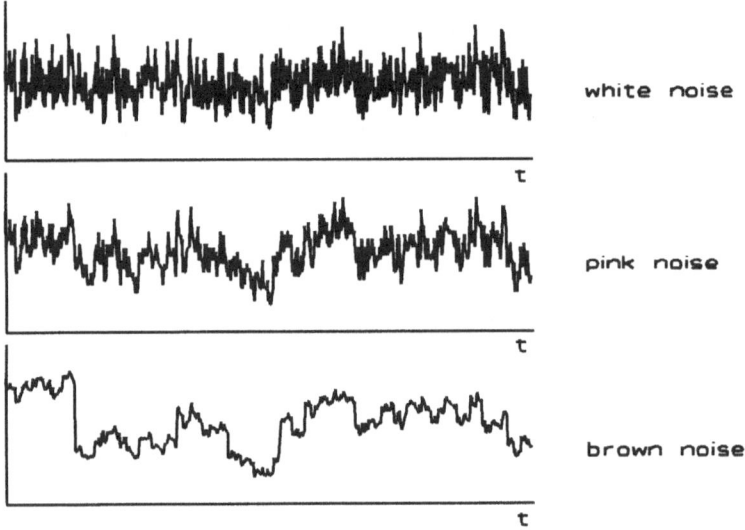

Figure 4.4.12 White, pink and brown noise.

4.5 Orthogonal functions and signal transforms

A. *Vectors and functions*

Via Fourier transformation an arbitrary function is reduced to a combination of elementary functions. It appears that this process shows similarities with the description of a vector by means of its coordinates. First we shall have a look at the properties of vectors and then delve into the similarities existing between the two representations.

A vector is a quantity with a particular 'value' and a direction. Such quantities often occur in physical descriptions (forces, speeds etc.). They are represented symbolically by means of arrows. The length of the arrow represents the value of the vector and its orientation the direction of the vector. Because two vectors generally have different directions the rules for addition are different from those for the addition of numbers. Addition of vectors takes place by means of the well-known parallelogram construction that we used already in chapter 3 when calculating the opposing force in a vibrating string (section 3.1). In fig. 4.5.1 the sum and the difference of two vectors (also vectors themselves!) are shown.

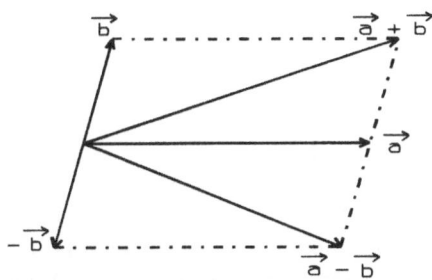

Figure 4.5.1 Addition of vectors.

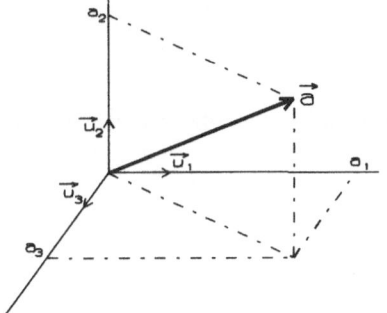

Figure 4.5.2 Decomposition of a vector in unit vectors.

Reversely a given vector may always be split into two (or more) components. We can for example reduce a vector in three-dimensional space to three basic vectors, the projections of the vector upon the three axes. (To make clear in a formula or calculation that we are dealing with vectors in this book a small arrow is printed above the symbol in question.)

Let us have a look at how a vector $\vec{a}$ can be described with the help of its coordinates in three-dimensional space. In fig.4.5.2 a vector $\vec{a}$ can be seen and also the projections of $\vec{a}$ upon the three axes. The three coordinates a_1, a_2 and a_3 can also be considered as weighting factors by which the 'unit vectors' $\vec{u}_1$, $\vec{u}_2$ and $\vec{u}_3$ must be multiplied before being added together to form the vector $\vec{a}$:

$$\vec{a} = a_1 \cdot \vec{u}_1 + a_2 \cdot \vec{u}_2 + a_3 \cdot \vec{u}_3 = \sum_{n=1}^{3} a_n \cdot \vec{u}_n$$

This principle can also be applied without any problem in spaces of more dimensions, and even in those of infinite dimensions. We cannot imagine such spaces but the mathematical definition is quite possible. Such spaces are useful in the analysis of problems with many variables. A vector in a N-dimensional space can be described as:

$$\vec{a} = \sum_{n=1}^{N} a_n \cdot \vec{u}_n \tag{4.30}$$

A system of N mutual perpendicular axes is called *orthogonal*. The length of a single vector (which is designated with the vector symbol between absolute-value stripes: $|\vec{a}|$), and also the angle between two vectors can be derived from the coordinates. Regarding the length, we know the relation in two dimensions as the theorem of Pythagoras:

$$|\vec{a}|^2 = a_1^2 + a_2^2$$

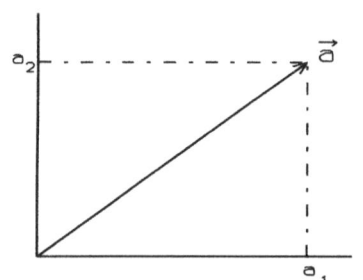

Figure 4.5.3 Pythagoras theorem.

This can be generalized for N dimensions by:

$$|\vec{a}|^2 = \sum_{n=1}^{N} a_n^2 \qquad (4.31)$$

Concerning the angle between vectors in a two-dimensional case (see fig.4.5.4) we work with the 'generalized' theorem of Pythagoras, also called the 'cosine rule':

$$|\vec{c}|^2 = |\vec{a}|^2 + |\vec{b}|^2 - 2\,|\vec{a}|\cdot|\vec{b}|\,\cos\phi$$

(If $\phi = 90°$ this becomes the 'normal' theorem of Pythagoras.) If we designate the coordinates of $\vec{a}$ and $\vec{b}$ with (a_1, a_2) and (b_1, b_2) we can derive the following expression for $|\vec{c}|^2$:

$$|\vec{c}|^2 = (a_1 - b_1)^2 + (a_2 - b_2)^2 = a_1^2 - 2a_1b_1 + b_1^2 + a_2^2 - 2a_2b_2 + b_2^2$$

$$= a_1^2 + a_2^2 + b_1^2 + b_2^2 - 2(a_1b_1 + a_2b_2)$$

$$= |\vec{a}|^2 + |\vec{b}|^2 - 2(a_1b_1 + a_2b_2)$$

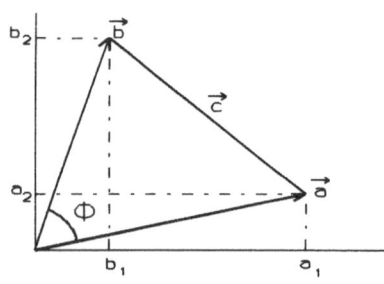

Figure 4.5.4 The cosine rule.

Comparison of the two expressions for $|\vec{c}|^2$ shows that

$$|\vec{a}|\cdot|\vec{b}|\,\cos\phi = a_1b_1 + a_2b_2$$

This can also be generalized for spaces of N dimensions:

$$|\vec{a}|\cdot|\vec{b}|\,\cos\phi = \sum_{n=1}^{N} a_n b \qquad (4.32)$$

(With $\phi = 0$ and thus $\vec{a} = \vec{b}$ we have again the rule for the length of a vector.) We abbreviate (4.32) to $\vec{a}\cdot\vec{b}$ and call this the 'dot product' of the two vectors $\vec{a}$ and $\vec{b}$. The dot product itself is a scalar.

If two vectors are perpendicular, their dot product equals 0, because the cosine of $90°$ is equal to 0. This gives the possibility of testing the orthogonality of vectors. In fig.4.5.2 the vector $\vec{u}_1$ has the coordinates 1, 0 and 0, and vector $\vec{u}_2$ the coordinates 0, 1 and 0. The dot product is: $1\cdot0 + 0\cdot1 + 0\cdot0 = 0$

A system of orthogonal vectors, each with a length 1 (such as the three unit vectors $\vec{u}_1$, $\vec{u}_2$ and $\vec{u}_3$ of fig.4.5.2), is called *orthonormal*. Two vectors, $\vec{a}$ and $\vec{b}$ are orthonormal if:

$$\sum_{n=1}^{N} a_n b_n = \begin{cases} 0 \text{ if } \vec{a} \neq \vec{b} \text{ (then } \cos \phi = \cos 90° \\ \\ 1 \text{ if } \vec{a} = \vec{b} \end{cases} \tag{4.33}$$

The dot product also makes it possible to determine the coefficients a_n. The coefficient a_k can be calculated via the dot product of vector $\vec{a}$ with the unit vector $\vec{u}_k$:

$$\vec{a} \cdot \vec{u}_k = a_1 (\vec{u}_1 \cdot \vec{u}_k) + a_2 (\vec{u}_2 \cdot \vec{u}_k) + \ldots + a_k (\vec{u}_k \cdot \vec{u}_k) + \ldots = a_k$$

The concepts of orthogonality and orthonormality can be applied also to functions. A set of functions $\phi_n(t)$ is called orthonormal in the interval (t_1, t_2) if for each pair of functions $\phi_i(t)$ and $\phi_j(t)$ in this interval the following relation holds:

$$\int_{t_1}^{t_2} \phi_i(t) \cdot \phi_j(t) \, dt = \begin{cases} 0 \text{ if } i \neq j \\ \\ 1 \text{ if } i = j \end{cases} \tag{4.34}$$

If one realizes that an integral is in fact an 'infinite sum', then apparently this definition has much in common with that of the orthonormality of vectors. In the same way as an arbitrary point in a space can be considered as a vector, a weighted addition of several elementary orthonormal vectors (the unit vectors), so an arbitrary function can be described as the 'weighted' sum of a number of elementary orthonormal functions ϕ_n. We thus have next to each other:

$$\vec{a} = \sum_{n=1}^{N} a_n \vec{u}_n \quad \text{and} \quad y(t) = \sum_{n=0}^{\infty} c_n \cdot \phi_n(t) \quad t_1 \leq t \leq t_2$$

As the coefficient a_k can be determined via the dot product of $\vec{a}$ and $\vec{u}_k$, so the coefficient c_k can be determined via:

$$\int_{t_1}^{t_2} \phi_k \cdot y(t) \, dt = \int_{t_1}^{t_2} \phi_k(t) \left\{ \sum_{n=0}^{\infty} c_n \cdot \phi_n(t) \right\} dt = \sum_{n=0}^{\infty} c_n \left\{ \int_{t_1}^{t_2} \phi_k(t) \, \phi_n(t) \, dt \right\} = c_k$$

or

$$c_k = \int_{t_1}^{t_2} \phi_k(t) \cdot y(t) \, dt \tag{4.35}$$

This holds for every set of orthonormal functions, such as Legendre polynomials, Laguerre functions, and the sets upon which the transforms are based which bear the names of their 'discoverers': Fourier and Walsh. (See also Ahmed et al. 1975.) The Fourier transform has been dealt with in detail. In problem 2.15 we saw that functions of the type $\sin 2\pi nft$ and $\cos 2\pi nft$ are orthogonal. We shall spend a short time on the study of the Walsh transform.

B. *The Walsh transform*

It was shown in 1923 by J. Walsh that the system of functions of which the first eight are shown below, is an orthonormal system.

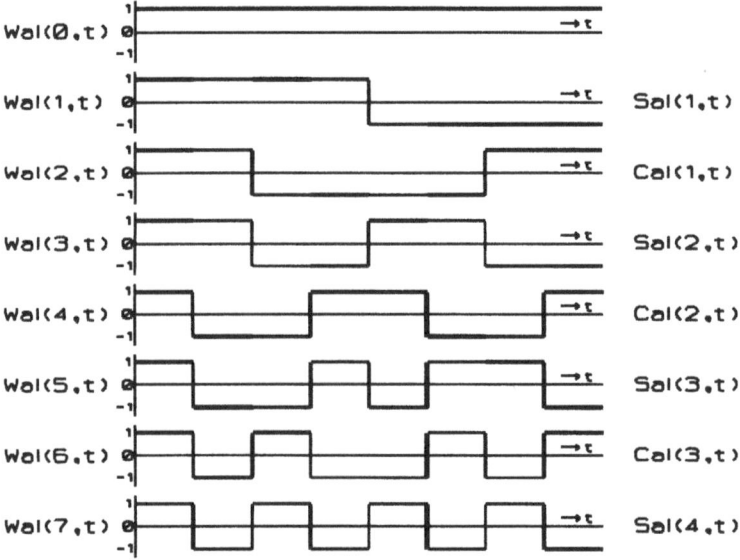

Figure 4.5.5 The first eight Walsh functions.

The definition of the functions is based on the interval $(0,1)$. The functions are given with Wal(n,t), in which t may take values from the interval $(0,1)$, and whereby n gives the rank order and also the number of zero-crossings of the particular functions. There is some similarity with the concept of '(half) frequency' of the Fourier transform. In Walsh transforms the term 'sequency' is made use of. There is no simple function rule for the Walsh functions, and therefore we can not directly check their orthogonality or orthonormality. With the help of the graphs above we can see that the product of two arbitrary Walsh functions has again the typical square wave shape and that the area above the zero line is equal to that under that line so that the total area (=the integral) is zero.

The product of a particular Walsh function with itself gives the constant value 1, so the area is 1 again. There exists a recursive definition for Walsh functions by which higher-order functions are expressed in lower-order functions. Figure 4.4.5 shows that Walsh functions with an even rank number are symmetric around $t=\frac{1}{2}$, and that those with an odd rank number are mirror-symmetric. We call such functions 'even' and 'odd' respectively. The Fourier functions have the same characteristic because $\cos 2\pi nft$ is symmetric around $\frac{1}{2}T$ and thus 'even', and $\sin 2\pi nft$ is mirror-symmetric around $\frac{1}{2}T$ and thus 'odd'. For this reason Walsh functions are divided into two groups according to their rank number, namely the even 'Cal'-functions and the odd 'Sal'-functions, where the C and S are borrowed from the Fourier functions. As with every orthonormal system the arbitrary function $y(t)$ can be written within the interval $(0,1)$ as

$$ y(t) = \sum_{n=0}^{\infty} d_n \cdot Wal(n,t) $$

In correspondence with what was stated about orthonormal functions above it holds for the coefficients d_n that:

$$ d_n = \int_0^1 y(t) \cdot Wal(n,t)\, dt $$

We can also work with the Cal- and Sal-functions:

$$ y(t) = a_0 \cdot Wal(0,t) + \sum_{n=1}^{\infty} \{ a_n CAL(n,t) + b_n SAL(n,t) \} $$

with $a_0 = d_o,\ a_n = d_{2n}$ and $b_n = d_{2n-1}.$

An attractive characteristic of Walsh functions is that the calculation of the coefficients d_n is very simple: the interval $(0,1)$ is subdivided into a number of segments of equal length (this number is a power of 2); in each segment the function is either multiplied by +1 (thus left unchanged) if $Wal(n,t)$ for those t-values is equal to +1, or multiplied by -1 (mirrored) if $Wal(n,t)$ is equal to -1. Then integration follows, which means that all integrals over all segments are added together. Thanks to its binary structure the Walsh transform is especially easy to implement with the help of a computer. Just as for the Fourier transform there is an efficient and fast algorithm, called the F(ast) W(alsh) T(ransform).

    If one calculates the Walsh transform of a sine function the following values are found for the first 16 coefficients:

$d_0 = 0.0$	$d_4 = 0.0$	$d_8 = 0.0$	$d_{12} = 0.00$
$d_1 = 0.635$	$d_5 = -0.275$	$d_9 = -0.051$	$d_{13} = -0.127$
$d_2 = 0.0$	$d_6 = 0.0$	$d_{10} = 0.0$	$d_{14} = 0.0$
$d_3 = 0.0$	$d_7 = 0.0$	$d_{11} = 0.0$	$d_{15} = 0.0$

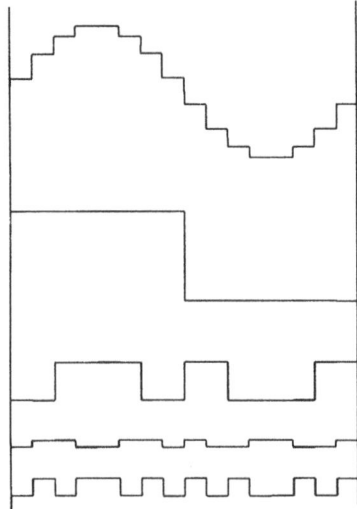

Figure 4.5.6 Walsh synthesis of a sine function.

Figure 4.5.6 shows the four Walsh functions Wal(1,t), Wal(5,t), Wal(9,t) and Wal(13,t), each multiplied by the corresponding d-coefficient taken from the table above. At the top the sum of the four Walsh functions can be seen, a staircase-like ('piecewise constant') approximation of the original sine function. It holds here as well that the approximation is more accurate when more functions are used in the addition.

Generating Walsh functions electrically is simple. In chapter 7 the construction of a Walsh synthesizer will be discussed. We shall see then that only a few elementary digital circuits are required. For that reason Walsh synthesis is used as a form of sound synthesis. For example the easiest way to construct a sine wave oscillator is to generate the four Walsh functions at the side and add them.

4.6 Problems

4.1   Give the octal, hexadecimal and binary equivalent of the following deci-
       mal numbers. Use a 16 bit, two's complement notation.
       a. 2748   b. -300   c. 592   d. -592

4.2   Calculate sum and product of the following pairs of numbers:
       a. 1100110_2 and 1011_2,
       b. 612_8 and 76_8,
       c. $1A7_{16}$ and 29_{16}.

4.3   Given a 16-bit AD-converter with a voltage range of 0 - 10 volt. Which
       numbers (in octal notation) will be produced by the converter when the
       input voltage is equal to:
       a. 0.56 V   b. 4.37 V   c. 8.89 V

4.4   Using the same convert as in problem 4.3, calculate the input voltage
       when the converter value is equal to
       a. 17_{10} b. 1417_{10} c. 2717_{10}

4.5   The proportion between the peak value and the RMS value of a signal is
       3. Calculate the Signal-to-Noise ratio after conversion by a 12 bit AD-
       converter.

4.6   What is the slew rate (= maximal speed with which the voltage can in-
       crease) of the output voltage of a deltamodulator when the stepsize is 5
       mV and the clock frequency is 450 kHz?

4.7   The deltamodulator of problem 4.6 receives a sinusoidal input signal
       with an amplitude of 5 volt. Calculate the maximal frequency of this sine
       wave that can be tracked by this modulator.

4.8   One period of a periodical signal is sampled. The 16 samples are given
       below. Calculate amplitude and phase of the fourth harmonic.
       1.72 2.14 -1.68 -0.32 2.59 1.46 -2.81 -4.32 -1.76 0.56 1.53 -2.87
       -4.48 3.25 2.45 1.47

4.9   Using Parseval's theorem calculate the RMS value of one period of a
       triangular signal fluctuating between -1 V and +1 V. Compare this result
       with that of problem 2.3.

4.10   Calculate the spectrum of the squarewave signal of fig.4.3.9.

4.11   Given a time-discrete signal fragment of 512 samples. No window has been applied, and the sampling frequency was 20 kHz.

a.   The spectrum has a peak with the following maximal amplitude coefficients: $C_{15} = 720$ and $C_{16} = 480$. Give an estimate of the frequency and amplitude of the signal component that caused this peak.

b.   Where and how do we find a frequency component of 750 Hz with an amplitude of 550 in this spectrum?

4.12   Calculate the Fourier coefficients of the following signal function (a 'rectified' sine wave).

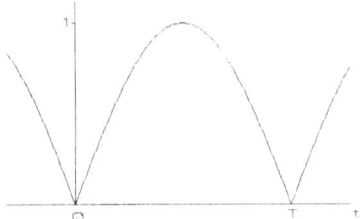

4.13   The period duration of a pulse signal is 8 ms, the pulse width is 1 ms.

a.   What is the frequency of the first harmonic?

b.   Which frequency components are missing in the spectrum?

c.   What is the level difference in dB between the first and the fourth harmonic?

4.14   Calculate the continuous spectrum of the single pulse signal; shown below. Calculate the location of the 'point of gravity' of this signal using rule (4.19).

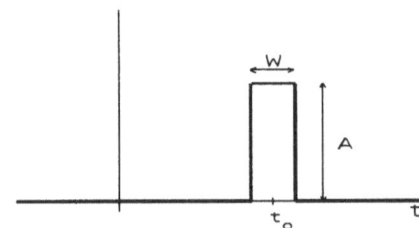

4.15   An AD-converter with a clock frequency of 40 kHz receives a sawtooth signal with a frequency of 7 kHz. Calculate eventual mirror frequencies which are caused by the first five harmonics due to aliasing.

4.16   Given a DA-converter with a zero-order hold filter. Calculate the aperture correction in dB which should be applied at the maximal signal frequency to compensate the attenuation in the pass band. Answer the same question for the case that fourfold oversampling has been applied.

4.17   Check with the same method as used in section 4.3.H for coefficient a_4, that it makes no difference whether a_1, b_1 and b_3 are calculated according to fig.4.3.31 or to fig.4.3.32.

4.18   Fig.4.3.13 shows one period of a periodical pulse signal. Calculate the Fourier coefficients a_0, a_n and b_n .

CHAPTER 5

System Theory

A signal processing system is a functional unit with one or more inputs to which we can supply signals, and one or more outputs where we can take off signals. An example of such a system is the harmonic oscillator that we got to know in chapter 3. In this chapter we will deal in the first place with the general properties of systems and we will treat the system as a 'black box' because we are only interested in the relation between the output and the input signals. Later in this chapter and particularly in the next chapter the 'inside' will be dealt with.

The black box approach has the advantage that very different systems (acoustical, electro-acoustical, electronic, analog and digital systems, and also systems like the speech and auditory organs) can be treated in the same way, and it does not matter whether we are dealing with acoustical, electrical or other signals. The following discussion is based on the diagam below with input signal $x(t)$ and output signal $y(t)$.

IN   OUT
$x(t)$ •   SYSTEM   • $y(t)$

5.1 Classification of systems

For the classification of systems several criteria can be used. As there are time-continuous and time-discrete signal functions, so there are time-continuous and time-discrete systems. This distinction often (but not always) corresponds with the classification *analog/digital*.

Based on certain properties of the relation between the output signal y and the input signal x systems are distinguished in *linear* and *non-linear* systems and besides that in *time-invariant* and *time-variant* systems and *causal* and *non-causal* systems. We will in the first place delve into some possible classifications.

A. *Analog and digital systems*

With analog systems the signal function has a physical 'carrier' (electrical volt-age, air pressure, magnetic field etc.) that fluctuates as a time-continuous func-tion. Upon this principle the construction of the system is based: an analog sys-tem accepts and produces such signals. On the other hand digital systems execute calculations with binary coded and electrically represented numbers derived from the signal function via sampling. They do so on the basis of a series of instruc-tions, a program. In the following I will not always deal separately with both possible systems. In those cases where the analog and the digital version behave in the same way it is sufficient to deal with one of them.

B. *Classification based on the input/output relation*

1. Linearity

The most important classification of systems is based on the distinction linear/non-linear (see also Gabel et al. 1973 and Poularikas et al. 1985). We call a system linear if both the *homogeneity* principle as well as the *superposition* principle are valid.
a) The homogeneity principle implies that multiplication of the input signal by a certain, constant factor leads to multiplication of the output signal with the same factor. If we symbolize the relation between the input signal $x(t)$ and the output signal $y(t)$ as follows:

$$x(t) \rightarrow y(t)$$

then the homogeneity principle is valid if

$$ax(t) \rightarrow a(t) \tag{5.1}$$

b) Set the input signal $x(t)$ equal to the sum of two signals or:

$$x(t) = x_1(t) + x_2(t)$$

If then the output signal consists of the sum of the two output signals $y_1(t)$ and $y_2(t)$ which belong to each of the input signals, we say that the superposition principle is valid.
Symbolically:

$$\text{if } x_1(t) \rightarrow y_1(t) \quad \text{and } x_2(t) \rightarrow y_2(t)$$

$$\text{then } x_1(t) + x_2(t) \rightarrow y_1(t) + y_2(t) \tag{5.2}$$

Both principles may be combined:

$$ax_1(t) + bx_2(t) \rightarrow ay_1(t) + by_2(t) \tag{5.3}$$

The superposition principle implies that no mutual influence or interaction of signals takes place. Compare e.g. an amplifier (linear) with a squaring circuit (non-linear):

amplifier

input/output relation: $y(t) = Ax(t)$

With input signals $x_1(t)$ and $x_2(t)$ we get $y_1(t) = Ax_1(t)$ and $y_2(t) = Ax_2(t)$

With input signal $x(t) = ax_1(t) + bx_2(t)$

we get $\qquad y(t) = A\{ax_1(t) + b\} = aAx_1(t) + bAx_2(t) = ay_1(t) + by_2(t)$

squaring circuit

input/output relation: $\quad y(t) = x^2(t)$

With the same input signal as in the previous case we now get the following output signal: $\qquad y(t) = \{x_1(t) + x_2(t)\}^2 = x_1^2(t) + x_2^2(t) + 2x_1(t)x_2(t)$

$$= y_1^2(t) + y_2^2(t) + 2x_1(t)x_2(t)$$

2. Time-invariance

We call a system time-invariant if the input/output relation does not change with a shift in time. The behaviour of the system depends in that case not on the point in time upon which it is used.
Symbolically:

$$\text{if } x(t) \rightarrow y(t) \text{ then } x(t - T) \rightarrow y(t - T) \tag{5.4}$$

3. The sine in/sine out-principle

It is not always easy to decide whether a given system is linear and time-invariant or not. The most simple situation is that the system equation, the mathematical relation between the input and the output signal, is known. If the system is linear, the same holds for the (differential or difference) equation, and if the system is time-invariant, the equation has constant coefficients. The equations of the harmonic oscillator that we discussed in chapter 3 have these properties and so the harmonic oscillator is a linear, time-invariant system.

It will become harder if the system is a real black box. Sometimes one may have recourse to testing it via the '*sine in/sine out principle*'. This term refers to the fact that, if to a linear, time-invariant system (a so-called LTI system) a

sinusoidal signal is connected, the output signal will be sinusoidal as well, with the same frequency but possibly another amplitude and/or initial phase angle.
Symbolically: $x(t) = A_1 \sin (2\pi f t + \phi_1)$ → $y(t) = A_2 \sin (2\pi f t + \phi_2)$ (5.5)

Proof:
Suppose that with the input signal $x(t) = \sin \omega t$ (setting $A_1 = 1$ and $\phi_1 = 0$ simplifies the calculation) corresponds the output signal $y(t) = f_o(t)$, so

$$\sin \omega t \rightarrow f_o (t)$$

then we have to prove that $f_o(t) = A \cdot \sin(\omega t + \phi)$
First we calculate what $y(t)$ is, if the input signal is a cosine function. From the time-invariance the following relation follows that is valid for all T-values:

$$\sin \omega (t + T) \rightarrow f_o (t + T)$$

Choose $T = \pi/2\omega$:

$$\sin \omega (t + \frac{\pi}{2\omega}) = \sin (\omega t + \frac{1}{2}\pi) = \cos \omega t \rightarrow f_0(t + \frac{\pi}{2\omega})$$

So: $$\cos \omega t \rightarrow f_0(t + \frac{\pi}{2\omega})$$

We can also work out the fact of the time-invariance as follows:

$$x(t + T) = \sin \omega(t + T) = \sin \omega t \cos \omega t + \cos \omega t \sin \omega t$$

With the superposition and homogeneity principle we can deduce that

$$y(t + T) = f_0(t + T) = f_0(t) \cos \omega T + f_0(t + \frac{\pi}{2\omega}) \sin \omega T$$

This holds for all values of t. Thus it is allowed to assume that $t = 0$:

$$y(T) = f_0(T) = f_0(0) \cos \omega T + f_0(\frac{\pi}{2\omega}) \sin \omega T$$

Both $f_o(0)$ and $f_o(\pi/2\omega)$ are constant. Let us replace them by respectively '*b*' and '*a*', and let us interchange the two terms. Then we get

$$y(T) = f_0(T) = a \sin \omega T + b \cos \omega T$$

Because this holds for all values of T, we may write t instead of T:

$$y(t) = f_0(t) = a \sin \omega t + b \cos \omega t$$

This can be elaborated in the known manner (rule(2.56)) to

$$y(t) = f_0(t) = A \sin(\omega t + \phi) \text{ with } A = \sqrt{a^2 + b2} \text{ and } \tan \phi = \frac{a}{b}$$

showing that the output signal is indeed the desired sine function.

Electrical systems can with help of the sine in/sine out principle be tested for linearity/time-invariance by connecting the input and output signal to respectively the X and Y input of an oscilloscope. The screen image should for all frequencies be either a straight line (then $\phi_1 = \phi_2$) or an ellipse (if there is a phase difference).

4. Causality

For a causal system it holds that the output signal which appears at a certain moment only depends upon the input signal supplied to the system before that time. This has as a consequence among other things that if $x(t) = 0$ for $t<0$ ($t=0$ marks the beginning of the x-signal), $y(t) = 0$ for $t<0$. This seems a trivial property, but this is not so. While working with non-real time systems (where the signal function is stored in some memory) it is possible to establish a non-causal input/output relation.

C. *Non-linear behaviour of practical linear systems*

Perfect linear systems only exist in theory. Every practical system will show deviations from the ideal linearity. We call this *non-linear distortion*. To indicate how near a system approaches the ideal, the amount of non-linear distortion should be indicated. That is not easy because non-linearity is not so much a property as the absence of it. In practice one manages by measuring the deviations of the sine in/sine out principle and/or the superposition principle. These deviations are defined by the *harmonic* distortion and the *intermodulation* distortion respectively.

1. Harmonic distortion

If the sine in/sine out principle is not completely valid, this mostly means that a sinusoidal signal with a somewhat deviating but periodic wave shape appears at the output. This means that to the signal harmonics are added and one specifies the amount of harmonic distortion via the ratio (in %) between the RMS value of the unwanted harmonics and that of the total signal.
The RMS value is the root of the average energy, and thanks to the theorem of Parseval the energy can be determined by the amplitudes C_n of the harmonics.

Thus we find for the total harmonic distortion (*THD*):

$$THD = \frac{\sqrt{C_2^2 + C_3^2 + C_4^2 + \dots}}{\sqrt{C_1^2 + C_2^2 + C_3^2 + \dots}} \cdot 100\% \approx \frac{\sqrt{C_2^2 + C_3^2 + C_4^2 + \dots}}{C_1} \cdot 100\% \qquad (5.6)$$

The simplification is justified by the fact that $C_2, C_3,\dots \ll C_1$.

One can measure *THD* by measuring the RMS values of the signal with and without (i.e. suppressed by a filter) the first harmonic. Sometimes the harmonic distortion per frequency component is specified:

$$d_n = \frac{C_n}{\sqrt{C_1^2 + C_2^2 + C_3^2 + \dots}} \cdot 100\% \approx \frac{C_n}{C_1} \cdot 100\% \qquad (5.7)$$

2. Intermodulation distortion.

If the superposition principle is not completely valid, interaction between the signals occurs. A consequence of this interaction could be that with two sinusoidal input signals the output signal contains new components with sum and difference frequencies. This can be seen as follows.

The most simple linear relation between input and output signal is given by the equation $y = ax$. In the case of a non-linear relation, this equation is replaced by the following power series: $\qquad y = a_1 x + a_2 x^2 + a_3 x^3 + \dots$.

With the input signal $\qquad\qquad x(t) = \sin \omega_1 t + \sin \omega_2 t$

appears as output signal:

$$y(t) = a_1(\sin \omega_1 t + \sin \omega_2 t) + a_2(\sin \omega_1 t + \sin \omega_2 t)^2 + a_3(\sin \omega_1 t + \sin \omega_2 t)^3 + .$$

The quadratic term can be worked out to:

$$a_2(\sin \omega_1 t + \sin \omega_2 t)^2 = a_2(\sin^2 \omega_1 t + 2 \sin \omega_1 t \sin \omega_2 t + \sin^2 \omega_2 t) =$$

$$= a_2\{\tfrac{1}{2} - \tfrac{1}{2}\cos 2\omega_1 t - \cos(\omega_1 + \omega_2)t + \cos(\omega_1 - \omega_2)t + \tfrac{1}{2} - \tfrac{1}{2}\cos 2\omega_2 t\}$$

Thus new frequency components ($\omega_1+\omega_2$ and $\omega_1-\omega_2$) appear. In the same way it can be shown that the cubic term results in the new frequency components $2\omega_1+\omega_2$, $2\omega_1-\omega_2$, $\omega_1+2\omega_2$, $\omega_1-2\omega_2$.

One can show in general that the *k*-power term leads to the frequencies

$$|m\omega_1 \pm n\omega_2| \quad \text{with } m + n = k \qquad (5.8)$$

These distortion products are more annoying than those caused by harmonic distortion because of the non-harmonic relation. The intermodulation distortion is specified in the same manner as the harmonic distortion via the proportion between the RMS value of the unwanted components and that of the total signal.

Non-linearity is only an unwanted property if the system concerned is supposed to be linear. For a number of applications non-linear systems are very important. This applies to e.g. measuring techniques, the modulation of signals and sound synthesis techniques.

Linear systems are used for the conversion, amplification and recording of signals. The theoretical treatment of both subjects is very different, because for the linear systems there is a elegant, generally applicable method.

For non-linear systems no uniform method exists, so that every system must be treated separately. The possibility of analysing such systems numerically with the aid of a computer has led to the fact that also from a theoretical point of view attention has substantially increased, and that for the analysis of these systems a theoretical foundation has been laid down. In chapters 6 and 7 several non-linear systems and operations are discussed.

5.2 The description of linear systems

For the analysis of linear, time-invariant systems there are besides the solution of the system equation (the method used in chapter 3) two more methods. The first focuses on the time domain, by tracing what the system 'does' with a given signal function. The second is a frequency domain method, based on the sine in/sine out principle. With this it is checked what the influence of the system is on the amplitude and phase of the spectral components. As we will see there is a close relationship between both methods, which is not astonishing, because it is a matter of different approaches to the same system.

A. *Time domain description; impulse response and convolution*

With a linear system it is possible for a given input signal $x(t)$ to predict what the output signal $y(t)$ will look like, if it is known how the system reacts to the input signal that consists of only one impulse. This reaction, this output signal we call the *impulse response* of the system. How this works can be shown most easily by a time-discrete system. As we shall see, it is a question of direct application of the superposition and homogeneity principles, and of time-invariance. The example is a digital system that consists of a time delay (= two memory cells P and Q), an adder and a multiplier as shown in fig.5.2.1.

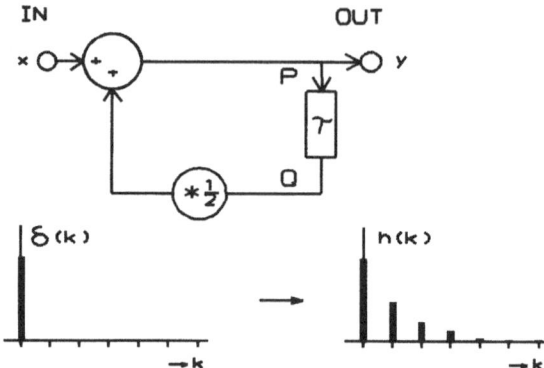

Figure 5.2.1 Impulse response of a time-discrete system.

Samples shift from P to Q. In Q we find numbers that were previously in P. These numbers form the output signal. In P the input samples are placed, after half of the value of the number in Q has been added. The difference equation for this system is: $y_k = x_k + \frac{1}{2}y_{k-1}$

This is a linear equation with constant coefficients. So the system is linear and time-invariant. The input signal consisting of one pulse is indicated by $\delta(k)$, The corresponding output signal, the impulse response, by $h(k)$. Below a simple computer program has been listed which we may look upon as the software version of this system, and which is of the same type as the programs we used in chapter 3 for the time-discrete harmonic oscillator.

	k	IN=δ(k)	P=IN+$\frac{1}{2}$Q =OUT=h(k)	Q
P = 0				
Q = 0				
IN = 1	0	1	1	0
5 Q = P	1	0	0.5	1
P = Q * 0.5 + IN	2	0	0.25	0.5
OUT = P	3	0	0.125	0.25
CALL WAITCL	4	0	0.0625	0.125
IN = 0	5	0	0.03125	0.0625
GOTO 5	.	.	.	.

If the start situation is that P and Q are equal to 0, we can trace what happens if we once supply the number '1' as an input signal. The first 6 output samples are shown in the table. The impulse response thus takes the values 1, 0.5, 0.25, . . . successively. Somewhat more accurately:

$$h(k) = 0 \quad \text{if } k < 0 \quad \text{and } h(k) = \frac{1}{2^k} \quad \text{if } k \geq 0$$

This impulse response is infinitely long. A system with an infinitely long impulse response is indicated by the letters IIR (Infinite Impulse Response). If the impulse response is finite, we use the term 'FIR system' (Finite Impulse Response). Imagine the case that we further supply to this system an (arbitrarily chosen) input signal which consists of 5 samples:

$$x(-2) = 3, x(-1) = 1, x(0) = 2, x(1) = 1.5 \text{ and } x(2) = 1.$$

To check what now appears at the output we regard this input signal as the sum of five input signals, each consisting of one single pulse. See fig.5.2.2.

The calculation of each of the five output signals which correspond to their input signals is easy. The only difference between these signals and that with which we have derived the impulse response is a shift in time and a change of amplitude. Because of the time-invariance the first means that the output signal is shifted in time correspondingly, while because of the homogeneity principle the amplitude factor is passed on directly to the output signal. So we see that the first pulse with amplitude 3 and index -2 leads to an output signal that begins with a sample with value 3 and index -2 and subsequently consists of a series of pulses decreasing by a factor 2.

The value of each sample of the complete output signal can be calculated by adding the five corresponding output samples. Let us for example see how the value of $y(1)$ can be calculated. In the figure we see that

$$y(1) = 0.375 + 0.250 + 1.000 + 1.500 + 0.000 = 3.125$$

Every number in this series is a sample of the impulse response multiplied by the value of the 'isolated' x-sample which is used as a scaling factor. For the first number $h(3)$ (=0.125) is multiplied by $x(-2)$ (=3.0), for the second $h(2)$ (=0.25) is multiplied by $x(-1)$ (=1.0) and so on. We may thus also write:

$$y(1) = 3.0 \cdot 0.125 + 1.0 \cdot 0.25 + 2.0 \cdot 0.5 + 1.5 \cdot 1.0 + 1.0 \cdot 0.0 = 3.125$$

or with the x- and h-samples:

$$y(1) = x(-2) \cdot h(3) + x(-1) \cdot h(2) + x(0) \cdot h(1) + x(1) \cdot h(0) + x(2) \cdot h(-1) = 3.125$$

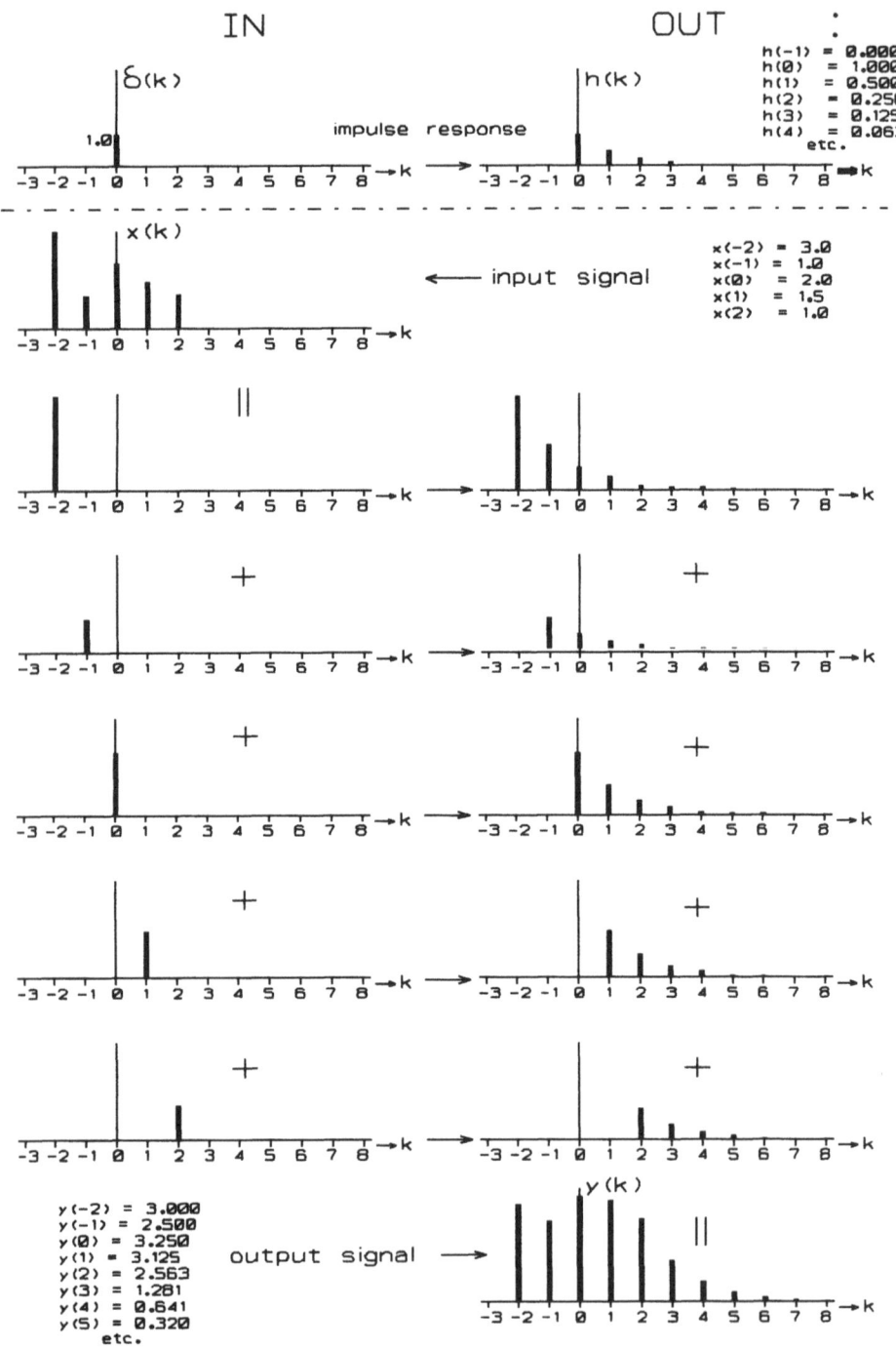

Figure 5.2.2 Convolution.

The index of the *h*-coefficient always follows from the 'distance' between the input and the output sample. We find the *h*-index by calculating the difference between the index of the output sample (here the '1' of $y(1)$) and that of the input sample:

h-index = *y*-index - *x*-index.

This leads us to the following expression for $y(1)$:

$y(1) = x(-2) \cdot h(1-(-2)) + x(-1) \cdot h(1-(-1)) +$
$\qquad + x(0) \cdot h(1-0) + x(1) \cdot h(1-1) + x(2) \cdot h(1-2)$

From this it follows what the general expression is for the output samples $y(k)$ at an arbitrary input signal *x* and an impulse response *h*:

$y(k) = .. \qquad + x(-5) \cdot h(k-(-5)) + x(-4) \cdot h(k-(-4)) + x(-3) \cdot h(k-(-3)) +$
$\qquad + x(-2) \cdot h(k-(-2)) + x(-1) \cdot h(k-(-1)) + x(0) \cdot h(k-0) + x(1) \cdot h(k-1) +$
$\qquad + x(2) \cdot h(k-2) + x(3) \cdot h(k-3) + \ldots$

Or in short:

$$y(k) = \sum_{n=-\infty}^{\infty} x(n) \cdot h(k-n) \qquad (5.9)$$

This expression is named the *convolution sum*. One also says that the output signal y is the convolution of the input signal x and the impulse response h and often the following shorter notation is used:

$$y(k) = x(k) * h(k) \qquad (5.10)$$

The convolution principle can also be applied to time-continuous signals, for such signals can also be looked upon as series of (infinitely densely superimposed) pulses. Instead of a sum we then get an integral:

$$y(t) = \int_{-\infty}^{\infty} x(\tau) \cdot h(t - \tau) \, d\tau = x(t) * h(t) \qquad (5.11)$$

The convolution principle here is, however, mainly of theoretical importance, while it is also applied practically to time-discrete signals. As an example I will describe the principle of the so-called 'transversal' digital filter (the exact meaning of this term will be explained in section 5.4).

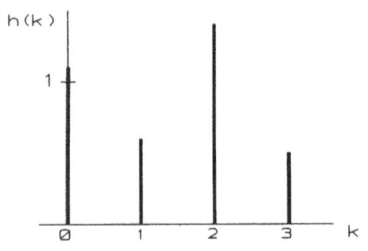

Fig.5.2.3
Desired impulse response.

Suppose that we want a system (filter), of which the impulse response consists of four samples:
$h(0) = 1.1$, $h(1) = 0.6$, $h(2) = 1.4$, $h(3) = 0.5$
(see fig.5.2.3). If we now write out the convolution sum

$$y(k) = \sum_n x(n) \cdot h(k-n)$$

we need only take into account the terms in which $h(0)$, $h(1)$, $h(2)$ and $h(3)$ occur, as all other h-values are equal to 0. We must thus determine the k-indices for which $k - n$ is equal to 0, 1, 2 or 3:

$$k - n = 0 \;\;\rightarrow\;\; n = k - 0 = k$$
$$k - n = 1 \;\;\rightarrow\;\; n = k - 1$$
$$k - n = 2 \;\;\rightarrow\;\; n = k - 2$$
$$k - n = 3 \;\;\rightarrow\;\; n = k - 3$$

Thus: $y(n) = x(k) \cdot h(0) + x(k-1) \cdot h(1) + x(k-2) \cdot h(2) + x(k-3) \cdot h(3)$

This operation can be realized with the circuit of fig.5.2.4, which consists of a series of (here: 4) memory cells or registers. Every time a new input sample appears the numbers in the four registers shift one place to the right. The number on the far right disappears. The contents of the four registers are, in the given manner, multiplied by the h-coefficients. The products are added. In this way the above sum is realized.

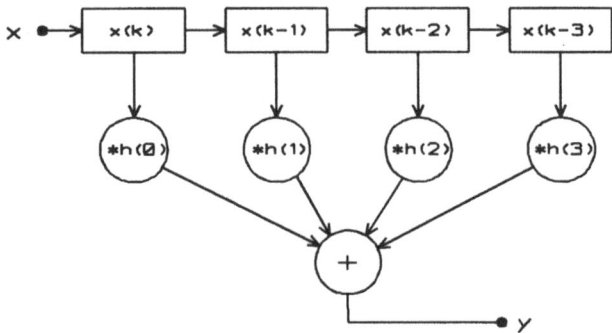

Figure 5.2.4 Transversal filter.

Such a FIR filter is called a *transversal* filter. It is easy to design and always stable. A 'long' impulse response however requires a proportionally large number of multiplications. If we send into this system a single '1' we see that the output signal indeed consists of the three consecutive impulses $h(0)$, $h(1)$, $h(2)$ and $h(3)$.

B. *Frequency domain description; the frequency response*

This description is based on the sine in/sine out principle by considering that
- every signal can be looked upon as a summation of sine vibrations,
- for a sine vibration the phase angle and amplitude is possibly changed by a linear, time-invariant system, but not the waveform and the frequency.

Symbolically (see (5.5)): $A_1 \sin (2\pi f t + \phi_1) \rightarrow A_2 \sin (2\pi f t + \phi_2)$

So if we know per frequency component how much the amplitude changes (e.g. in the form of the ratio $A_2(f)/A_1(f) = H_A(f)$, the so-called *amplitude response,* usually calibrated in dB's) and if we also know how much phase shift is introduced (the difference $\phi_1(f) - \phi_2(f) = \Delta\phi(f)$, of the *phase response*) then we are able to calculate from the spectrum $X(f)$ of the input signal $x(t)$ the spectrum $Y(f)$ of the output signal $y(t)$.

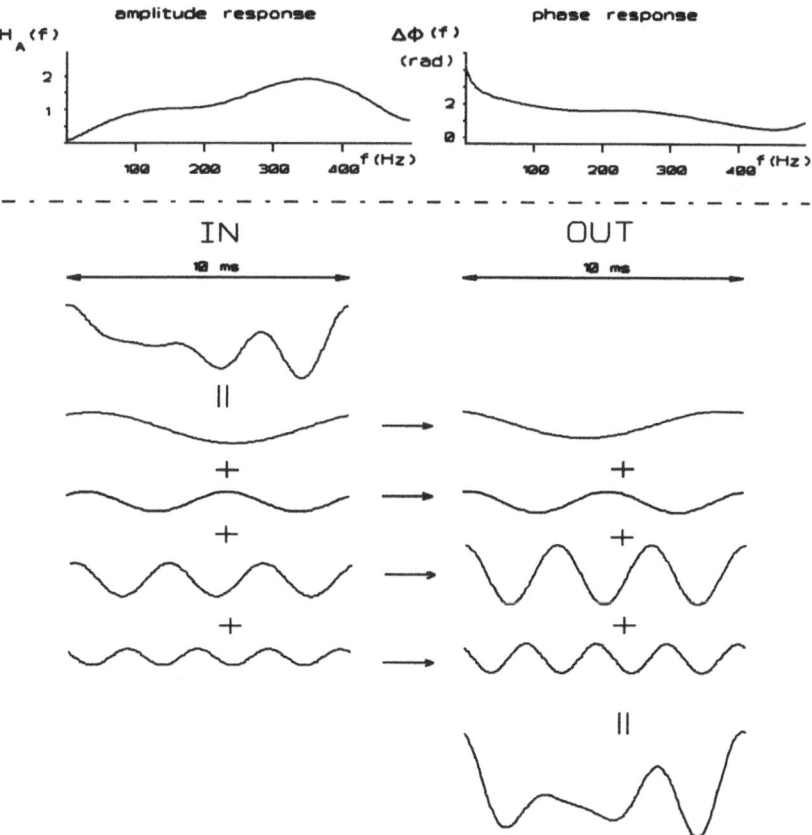

Figure 5.2.5 Construction of the output signal using the frequency response.

Because $A_2(f) = \dfrac{A_2(f)}{A_1(f)} A_1(f) = H_A(f) \cdot A_1(f)$ we find the following relation for

the amplitude spectrum: $Y(f) = X(f) \cdot H_A(f)$ (5.12)

The amplitude and phase response of a system can be represented graphically (see fig.5.2.5); together they form the *frequency response* $H(f)$ of the system.

   Fig.5.2.5 is the frequency domain version of fig.5.2.2. On the left, just below the dashed line an arbitrary input signal is shown. Below the sinusoidal components are depicted. Using the amplitude and phase response for each of these components it is determined how they appear at the output. This is shown on the right-hand side. Finally the complete output signal is constructed by adding these output components.

C. Determination of impulse and frequency response of a system

Just as with the Fourier transform of signal functions we can also determine the quantities concerned via calculation or by means of measurement.

- *Calculation*

If the system equation is known one can try to solve it with either a pulse or a sine function with an arbitrary frequency as input signal. The solution in the first case is the impulse response $h(t)$, and in the second case the frequency response $H(f)$. We have applied both methods to the harmonic oscillator, where we have determined $h(t)$ (the damped sinusoidal vibration) and $H(f)$ (the resonance curve). With time-discrete systems h is often simply determined by checking what happens if we send an '1' into the system. See for example the sections 3.2, 3.4 and 5.2.A.

- *Measurement*

a. It is obvious that the impulse response can be measured by recording the output signal if a short pulse is supplied to the system as input signal.
   Examples: Sending a short electrical pulse to a loudspeaker, or firing a pistol in a concert hall.

b. The frequency response can be determined by supplying a sinusoidal signal to a system and by measuring the change of the amplitude and the phase shift for a large number of frequencies. Example: the adjustment of a tape recorder by means of a test tape.

- *Conversion (h → H and H → h)*

It is not necessary to measure both h and H, because, as will be shown below, one quantity can be derived from the other. The spectrum of a single pulse (see fig. 4.3.17) has zero points. Their location on the frequency axis depends on the pulse width W as we have seen. The narrower the pulse, the higher the zero point frequencies. With an infinitely narrow pulse ($W=0$) the first zero point, which we meet at frequency $f = 1/W$, lies infinitely far away, which means that the continuous spectrum of a single, infinitely narrow pulse is completely flat. It contains 'all' the frequencies with the same amplitude (fig.5.2.6). By supplying such a pulse to the system it is as if we execute the above described measurement at one blow. Suppose that at the output we then meet the in fig.5.2.7 reproduced signal, to which corresponds the spectrum also reproduced in fig.5.2.7. This spectrum reproduces what happened to all the originally equally strong frequency components. In other words:

The spectrum of the impulse response is identical to the frequency response of the system.

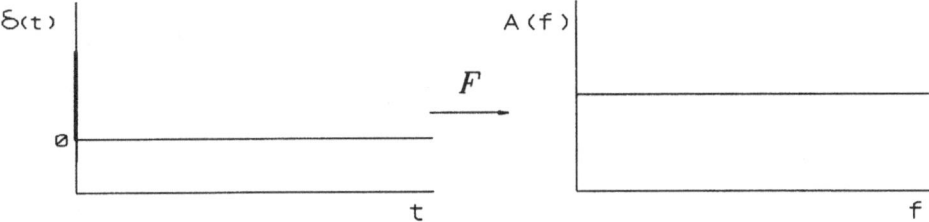

Figure 5.2.6 Time function and spectrum of a single, very narrow impulse.

Impulse response and frequency characteristic form a Fourier pair. This can be shown by results found earlier. In the figures 3.3.2 and 3.5.3 the impulse response and frequency characteristic of the harmonic oscillator are reproduced. In section 4.3.G we have looked at the continuous spectrum of a damped sinusoidal signal. Indeed this spectrum is identical to the resonance curve.

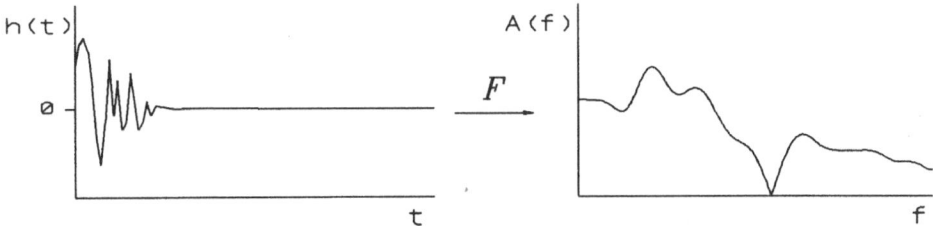

Figure 5.2.7 Impulse response and spectrum.

In fig.5.2.8 the above is recapitulated. One sees a system with impulse response $h(t)$ and frequency characteristic $H(f)$, an input signal $x(t)$ with spectrum $X(f)$ and the corresponding output signal $y(t)$ with spectrum $Y(f)$. If * indicates the convolution operation, F the forward and F^{-1} the inverse Fourier transform, we can state the relations between the functions as follows:

$$
\begin{aligned}
y(t) &= x(t) * h(t) & X(f) &= F\{x(t)\} \text{ or } x(t) = F^{-1}\{X(f)\} \\
Y(f) &= X(f) \cdot H(f) & Y(f) &= F\{y(t)\} \text{ or } y(t) = F^{-1}\{Y(f)\} \\
& & H(f) &= F\{h(t)\} \text{ or } h(t) = F^{-1}\{H(f)\}
\end{aligned}
\tag{5.13}
$$

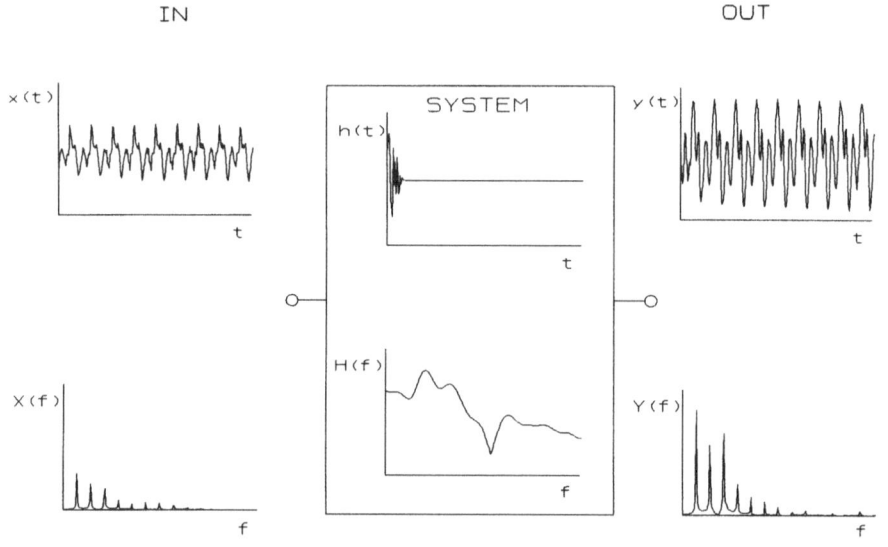

Figure 5.2.8 Relation between input signal, system properties and output signal.

To be able to analyse the system one should know one component of two of the three Fourier pairs (x,X), (y,Y) or (h,H). We are then able for example
- to extract the system properties from input and output signal,
- to calculate the output signal from the input signal and the system properties ('filtering'),
- to calculate the input signal from the output signal and system properties ('inverse filtering').

We can follow different trajectories; for example:

given $x(t)$ and $H(f)$, calculate: $h(t) = F^{-1}\{H(f)\}$ and $y(t) = x(t) * h(t)$

or calculate: $X(f) = F\{x(t)\}$, then $Y(f) = X(f) \cdot H(f)$ and $y(t) = F^{-1}\{Y(f)\}$.

These operations are really performed in practice. In this way via inverse filtering one has been able to remove the effect of the sound funnel used in old acoustic recordings and to improve the sound quality of historical recordings.

Sometimes it is possible to calculate from just one of the three components (the output signal $y(t)$), both $h(t)$ and $x(t)$. However then there must be extra information about the signal, for example the fact that the signal is a speech signal. Examples of such techniques are treated in chapter 7.

5.3 Distortion-less linear systems

If we do not wish a linear system to influence the signal some way or other we must make sure that the output signal function is identical to the input signal function. Two deviations can be tolerated: an possible multiplication of the function by a constant scale factor p, and an possible time delay t_o of the total signal. See fig.5.3.1.

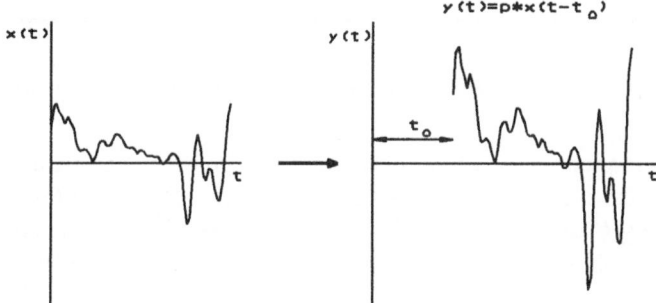

Figure 5.3.1 Distortionless transmission.

The impulse response of such a system is a by p multiplied and over t_o shifted impulse (fig.5.3.2), thus: $h(t) = p \cdot \delta(t - t_o)$

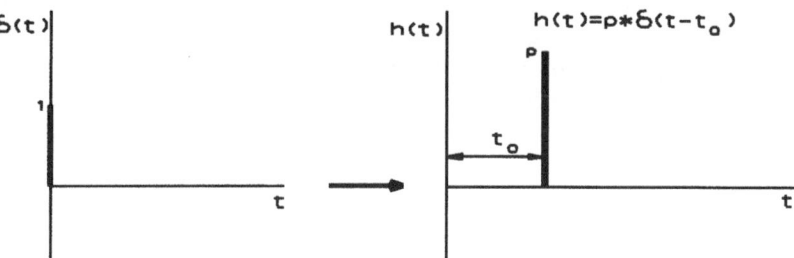

Figure 5.3.2 Impulse response with distortionless transmission.

We are now able to determine the amplitude and phase response of the system via the Fourier transform of the impulse response, but also using the argument that since each signal gets the same amplitude factor p and the same delay t_o this also holds for a sinusoidal signal with an arbitrary frequency, amplitude 1 and initial phase angle 0. According to the sine in/sine out principle it generally holds that: $A_1 \sin (2\pi f t + \phi_1)$ → $A_2 \sin (2\pi f t + \phi_2)$

And here: $\sin 2\pi f t$ → $p \sin 2\pi f(t - t_0) = p \sin (2\pi f t - 2\pi f t_0)$

The amplitude response is now for all frequencies: $H_A(f) = A_2(f)/A_1(f) = p$ (5.15)

The phase response $\Delta\phi(f) = \phi_1(f) - \phi_2(f)$ is: $\Delta\phi(f) = 2\pi f t_0$ (5.16)

Shown graphically (fig.5.3.3):

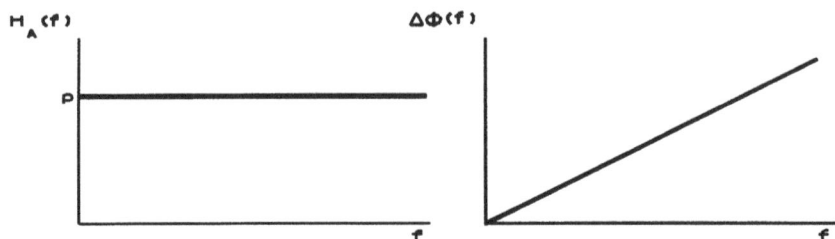

Figure 5.3.3 Amplitude and phase response with distortionless transmission.

The system is characterized by a straight horizontal amplitude and a linear phase characteristic. The phase shift increases linearly with the frequency and the faster the larger t_o. Reversely t_o can be derived from the steepness of the phase characteristic by reading the phase shift $\Delta\phi$(in degrees or radians) at an arbitrary frequency. The time delay t_o can be calculated from: $t_0 = \Delta\phi/2\pi f = \Delta\phi/360°\cdot f$.

That the phase shift should be proportional to f can be seen by having a look at the following example: for a time delay of 10 ms a frequency component of 100 Hz (period duration: 10 ms) must be shifted over one period $= 2\pi$ radians or $360°$. A component of 200 Hz (period duration: 5 ms) must be shifted over 2 periods $= 4\pi$ radians $= 720°$.

Usually we want sound processing systems to be linear and distortion-less over the range of 'audible' frequencies, thus from ca. 20 Hz till 20000 Hz. Due to the assumed phase insensitivity of the ear, not much attention has been paid to the phase characteristic. In the last few years this has changed somewhat because we have learned that the ear is not 'phase deaf' and that phase distortion can be audible, in particular in non-stationary signals. For example there are now linear phase loudspeakers on the market. Other systems that belong to the category of linear/distortionless transmission systems:

    amplifiers
    transducers (loudspeakers and microphones)
    time delays
    recording systems (records, tape; analog and digital)
    AD- and DA-converters

The degree in which a particular linear system deviates from the ideal distortion-less system (one speaks here of linear distortion, wave shape distortion and also of amplitude and/or phase distortion) can be specified exactly by means of the amplitude and phase characteristic. One can show the complete characteristics; often only the tolerances are specified ("amplitude characteristic straight within 2 dB between 20 and 20000 Hz").

5.4 Filters

Under this term fall all systems that are linear (the superposition principle holds, and if the system is time-invariant the sine in/sine out principle holds as well), but that cause changes in the waveform. The impulse response is not a single (delayed, enlarged or attenuated) impulse, or equivalently, the amplitude characteristic is not horizontal and flat, and/or the phase characteristic is not linear.

The manner in which a filter is specified follows directly from the above: for a time-domain specification one describes the impulse response, and for a frequency domain specification one gives the amplitude and/or phase characteristic.

A.   *The frequency domain specification of filters*

1. The amplitude response
The frequency domain specification is the most common one, at least certainly for analog filters; and this is the reason we will now discuss this first. One can show the complete characteristics but often it is enough to describe them with the aid of *cutoff frequencies* and *flank steepness*. Cutoff frequencies are those frequencies at which the amplitude characteristics changes from a horizontal into a non-horizontal line. To put it more accurately: those frequencies at which the signal is attenuated with 3 dB. The term '3 dB-points' is used as well. The further one goes from the cutoff frequency the stronger the attenuation. Often the relationship between the attenuation (expressed in dB) and the frequency(-interval) distance is linear which means that the flank steepness can be expressed in (for example) dB/octave. Below some filter characteristics are shown with a cutoff frequency of 1000 Hz and a steepness of 12 dB/octave. With a characteristic as in fig.5.4.1a one speaks of a *low pass filter* and in fig.5.4.1b of a *high pass filter* for obvious reasons.

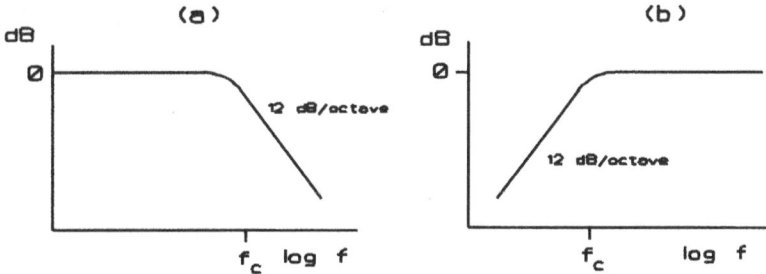

Figure 5.4.1 Lowpass and highpass filter.

Besides this there is the important group of band filters that have two cutoff frequencies and which let through or suppress the range in between (*band pass filter* and *band suppress filter* or *notch filter*), and that usually have large

flanksteepness, see fig.5.4.2a and fig.5.4.2b. Instead of the two cutoff frequencies one also gives the centre or resonance frequency and bandwidth. We have
seen this with the harmonic oscillator.

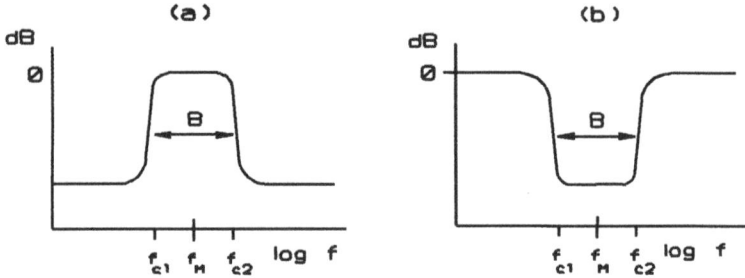

Figure 5.4.2 Bandpass and bandsuppress filter.

We further distinguish the following filter types:
- the filter bank, a series of independent and parallel band filters with adjacent
  cutoff frequencies (see also fig.4.3.34).
Usually there is a constant proportion between the frequencies of the upper and
lower cutoff frequencies. If this proportion is a factor 2 we speak of an *octave
filter*, with a factor 1.5 of a *half-octave* (= fifth) *filter*, and with 1.2 of a *third-
octave filter*. If furthermore the attenuation per passband can be adjusted then we
speak of a *spectrum shaper* because of the possibility of setting a global frequency characteristic, and also of an *equalizer* because of the possibility of
realizing certain frequency corrections.
- the comb filter, a filter with a periodic sequence of passbands (section 6.3,
  fig.6.3.3).

2. The phase response (Preis 1982)
To specify the phase response one can choose from three possibilities:
a. The phase shift $\Delta\phi(f)$.
   This is the value of $\Delta\phi(f) = \phi_{in}(f) - \phi_{out}(f)$ (5.17)
   in degrees or radians as a function of the frequency. This method was used
   before.
b. The 'phase delay' τ_f. This is the time delay of each sinusoidal component. It
   can be derived from (5.17) as the phase shift with regard to the input signal,
   thus $\phi_{out} - \phi_{in}$, is equal to $-\Delta\phi$:

$$\sin(2\pi ft - \Delta\phi(f)) = \sin(2\pi ft - 2\pi f\frac{\Delta\phi(f)}{2\pi f}) =$$

$$= \sin 2\pi f(t - \frac{\Delta\phi(f)}{2\pi f}) = \sin 2\pi f(t - \tau_f)$$

Thus: $$\tau_f = \frac{\Delta \phi(f)}{2\pi f} \qquad (5.18)$$

- The 'group delay' τ_{gr}. This measure is based on the principle of stationary phase (see section 4.3.G, fig.4.3.24). There we saw that using the principle of stationary phase the centre of gravity of a signal function can be derived from the phase spectrum with $t_p = (-1/2\pi)\cdot d\phi/df$. The time delay τ_{gr} of this centre of gravity can then be calculated from

$$\tau_{gr} = t_{p,out} - t_{p,in} = -\frac{1}{2\pi}\frac{d(\phi_{out} - \phi_{in})}{df} = \frac{1}{2\pi}\frac{d}{} \qquad (5.19)$$

It is important to see that τ_f and τ_{gr} do not need to be equal. Therefore two examples (derived from Blauert (1972)):

Example 1
For a linear, distortion-less system (with amplitude factor 1 and time delay t_o) it holds that as fig.5.4.3 shows, a pulse-like input signal only undergoes a time delay. This means for

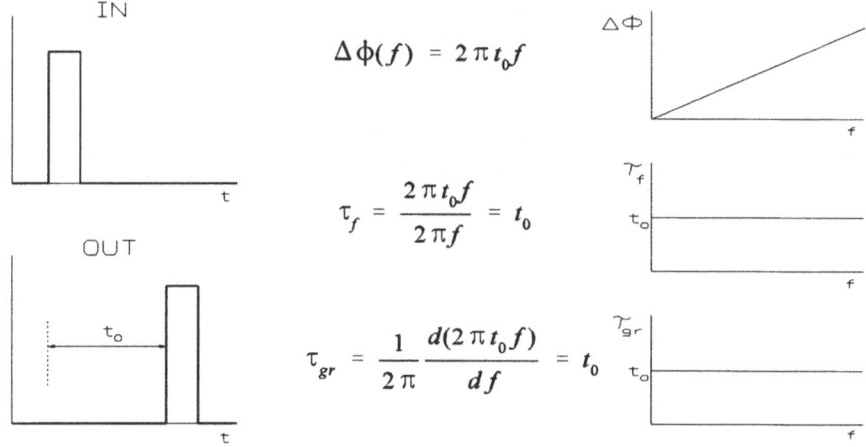

$$\Delta \phi(f) = 2\pi t_o f$$

$$\tau_f = \frac{2\pi t_o f}{2\pi f} = t_0$$

$$\tau_{gr} = \frac{1}{2\pi}\frac{d(2\pi t_o f)}{df} = t_0$$

Figure 5.4.3 Phase response, phase delay and group delay of a distortion-less transmission system.

The phase delay is constant, thus all sine components are equally delayed and this of course holds as well for their sum, the signal itself.

Example 2

Let us have a look at a system where the output is the mirrored version of the input signal. To mirror a signal we must mirror all the frequency components, or in other words give them a phase shift of 180°, thus:

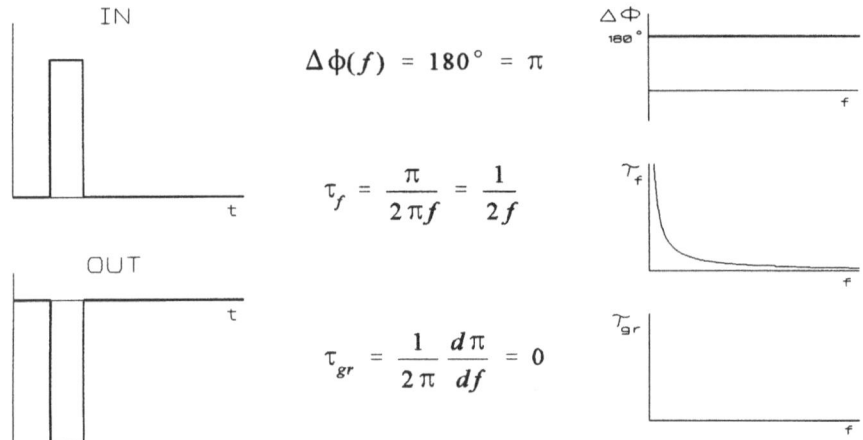

$$\Delta \phi(f) = 180° = \pi$$

$$\tau_f = \frac{\pi}{2\pi f} = \frac{1}{2f}$$

$$\tau_{gr} = \frac{1}{2\pi} \frac{d\pi}{df} = 0$$

Figure 5.4.4 Phase response, phase delay and group delay with mirrored signals.

Thus with a constant phase shift a time delay corresponds that increases with lower frequencies (at 1000 Hz → 180° = 0.5 ms; at 100 Hz → 180° = 5 ms; at 10 Hz → 180° = 50 ms), but the 'centre of gravity' of the signal is not shifted. Indeed: $\tau_{gr} = 0$.

B. The time domain specification of filters

Specification of typical time-domain filters is not standardized and a classification system does not exist for such filters except for the distinction between FIR and IIR filters. It is yet becoming more common to specify the impulse response, especially with linear phase systems such as CD-players. That is because phase linearity causes the impulse response to be symmetric. To show that here is the following argument:

The impulse response of a system can be determined via the inverse Fourier Transform of the frequency response (rule (5.13)). Suppose we have a system that has an arbitrary amplitude response $H_A(f)$ and a linear phase response $\Delta\phi(f)$ (thus $\Delta\phi(f) = -2\pi f t_o$). Using the formula for the Fourier integral (4.16) we may write for $h(t)$:

$$h(t) = \int_0^\infty H_A(f) \cos(2\pi ft - \Delta\phi(f)) \, df = \int_0^\infty H_A(f) \cos(2\pi ft - 2\pi ft_0) \, df$$

$$= \int_0^\infty H_A(f) \cos 2\pi f(t - t_0) \, df$$

This function is symmetrical around t_0 (and has its maximum at that time point). To prove this we show that $h(t_0 - t) = h(t_0 + t)$:

$$h(t_0 - t) = \int_0^\infty H_A(f) \cos(-2\pi ft) \, df, \quad h(t_0 + t) = \int_0^\infty H_A(f) \cos(+2\pi ft) \, df$$

These two expressions are equal to each other because $\cos(-2\pi ft) = \cos(+2\pi ft)$. Conclusion: a linear phase system has an impulse response that is symmetrical around t_0, the time delay of the system.

The reverse (a symmetrical impulse response implies phase linearity) can be proven as well. If the coefficients of a transversal filter (fig.5.2.4) are chosen so that the impulse response of the filter is symmetric, then the filter has a linear phase response. This is an extra bonus making this type of digital filter even more attractive. Linear phase filters introduce no phase shift between frequency components; the only time effect is a delay of the total signal.

Finally sometimes the term *minimum phase system* is used in this context. A short remark about this type of system: for causal systems it can be proven that there is a relation between the amplitude response $H_A(f)$ and the phase response $\Delta\phi(f)$. This means that with such systems there is a theoretically predictable, minimal phase shift. The relationship between H_A and $\Delta\phi$ is given by the Hilbert Transform:

$$\Delta\phi_{min}(f) = \frac{1}{\pi} \int_{-\infty}^\infty \frac{\ln H_A(f')}{(f - f')} \, df'$$

If the actual phase shift which occurs is equal to the theoretically predictable minimal phase shift, we speak of a 'minimum phase system'. Clearly this phase shift can lead to phase distortion, to non-linearity of the phase response.

5.5 Problems

A time-discrete, linear and time-invariant system, using a clock frequency of 20 kHz, has the impulse response shown below ($h(0) = h(2) = 0.5$, $h(1) = 1$) Answer the questions 5.1 to 5.8.

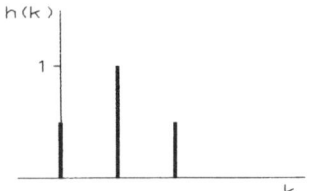

5.1   Is this a FIR or a IIR system? How can it be constructed?

5.2   Is this a system with or without linear distortion?

5.3   Is this system phase linear or not?

5.4   Derive the convolution sum for this impulse response.

5.5   Calculate the output signal $y(k)$ when the input signal $x(k)$ is equal to:
      $x(k) = \sin s \cdot k$

5.6   Using the result of problem 5.5 sketch the amplitude response of this system. What kind of filter is it?

5.7   Calculate the flank steepness in dB/octave by deriving the amplitude factors at 3500 Hz and 7000 Hz.

5.8   What is the time delay of the system? Calculate the phase shift at 1000 Hz and sketch the phase response.

5.9   With strong overmodulation a sinusoidal signal changes into a square wave. Calculate the total harmonic distortion (*THD*).

5.10 Which device does not belong in this list:
           microphone - pre-amplifier - fader - presence filter - line-amplifier.
      Why not?

5.11 At which frequency will be the group delay of a harmonic oscillator maximal?

5.12 Given a transversal filter with 21 cells, with all *h*-values equal to 1.21. The clock frequency is 10 kHz.
  a. Show that this filtering is equivalent to replacing each sample by the average value of that sample and the 20 preceding samples.
  b. This 'moving average' is a filtered version of the original signal. Describe this filter.
  c. Is this a linear phase filter? If so, what is the time delay?
  d. Show that due to this filtering the sampling frequency may be decreased tot ca. 1000 Hz. What is the easiest way to achieve this?

5.13 Prove that the principle of homogeneity (5.1) follows from the superposition principle when the factor *a* is a rational number.

CHAPTER 6

Systems for Sound Signal Processing

Based on the theory dealt with in the last chapter we shall now deal with practical systems and operations. In this I have consciously limited myself to electric and electronic systems, analog and digital. For the treatment of subjects from electro-acoustics I refer to the vast supply of literature in this field. The subdivision of this chapter is based upon the classification of systems dealt with in chapter 5. I shall thus first pay attention to linear systems (distortion-less systems and filters) and then to non-linear systems. Before I begin to deal with the treatment of concrete electrical systems I would like first to summarize some elementary facts and formulae for such systems.

6.1 Elementary electrical quantities, concepts and circuits

A. *Voltage, current, resistance and power*

In an electrical system the physical carrier of the signal function is an electrical voltage or current. The value of these quantities can be proportional to the signal function (the analog representation), or one works with an independent series of voltage (or current) carriers by which only two values are differentiated (the binary codification of the digital signal representation).

An electrical voltage is the result of a different concentration of electrical charge carriers. There are materials in which the mobility of the charge carriers is large (conductors; most metals can conduct electricity). With such conductors electrical voltages can be transported.

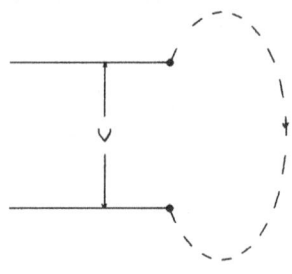

Figure 6.1.1
Voltage and current.

We shall mainly concentrate on the situation that there is a voltage difference between two conductors, for example between two copper wires. If between two such wires a more or less conductive connection is made, the charge concentration difference will decrease, because the charge carriers will move through the connection. We speak in short of an electrical current. For the measurement of voltages and currents appropriate units are required: *volt* V and *ampere* A. With the help of such measurements it can be observed that in the situation

sketched above with a given voltage V a current i will flow, so that i is proportional to V. This relation (which has its exceptions) is called *Ohm's law*. If two
quantities are proportional they can be made equal to each other by multiplying
one by a proper factor of proportionality. This leads to the well-known formula
of Ohm's law:

$$V = R \cdot i \qquad (6.1)$$

The factor of proportion R is called the (electrical) *resistance*. The resistant
connection through which a current of 1 A flows if the voltage difference
amounts to 1 V has per definition the unit of resistance, *ohm* (symbol: Ω). Good
conductors have a low resistance, which means that with a given voltage a strong
current will flow. There are also materials that are poor conductors: isolators. A
connection of this material will have a high R-value. Resistance values can in
practice vary from less than 0.001 Ω to 10^{10} Ω and more.

 If because of a voltage an electrical current begins to flow, energy comes free
for example as heat or mechanical power. This energy is provided by the source
of the electrical voltage. The amount of energy depends upon the voltage V and
the current i, and is expressed in the universal unit of energy, *Watt*:

$$E = V \cdot i \qquad (6.2)$$

With the help of Ohm's law two other versions of this relation can be derived:

-     with $V = R \cdot i$: $E = R \cdot i^2$ $\qquad (6.3)$

-     with $i = V / R$: $E = V^2 / R$ $\qquad (6.4)$

Electrical voltages can be generated in various ways, but it always concerns
processes in which non-electrical energy is converted into electrical energy. This
can occur for example via chemical (battery), mechanical (dynamo) and electromagnetic (solar cells) processes. The result of such a process is a voltage difference between two conducting connections, for example the poles of a battery. In
the batteries shown in fig.6.1.2 there is a voltage of 1.5 volts between the points
A and B, and one of 4.5 volts between C and D, but when there is no connection
between the batteries there is no voltage difference between A or B on the one
hand and C or D on the other. This changes only if a conductive connection between the two batteries is made. Before we can investigate what happens in such
a case, we have to realize that an electrical current flowing through a conductor
can have two directions. To know the situation at the given voltage source one
marks the two poles with a plus and a minus sign. If we know from the example
that A and D form the positive poles, and B and C the negative, we can see what
will occur if we connect the two batteries.

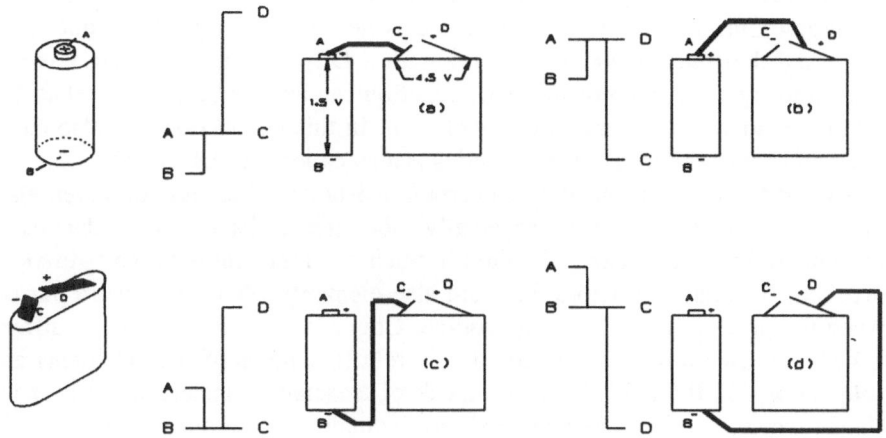

Figure 6.1.2 Electrical Voltages.

(a) $V_{AC} = 0$ (poles connected by means of a conductor)
        $V_{BD} = 6$ volts (D positive with regard to B)
(b) $V_{AD} = 0$, $V_{BC} = 3$ volts (B positive with regard to C)
(c) $V_{BC} = 0$, $V_{AD} = 3$ volts (D positive with regard to A)
(d) $V_{BD} = 0$, $V_{AC} = 6$ volts (A positive with regard to C)

Voltage poles that are not in one way or another galvanically connected are called 'floating'. The connections shown in fig.6.1.2 can be made directly, but they can also be indirectly formed using a central conductor ('ground'). If the voltage source produces a constant voltage we speak of a DC voltage ('Direct Current'). There are also voltage sources where the voltage between the poles is not constant as with batteries, but fluctuates in time. This is for example the case with the poles of a mains outlet.

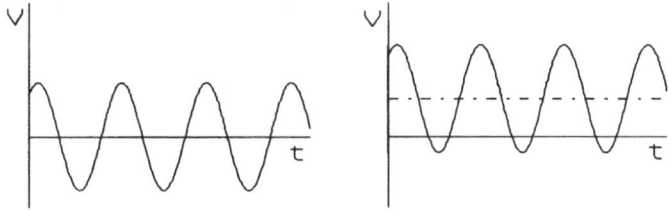

Figure 6.1.3 Alternating voltage with and without DC-component.

Here it can occur that the polarity of the poles (and thus also the direction of the current) alters. If this is the case in such a way that the average value of the voltage is 0 volts (as in fig.6.1.3, left side) then one speaks of a (true) AC voltage (Alternating Current). If the average value is not 0 (as in fig.6.1.3, right side) then we say that there is an AC voltage with a DC component, because as we have seen in fig.4.3.3, a non-symmetric function can always be considered as the sum of a symmetrical one and a constant term. In this connection it is also clear why the constant term of the Fourier series is also called the 'DC term'.

Every conductive connection has a certain resistance. There are, however, also conductive components that are especially fabricated to have a particular resistance value to be used as parts of a circuit. Such a component is called a *resistor*. Apart from 'ordinary' resistors there are also elements where if they are placed between the poles of an AC voltage source, Ohm's law holds but a phase difference occurs between the voltage and the current (examples of such elements are the capacitor and the coil). If so we speak of 'reactance' instead of resistance, while the word *impedance* as a general term contains both groups. Although this fact is important in particular for analog circuit techniques, because it introduces frequency dependency, I want to spend as little attention as possible to it in this elementary introduction.

Time dependent voltages (and currents) can be considered as examples of signal functions. Everything stated about these (such as the concept of RMS-value, level comparison via the dB, etc.) is thus also applicable to these voltages. In considering the dB it must be kept in mind that the electrical energy at various places in a circuit is not only dependent upon V^2 (or i^2) but also upon R. We shall encounter examples of this complication.

B. *Series and parallel circuits of resistors*

Ohm's law enables us to calculate what occurs if we connect more than one resistor between the poles of a voltage source. With two resistors there are two possibilities: series connection (fig.6.1.4, the rectangles indicate the resistors) and parallel connection (fig.6.1.5).

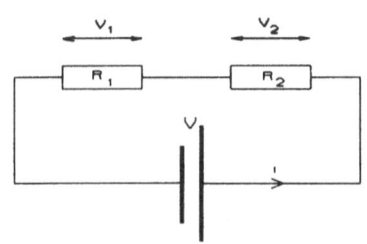

Figure 6.1.4 Series circuit.

a)   Series connection.
Which current will flow through this circuit? Because the current that flows through R_1 also flows through R_2, the voltage V will divide itself over the two resistors such that this is the result:

$$V = V_1 + V_2 = i \cdot R_1 + i \cdot R_2 = i(R_1 + R_2)$$

The current is that which would flow through one resistor with the value $R_1 + R_2$. With a series circuit the resistance values can thus be added. We can derive that for the relation between V and V_2 the following holds:

$$\frac{V_2}{V} = \frac{i R_2}{i (R_1 + R_2)} \quad \text{or} \quad V_2 = \frac{R_2}{R_1 + R_2} V \tag{6.5}$$

We speak of a *voltage divider*.

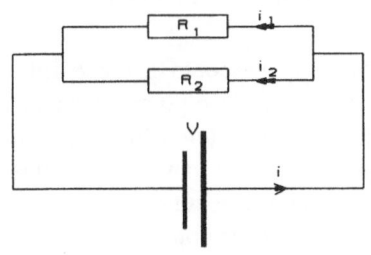

Figure 6.1.5 Parallel circuit.

b)   Parallel circuit. Over both resistors now occurs the same voltage V.
Through R_1 flows current $i_1 = V/R_1$
Through R_2 flows current $i_2 = V/R_2$
The total current i is the sum of these two currents:

$$i = i_1 + i_2 = \frac{V}{R_1} + \frac{V}{R_2} = V(\frac{1}{R_1} + \frac{1}{R_2})$$

If we wish to replace the two resistors by one without changing the total current we must give the resistor the value R so that $1/R = (1/R_1 + 1/R_2)$. The total resistance of the two parallel connected resistors is thus

$$R = \frac{R_1 \cdot R_2}{R_1 + R_2} \tag{6.6}$$

C. *Voltage source, input and output impedance*

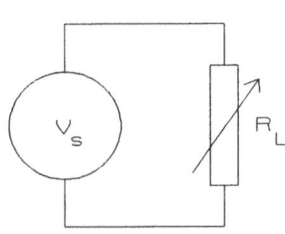

Figure 6.1.6
Voltage source with a variable
load resistance.

If a resistor is placed between the poles of a voltage source, a current will flow and energy will be withdrawn from the source. Because the available amount of energy is limited, of course, we must see what occurs if we withdraw more energy from the source.
   To this end we look at the circuit to the left (fig.6.1.6). The rectangle with the arrow indicates a variable resistor. The energy is converted into heat. The amount is:

$$E = \frac{V_s^2}{R_L}$$

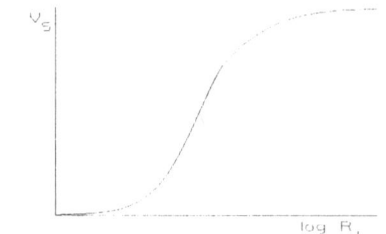

Figure 6.1.7 Relation between load resistance and output voltage.

Assuming V_s to be constant, according to this formula, we could make E infinitely large by reducing R_L. This is, of course, impossible. If we go on to reduce R_L, V_s will also start to diminish at a certain moment. The relationship between R_L and V_s is shown in the diagram above. It appears that we can describe this behaviour by acting as if the poles of the voltage source are connected via an 'internal' resistor with an 'ideal' voltage source (with an ever constant voltage).

In fig.6.1.8 this situation is sketched. The internal resistance R_i is drawn as a 'normal' resistor, thus as a rectangle, but this resistor is not a true, physical component. It is a conceptual resistance, a help with which to describe the behaviour of the circuit (and with that a property of voltage sources) as it depends on the load resistance R_L. R_i and R_L form a series circuit of two resistors and V_s now follows from the voltage-divider formula:

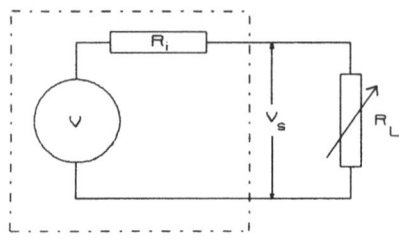

Figure 6.1.8 Internal resistance.

$$V_s = \frac{R_L}{R_i + R_L} V$$

It is this relation that is shown in fig.6.1.7 with R_L as variable. We see that if
- $R_L \gg R_i$ → $V_s \approx V$ (voltage constant),
- $R_L \ll R_i$ → $V_s \approx 0$,
- $R_L = R_i$ → $V_s = \frac{1}{2}V$.

We can characterize every voltage source by means of its internal resistance that indicates what the effect is of connecting a load resistor.

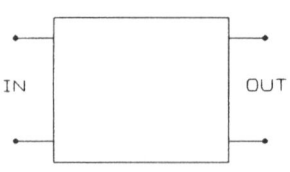

Figure 6.1.9 Four-pole.

An electrical system with one (voltage-) input and one (voltage-) output can be represented symbolically as a 'four-pole' (fig.6.1.9) with an input that accepts a voltage and an output that produces one and is thus a voltage source. If we connect a voltage to the input a (possibly very small) current begins to flow. This means that we can also describe the input behaviour with a resistance, the *input resistance* R_{in}. Fig. 6.1.10 shows the complete block diagram of an arbitrary system.

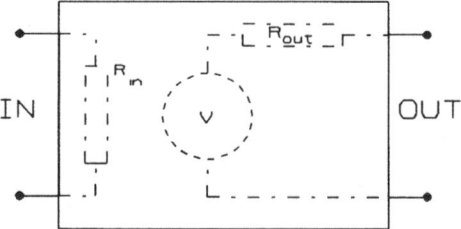

Fig.6.1.10 Input and output resistance.

We call the internal resistance of the output the *output resistance*, indicated by R_{out}. Once again: the components drawn do not have to be physically present but they serve to describe the behaviour of the system regarding voltages and currents at input and output. If for example several inputs are simultaneously connected to one output, then this output is loaded with a resistance the value of which is equal to that of the parallel connection of all the input resistances. It can be stated that in general it is easier to connect systems with one another when the output resistance of the source is small compared to the input resistance of the receiver.

Maximum energy transfer takes place only if the input resistance of the second system is equal to the output resistance of the first. We then speak of the proper *matching*. This can be proven by considering R_L in fig.6.1.8 as a variable resistor and by calculating how much energy is withdrawn from the source for different values of R_L. E is thus treated as a function of R_L:

$$E(R_L) = \frac{V_s^2}{R_L} = \frac{R_L}{(R_1 + R_L)^2} V^2 \tag{6.7}$$

This relation between E and R_L is shown in fig.6.1.11. The point where the graph reaches its maximum can be found by differentiating the function $E(R_L)$ to R_L and then solving the resulting equation $E'(R_L) = 0$:

$$\frac{dE}{dR_L} = \frac{V^2(R_i + R_L)^2 - 2(R_i + R_L)R_L V^2}{(R_i + R_L)^2} = 0$$

The whole expression is zero if the numerator is equal to zero:

$$(R_L + R_i)^2 - 2(R_L + R_i)R_L = R_L^2 + R_i^2 + 2R_L R_i - 2R_L^2 - 2R_L R_i = R_i^2 - R_L^2 = 0$$

thus $\quad R_L = R_i$

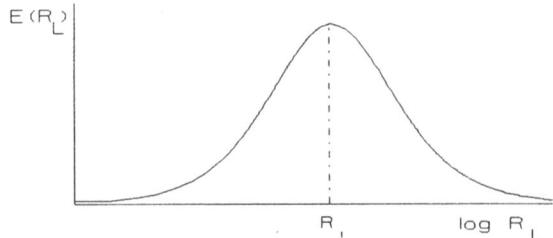

Figure 6.1.11 Relation between energy and load resistance.

With this value of R_L the power transferred is maximal. This quantity is:

$$E_{max} = \left\{ \frac{R_L}{(R_i + R_L)^2} V^2 \right\}_{R_L = R_i} = \frac{V^2}{4R_i} \qquad (6.8)$$

I have already mentioned that when the input and output terminals of a system are totally (galvanically) isolated from each other, we speak of a 'floating' input and output. It is then indeed possible (but not necessary) to connect an arbitrary input terminal with an arbitrary output terminal and to relate in this way the voltages to each other in the same way as in the example with the batteries (fig.6.1.2). Usually there is an internal connection between an input terminal and an output terminal ('common', '0', 'earth').

Connecting analog or digital systems can cause problems. The difference is that analog systems are frequently built up from separate building blocks that must be connected by the user, whereas digital systems are usually given as complete systems where necessary connections are made via standardized inputs and outputs with buffers.

Finally: in analog systems we distinguish passive and active systems. The latter contain components (such as transistors) by which energy can be introduced into the system. In this way the signal energy can be increased (amplification). For this purpose the systems are provided with a power supply that delivers the energy required. In passive systems the energy eventually needed is withdrawn from the signal itself. Digital systems are practically without exception active.

6.2 Linear systems with distortion-less transmission

Distortion-less transmission means (as we have seen in section 5.3) that the wave shape remains intact and that the only possible changes in the signal function that the system may cause, are a multiplication with a constant scaling factor p and/or a delay of the signal with a time Δt. Such systems have a straight, horizontal amplitude response and a linear phase response. There are two main groups in this category of systems: the systems to control the scaling factor and/or time delay (attenuators, amplifiers, delay units) and the systems to convert the signal function from one representation into another while keeping the wave shape intact as well as possible. To this latter group belong electro-acoustic transducers (loudspeakers, microphones), recording equipment (to register a signal function on tape or record) and also AD- and DA-converters. For the reason named in the introduction only the converters will be dealt with in this section. From the first group the analog *amplifiers* and *attenuators* give the most material for discussion, because once the signal is digitized, the introduction of a scaling factor boils down to a multiplication that is very easily realized with digital equipment. *Time delay units* require some form of memory for which digital systems may be used.

A. *Passive, analog systems for changing the scale factor*

1. The potentiometer

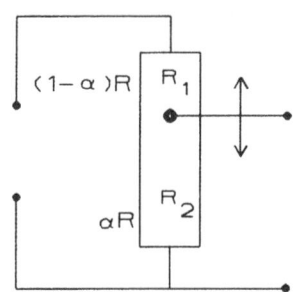

The most simple and most applied attenuator is the *potentiometer*. This is based upon the series circuit of resistors dealt with in the previous section. We saw that with this circuit voltage 'division' was possible (see formula (6.5)). A potentiometer does not work with two resistors, but with one resistor strip provided with a moveable tap (fig.6.2.1). The normal symbol for a potentiometer is:

Figure 6.2.1 Potentiometer.

By means of the tap the resistor strip is divided into two parts so that we in fact still have two resistors R_1 and R_2. The resistance values are however not fixed. R_2 is variable between 0 and the total resistance of the strip R. Instead of R_2 we can thus write αR, in which $\alpha < 1$.

According to the voltage divider formula we now find for the output voltage V_p:

$$V_p = \frac{\alpha R}{R} V = \alpha V$$

V_p can thus be set between 0 and V. Although simple in construction, a potentiometer is a system with an input and an output and thus also with a R_{in} and a R_{out}, with a straight amplitude response and a negligible time delay.

It is easily shown that R_{in} not only depends upon R but also upon the input resistance of the apparatus following the potentiometer and that R_{out} depends not only upon R but also upon the output resistance of the apparatus preceding the potentiometer (see problem 6.4).

2. The transformer.

Another passive system for changing the scaling factor is the *transformer*. This consists of two coils of which the magnetic fields coincide. In this way a changing voltage connected to one of the two coils (the *primary* coil) is transferred to the other, the *secondary* coil. Because the strength of the magnetic field and the value of the induction voltage depends upon the number of windings in the coil, it holds that the scaling factor, the proportion between the function values of the output and the input signal is equal to the proportion between the number of windings of the secondary coil (N_S) and that of the primary coil (N_P):

$$p = \frac{N_s}{N_p} \tag{6.9}$$

This scaling factor can thus be either more or less than 1.

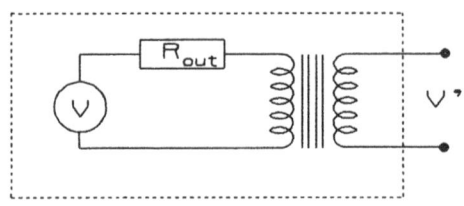

Figure 6.2.2 Transformer.

In an ideal transformer no energy is lost. If the signal voltage of some voltage source is transformed downwards, the output resistance is also decreased and to such an extent that the amount of energy remains the same. See fig.6.2.2. According to (6.8) the maximum power that can be delivered by the voltage source is $V^2/4R_{out}$.

At the output of the transformer we find a voltage $V' = (N_S/N_P)V$, and an output resistance R'_{out}, and thus a maximum power $V'^2/4R'_{out}$. As these amounts of energy should be equal we find:

$$\frac{V^2}{4R_{out}} = \frac{V'^2}{4R'_{out}} = \frac{\dfrac{N_s^2}{N_p^2}V^2}{4R'_{out}} \quad \text{or} \quad \frac{R'_{out}}{R_{out}} = \frac{N_s^2}{N_p^2} \tag{6.10}$$

Advantages of transformers:
- They bring galvanic separation and therefore floating inputs and outputs. In this way various transmission and connection problems can be avoided.
- Impedances can be matched to each other.
- The scaling factor can be changed without loss of energy.

Disadvantages:
- Especially with low frequencies and strong currents transformers are large, heavy and expensive.
- Losses and nonlinear distortion occur, for example, because of saturation.
- They are sensitive to interference.
- They can introduce unwanted phase shifts.

B. *Active, analog systems for changing the scaling factor*

1. The amplifier
These systems, named 'amplifiers', always contain active elements in the form of transistors or integrated circuits (ICs), and a power supply. We shall not go in depth into the electronic aspects, but regard the amplifier as a black box, a system with an input (with an input resistance R_{in}) and an output with output resistance R_{out}. The relation between input voltage V_i and output voltage V_u is:

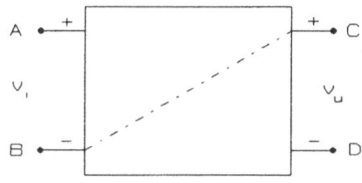

Figure 6.2.3
Inverting amplifier.

$$V_u = G \cdot V_i$$

Here G is the amplification factor, the gain. If the input and output are floating we cannot relate both voltages to each other. We can only state that V_u is a certain factor larger (or smaller) than V_i, and that if A is positive in relation to B (see fig.6.2.3), that then for example C is positive in relation to D. We may then indeed freely connect an input terminal with an output terminal. If we choose B and C for this, then A is not only positive in relation to B but also D is negative in relation to B. Making A more positive leads to making D more negative. We then speak of an *inverting* amplifier. If we connect B with D the voltages run coincidentally on A and C and we have a *non-inverting* amplifier. With some types of amplifiers we have indeed this choice; with others the connection concerned is internally already present (and serves usually as reference electrode for the whole circuit).

As a symbol for an amplifier a triangle is used pointing like an arrow in the direction of the output. Reference electrodes are often not drawn; inverting and non-inverting inputs are indicated as in fig.6.2.3 with + and -. The active components applied in an amplifier show clearly nonlinear characteristics. It is thus necessary to take measures to assure that the system as a whole behaves in a proper linear way. A much applied technique is the so-called *feedback*. We can study the principle without being obliged to go into the details of the circuit.

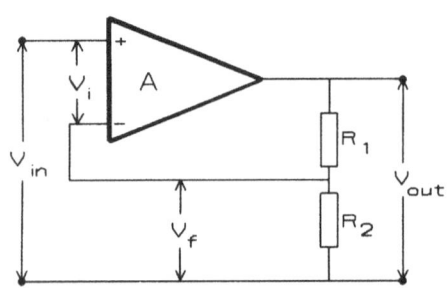

Figure 6.2.4 Feedback.

Fig.6.2.4 shows an example of a typical feedback circuit. An attenuated version V_f of the output signal (V_y) is returned to the inverting input of the amplifier:

$$V_f = \frac{R_2}{R_1 + R_2} V_{out} = \beta V_{out}$$

If the gain factor is equal to A the relation between V_y, V_x and the 'true' input voltage V_i can be derived as follows:

$$V_{out} = A V_i = A(V_{in} - V_f) = A(V_{in} - \beta V_{out}) = A V_{in} - A\beta V_{out}$$

$$V_{out} + A\beta V_{out} = A V_{in}$$

or:
$$V_{out} = \frac{A}{1 + A\beta} V_{in} = A' \cdot V_{in} \qquad (6.11)$$

The 'closed-loop' gain A' ($= A/(1 + A\beta)$) which is the result of the feedback connection to the inverting input of the amplifier, is smaller than the 'open-loop' gain A.

If $A\beta \gg 1$ (which is often the case) then $A' \approx \dfrac{1}{\beta} = \dfrac{R_1 + R_2}{R_2}$

This means that the gain is completely determined by the values of R_1 and R_2, and is thus independent of A. The amplification factor is constant, the amplitude response is a straight, horizontal line. At the end of this section an overview shall be given of the other consequences of feedback. With feedback we have a form of 'error correction': by subtracting V_f from V_x, V_i is distorted in a way that is opposite to the distortion that is caused by the amplifier. These distortions cancel each other through which the final total distortion can be very small. It is an effective and simple method with which to improve the linearity of an amplifier, but as with every system where the output signal is returned to the input there is always the risk of instability. It is actually more appropriate to speak of negative feedback because the feedback signal is subtracted from the input signal.

Positive feedback by which (a part of) the output voltage is added to the input voltage is also applied, for example in the construction of oscillators. The cumulative effect of positive feedback leads to large voltage variations, to oscillations. If the phase shift between output and input comes in the neighbourhood of 180°, the output signal is the mirrored version of the input signal and subtraction has then in fact become addition (- (-x) = + x). Negative feedback becomes positive feedback, the circuit will oscillate. With feedback one must thus always take care that the open loop gain at the frequencies where this phase shift occurs, is small.

Another method for improving the linearity without the risk of instability is the feed-forward technique, see fig.6.2.5.

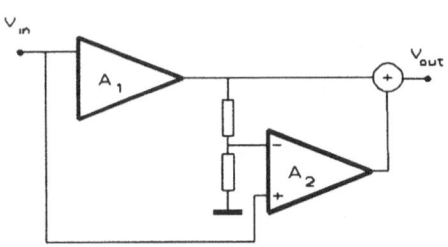

The signal amplified by A_1 is equally attenuated by a voltage divider. This signal is connected with the original input signal at the + and - input of an amplifier, which thus amplifies the difference between these two signals ('difference amplifier'). If the output signal of A_1 is an exact enlarged copy of the input signal the two input voltages of A are equal and the difference between them 0.

Figure 6.2.5 Feed forward correction.

Then the output signal is also 0. If there is really a difference it is detected in this way and afterwards once more added to the signal. This circuit is more complex but indeed unconditionally stable.

2. Amplifier specifications
In the following list specifications a) and b) are typical amplifier specifications. Specifications c) up to e) hold in part for other linear systems, and specifications f) until to k) for practically all electrical systems.
a. The amplification factor
    This is indeed given as a factor if one is concerned primarily with the enlargement of the amplitude. If the increase of the signal energy is also important, the amplification is also given in dBs. Let us take an example.
    Imagine an amplifier with a R_{in} of 100000 Ω that with an input voltage of 1 volt gives an output signal of 2 volts over a resistance of 3 Ω. The proportion of the voltages is then 2 which can be wrongly translated into 6 dB with the formula 20log V_1/V_2.

The proportion of the energies is (using (6.4)): $\dfrac{\dfrac{2^2}{3}}{\dfrac{1^2}{10^5}} = 102.5$ dB

Sometimes both the open loop gain and the closed loop gain are specified. The difference (in dB) between these numbers is the amount of feedback.

b. Linear distortion

This distortion is specified by stating how much the amplitude response deviates from the ideal straight horizontal line and/or the phase response from a straight line through the origin. For this the frequency range for which the specification holds is given, for example 'straight within 0.5 dB between 20 and 20000 Hz'. There are amplifiers of which the gain factor remains constant at low frequencies up to 0 Hz. We then speak of a *DC amplifier*.

c. Non-linear distortion

Specification of the harmonic distortion and the intermodulation distortion in % (see the formulae (5.6) and (5.7)).

d. Crossover distortion

In some amplifiers separate components are used for the positive and for the negative parts of the signal function. This can lead to distortion of the wave shape at crossings of the zero-line. This is especially disturbing with small signal amplitudes.

e. Transient Intermodulation Distortion

The components at the input side of an amplifier can often follow fast signal changes better than components at the output side which must deliver more power. With this it can occur that with a sudden voltage jump at the input the feedback voltage comes 'too late', so that within a short period of time the amplifier works in open loop mode. This can lead to serious overload distortion (see specification i. below) and saturation symptoms that do not decrease until after a relatively long time. The trick to improve the performance of a 'bad' amplifier by applying a huge amount of feedback does work for the static aspects, but leads to this type of distortion, and thus to a bad dynamic behaviour. To avoid this type of distortion it is in general advisable to give the separate stages of an amplifier their own local feedback.

f. The input resistance $R_{in}(\Omega)$

The resistance that can be measured between the input jacks.

g. The output resistance $R_{out}(\Omega)$

The internal 'pseudo'-resistance that determines what the voltage source-characteristic of the output will be.

h. The minimum load resistance $R_L(\Omega)$

R_{out} and R_L determine together the output voltage and the energy. A maximum energy transfer takes place if $R_L = R_{out}$, but the design of the circuit sometimes limits the minimal R_L that may be connected.

With amplifiers designed to drive an electrodynamic loudspeaker the proportion R_L/R_{out} is called the *damping factor*. If such a loudspeaker performs a 'free' vibration that does not correspond with the electrical signal (for example as a transient, see section 3.5) it functions as a dynamo and produces an induction voltage. The loudspeaker is then voltage source that is loaded by the internal resistance R_{out} of the amplifier. The larger the damping factor, the stronger the induction current by which in a short time all the energy is withdrawn from this voltage source and the movement is damped.

i. The maximum output voltage (V)

The voltages in an amplifier cannot be larger than ca. 80% of the supply voltage (except with the use of a transformer). If the product of the input voltage and the gain factor exceeds this value the output signal keeps at this value. Then 'clipping' occurs: the peaks of the signal are cut off. This of course is a serious form of nonlinear distortion. Sometimes the *sensitivity* of the amplifier is given. This is the input voltage required to give the maximum output voltage. It can be calculated by dividing the output voltage by the gain factor.

j. The maximum power (W).

The maximum output voltage V_{max}, R_L and R_{out} determine together the power that is delivered. If $R_L \gg R_{out}$ this power is equal to V^2_{max}/R_L.

k. The signal-to-noise ratio, SNR (dB).

Statistical fluctuations in voltages and currents that are the result of the corpuscular nature of electrical charge lead to a weak noise signal at the output. The ratio in dBs between the RMS-value of the maximum output signal and that of this noise signal is called the *signal to noise ratio*.

A few final remarks:

The effect of feedback on the properties of amplifiers can be summarized as follows:

a. with feedback to the input circuit of the amplifier R_{in} is usually increased;
b. with feedback from the output circuit of the amplifier R_{out} is usually decreased;
c. the signal-to-noise ratio is improved:
d. linear, nonlinear and crossover distortion are reduced;
e. the transient intermodulation distortion increases.

It is practical to subdivide amplifiers based on the power delivered into preamplifiers, line amplifiers and power amplifiers. For weak signals pre-amplifiers (power < ca. 100 mW) are especially suitable, thanks to a good SNR, as buffer and matching amplifiers and as basis for the construction of, for example, active filters for tone control and frequency correction (see section 6.3.B). With a line amplifier the power lies between 100 mW and 1 W. These amplifiers are used for circuits with low impedances and for example for sending a signal over a large number of parallel channels. Power amplifiers (1 - 2000 W) are mainly used for driving loudspeakers.

C. *Operational amplifiers* (Graeme et al. 1971)

These amplifiers, which owe their name to the purpose for which they were developed, the (analog) execution of particular mathematical operations such as additions with electrical voltages, form a separate group. This special application leads to priorities different from those of amplifiers for sound signals.
An operational amplifier is a DC amplifier with
- a very large open loop gain,
- a very high input resistance and a very low output resistance,
- a differential input.

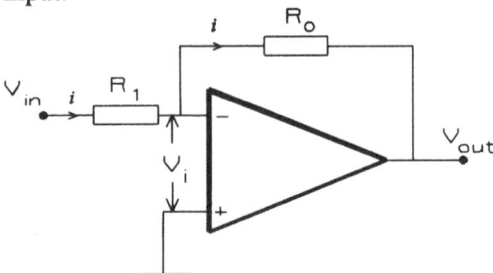

Figure 6.2.6 Operational amplifier.

The basic circuit is shown in fig.6.2.6. Via R_2 feedback is accomplished from the output to the inverting input. The consequence is that the actual input voltage V_i is very small. This is necessary as can be shown with the following example: imagine that the maximum variation of the output voltage is 10 volt, and that the open loop gain is 100000, then to prevent overload V_i must be smaller than $10/100000 = 10^{-4}$ Volts, . For calculations we can ignore such extremely small voltages and thus assume that $V_i = 0$ Volt (virtual earth).

If V_x is positive, V_y will be negative and a current i will flow from the input to the virtual earth point and from there to the output because through the high value of R_{in} there is no other possible route. We find therefore for i:

$$i = \frac{V_{in}}{R_1} = -\frac{V_{out}}{R_0} \text{ or } V_{out} = -\frac{R_0}{R_1}V_{in} = -R_0 i \qquad (6.12)$$

The amplification factor is thus R_2/R_1; the minus sign indicates the inversion of the signal. If $R_1 = R_2$ the circuit functions just as an inverter. The circuit can be extended to an adder (fig.6.2.7). It now holds that $i_1 + i_2 + i_3 = i$ thus also

$$\frac{V_1}{R_1} + \frac{V_2}{R_2} + \frac{V_3}{R_3} = -\frac{V_{out}}{R_0} \text{ or } V_{out} = -\left(\frac{R_0}{R_1}V_1 + \frac{R_0}{R_2}V_2 + \frac{R_0}{R_3}V_3\right) \qquad (6.13)$$

This circuit performs a 'weighted' addition. The number of inputs can be increased. If $R_1 = R_2 = R_3 = R_0$ we have 'normal' addition (+ inversion); by using a DC voltage as an input voltage, the output signal gets a DC component ('offset').

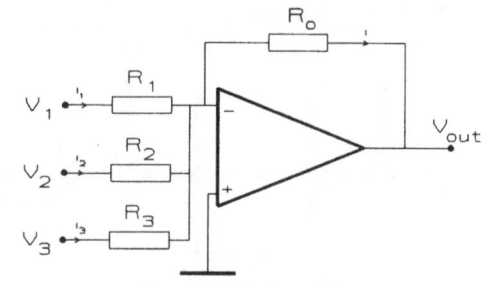

Figure 6.2.7 Adder.

Thanks to the large gain factor the operational amplifier can also be used for comparing two voltages. Fig.6.2.8 shows this *comparator*. If V_1 is larger or less than V_2 (the difference being greater than ca. 10^{-4} Volts) the output voltage will be equal to the maximum or the minimum value. There is a very small transition range when $V_1 \approx V_2$ (see fig.6.2.9). This is by the way a very nonlinear behaviour!

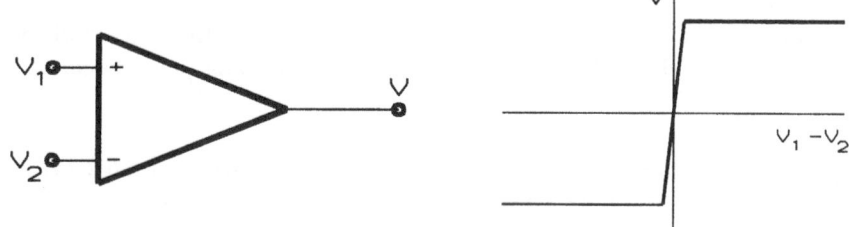

Figure 6.2.8 Comparator.

Figure 6.2.9 Relation between the input and output signal of a comparator.

D. *Time delay unit*

Such a system can be realized by storing signal samples in a digital memory and by recalling them after a while, for example as in fig.6.2.10. The write address starts at 0, and is increased by one every time a sample is stored. When the highest address is reached, the memory is full, and the write-address pointer is reset to 0. The process starts over again. The 'old' samples are replaced by new ones but before this happens they have already been fetched from memory with the help of a read-address pointer. The read address is derived from the write address by subtracting a (adjustable) number k from the latter. The time delay thus amounts to k/f_s seconds (f_s: clock frequency).

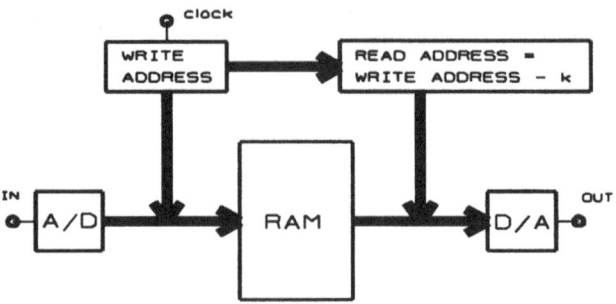

Figure 6.2.10 Digital delay unit.

E. *AD- and DA-converters*

The primary function of these devices is discussed in section 4.2. We shall now delve into the working principle and so begin with the

1. DA-converter
The task of this circuit is to convert a binary electrical number into an electrical voltage proportional to this number. Let us take as an example the 4-bits number 1101. The value of this number is

$$1 \cdot 2^3 + 1 \cdot 2^2 + 0 \cdot 2^1 + 1 \cdot 2^0 = 13_{10}$$

This formula suggests a possible construction that is shown in fig.6.2.11. If the position of the switches depends upon the binary code (1 = closed, 0 = open) we shall find with this number a voltage of 11 Volts at the output, and with the number 1001 a voltage of 9 Volts etc.

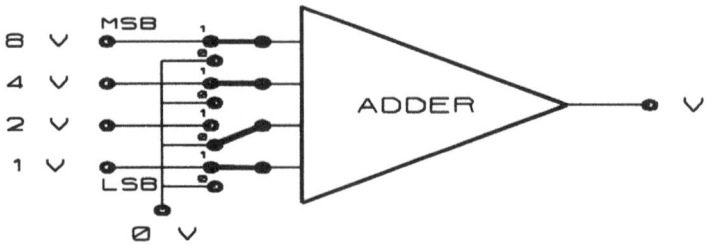

Figure 6.2.11 Principle of a DA-converter

The practical construction of the DA-converter is indeed based upon this idea. See fig. 6.2.12. As an adder we use an operational amplifier. We then need only one reference voltage V_{ref}, because by increasing every consecutive input resistance by a factor 2 the contribution of every next input is decreased by that same factor.

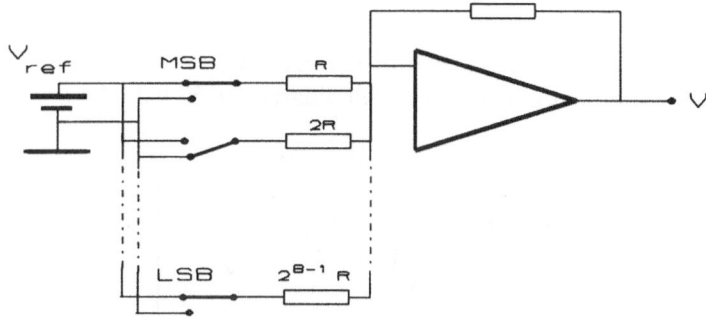

Figure 6.2.12 Binary weighted DA-converter

This converter is therefore called a 'binary weighted' DA-converter. The switches are not mechanical (too slow, too large) but electronic. An electronic switch is a semi-conductive element (Field Effect Transistor or FET) of which the electrical resistance can be made very large or very small with the help of a voltage connected to the control input of the element. In this way practically the same effect as with a 'true' mechanical switch can be achieved. The bit at the far left (MSB for Most Significant Bit) serves the upper switch, the bit at the far right (LSB for Least Significant Bit) the lowest switch.

The problem with this circuit is the large difference in the values of the resistors needed (a factor 2^{B-1} for a B-bits converter). As the accuracy and stability of resistors are usually proportional to the resistance value there is in absolute sense a large difference in this respect between the large and the small resistors. There is another DA-circuit where this problem can be avoided at the cost of double the number of resistors. This converter uses the R-2R ladder network that we had a look at in section 2.4 (see fig.2.4.2). Fig.6.2.13 shows the circuit diagram. We have seen that at each consecutive junction in this network the voltage is smaller

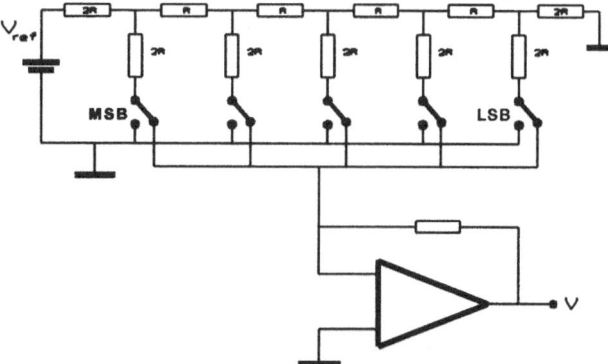

Figure 6.2.13 DA-converter with R-2R ladder.

by a factor 2. We thus achieve the desired effect with resistance values that only differ a factor 2.

2. AD-converter
AD-conversion can also be realized in several ways. A well-known method is 'successive approximation' (see the circuit diagram in fig.6.2.14).

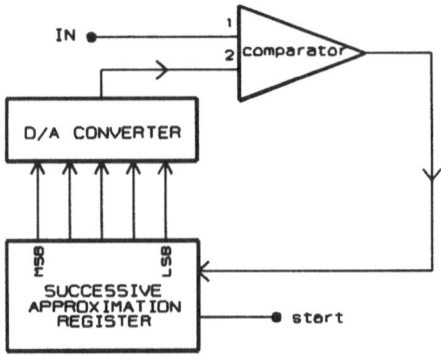

Figure 6.2.14 Successive Approximation AD-converter.

The converter consists of three parts: a comparator (see fig.6.2.8), a DA-converter (in this example a 10-volt type), and a digital circuit that is called the 'successive approximation register' (s.a.r.) that works as follows: if a start command is given to the converter, the MSB of this register is set to '1'. The DA-converter gives 5 volts. This voltage is compared with the input voltage. If it is larger the MSB is reset to '0'. If not it stays at '1'. This process is repeated with the next bits. With an input voltage of 6 volts and a 5-bit converter the consecutive stages are those shown in the table:

	S.A.R.	D.A.C.	
1	10000	5.0	V
2	11000	7.5	V
	reset		
	10000	5.0	V
3	10100	6.25	V
	reset		
	10000	5.0	V
4	10010	5.62	V
5	10011	5.91	V

The whole process takes time, of course; still here the combination of good resolution and comparatively high speed is still possible. During the conversion process the input voltage should remain constant. For this a Sample and Hold circuit is used (see fig.6.2.15). Every time the switch is closed for a moment as much charge flows from or to the capacitor to make its voltage equal to that of the preceding amplifier. The following buffer amplifier assures that the voltage on the capacitor is practically constant for some time. All AD-converters are preceded by such a Sample and Hold circuit.

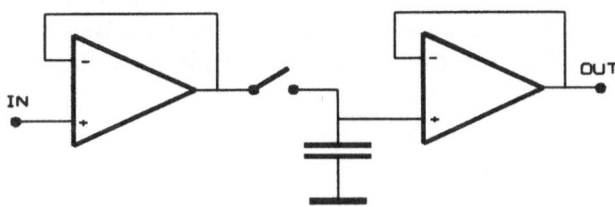

Figure 6.2.15 Sample and Hold circuit.

Other AD-converters are the 'double ramp' converter, where it is measured how much time is needed for a linearly rising voltage to reach (starting from 0 volt) the input voltage. The number of clock pulses is then directly the digital code for the voltage level concerned. For more accuracy one lets the voltage first rise rather quickly till a particular distance from the input voltage after which a slower rise up to the level of the input voltage follows. This principle was initially used only with digital voltmeters. Here for example once per second a conversion takes place and so speed in not critical. Nowadays one finds this principle also applied to faster converters. The fastest AD-converter is the 'flash' converter that contains one comparator for every voltage level of the quantizer. With a 5-bits converter simultaneously 32 comparisons take place. It is practically immediately determined which comparator-voltage corresponds most with the input voltage, after which the binary code is derived from the rank number of the comparator. The problem is that for 16-bits conversion about 65000 comparators are required.

6.3 Filters

A. *The principle of filtering; analysis and design of filters*

In chapter 3 we analysed a filter system, the harmonic oscillator, by solving the system equation. This method completely imparts the properties of the system, but can only be applied to simple systems. With more complicated filter systems other techniques (transform methods, pole-zero diagrams etc.) must be used. Therefore I restrict the discussion in this section to the analysis of the two simple filters shown in fig.6.3.1 a and b. The coefficient q in these diagrams should have a value ≤ 1.

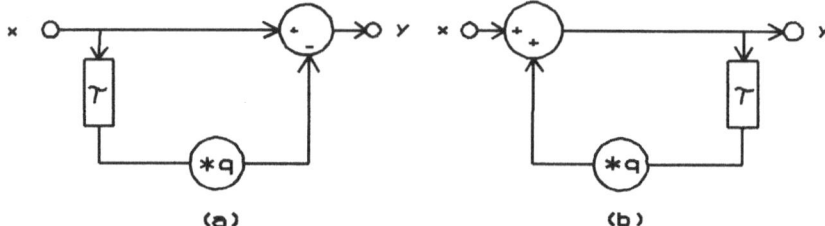

Figure 6.3.1 Two simple filters.

We have encountered filter (b) before, during the discussion of the convolution sum in section 5.2. The systems equations are:

$$y(t) = x(t) - q \cdot x(t - \tau) \qquad \text{and} \qquad y(t) = x(t) + q \cdot y(t - \tau)$$

The equation on the left is its own solution and thus trivial. The solution of the right-hand equation is:

$$y(t) = \sum_{n=0}^{\infty} q^n x(t - n\tau)$$

Proof via substitution:

$$\sum_{n=0}^{\infty} q^n x(t - n\tau) = x(t) + q \sum_{n=0}^{\infty} q^n x(t - \tau - n\tau)$$

$$= x(t) + \sum_{n=0}^{\infty} q^{n+1} x\{t - (n+1)\tau\}$$

$$= q^0 x(t - 0 \cdot \tau) + \sum_{n=1}^{\infty} q^n x(t - n\tau) = \sum_{n=0}^{\infty} q^n x(t - n\tau)$$

Now we can determine the impulse response and frequency response by using an impulse signal $\delta(t)$ or a sinusoidal signal $\sin 2\pi f t$ respectively for $x(t)$.

Impulse response (see fig.6.3.2 a and b):

$$h(t) = \delta(t) - q\,\delta(t - \tau) \qquad h(t) = \sum_{n=0}^{\infty} q^n \delta(t - n\tau)$$

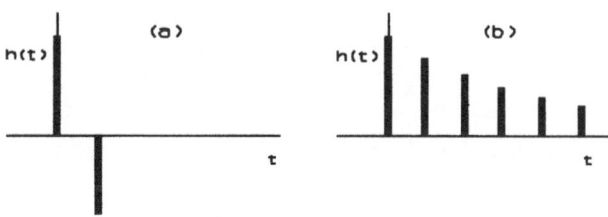

Figure 6.3.2 Impulse responses of the two filters.

Frequency response

This requires slightly more work to determine. First filter (a):

$$
\begin{aligned}
y(t) &= \sin 2\pi f t - q \sin 2\pi f t(t - \tau) \\
&= \sin 2\pi f t - q \sin 2\pi f t \cos 2\pi f \tau + q \cos 2\pi f t \sin 2\pi f \tau \\
&= (1 - q \cos 2\pi f \tau) \sin 2\pi f t + q \sin 2\pi f \tau \cos 2\pi f t \\
&= H_A(f) \cos (2\pi f t + \Delta\phi(f))
\end{aligned}
$$

The amplitude response of this filter is thus:

$$
\begin{aligned}
H_A(f) &= \sqrt{(1 - q\cdot\cos 2\pi f\tau)^2 + (q\cdot\sin 2\pi f\tau)^2} \\
&= \sqrt{1 - 2q\cdot\cos 2\pi f\tau + q^2}
\end{aligned}
\tag{6.14}
$$

For filter (b) a direct calculation is also possible. We then have to apply the sine in/sine out-principle and assume that with input signal $x(t)$ we get as output signal $y(t) = A \sin(2\pi f t + \phi)$. We then derive the expressions for H_A and $\Delta\phi$. An easier procedure is to determine both responses via the Fourier transform of the impulse response. Of course, both methods give the same results and we find for the amplitude response:

$$
H_A(f) = \frac{1}{\sqrt{1 - 2q\cdot\cos 2\pi f\tau + q^2}}
\tag{6.15}
$$

The two amplitude responses are shown in fig.6.3.3. Both filters are comb filters. This is no surprise as the addition of delayed sinusoidal signals leads to constructive or destructive interference, depending upon the phase shift. In many respects filter (b) is the counterpart of filter (a). See the following table:

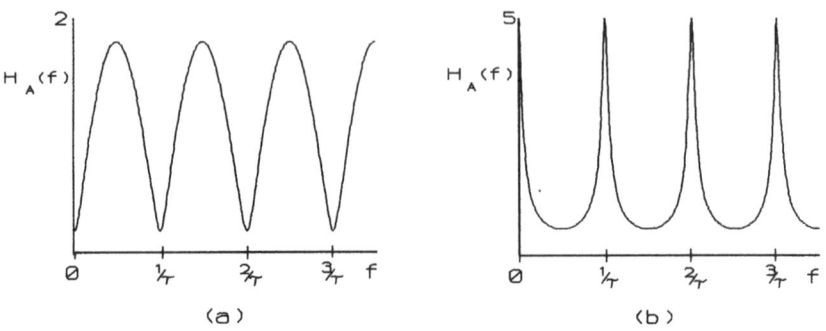

Figure 6.3.3 Amplitude responses of both filters.

filter (a)	filter (b)
- non-recursive	recursive
- impulse response finite (FIR-filter)	impulse response infinite (IIR-filter)
- maxima in amplitude response when	

$$\cos 2\pi f\tau = -1 \qquad\qquad \cos 2\pi f\tau = +1$$

thus at the frequencies $f = (2n+1)/2\tau$ $\qquad$ $f = n/\tau$ $(n = 0, 1, 2, \ldots)$

- value of the maximum

$$H_{A,max} = 1 + q \qquad\qquad H_{A,max} = 1/(1 - q)$$

thus with $q = 1$: $\quad H_{A,max} = 2$ $\qquad\qquad H_{A,max} = \infty$ (pole)

- minima in amplitude response when

$$\cos 2\pi f\tau = +1 \qquad\qquad \cos 2\pi f\tau = -1$$

thus at the frequencies $f = n/\tau$ $\qquad\qquad f = (2n + 1)/2\tau$

- value of the minimum

$$H_{A,min} = 1 - q \qquad\qquad H_{A,min} = 1/(1 + q)$$

thus when $q = 1$:

$$H_{A,min} = 0 \ \text{(zero)} \qquad\qquad H_{A,min} = \frac{1}{2}$$

Both filters can be realized as an analog and as a digital circuit. When in the digital version the time-delay τ is equal to the clock-period $1/f_s$, then with filter (a) the first maximum lies at $f = 1/2\tau = \frac{1}{2}f_s$. Higher frequencies are not allowed in a digital system and this means that filter (a) is in fact a high-pass filter. If furthermore $q = 1$, the output signal consists of the subsequent differences between the input samples. The system then performs numerical differentiation as described in section 2.3.D. When the time-delay equals a multiple of $1/f_s$, the filter changes into a comb filter. In the same way it can be argued that the digital version of filter (b) is a lowpass filter, and acts as a numerical integrator when $\tau = 1/f_s$ and $q = 1$. The digital versions of the filters (a) and (b) are the building-stones with which more complicated filters can be constructed. The digital filter discussed in section 3.6 is for example the 'double' version of filter (b), while filter (a) is the simplest version of the transversal filter we encountered in section 5.2. We shall also meet them in several filter and oscillator circuits in this section, and as analysis and synthesis filters in LPC-systems (section 7.3).

Both filters can be combined in a series circuit. Due to linearity it does not matter in which order this is done, but placing filter (b) before filter (a) has the advantage that only one time delay is required; see fig.6.3.4.

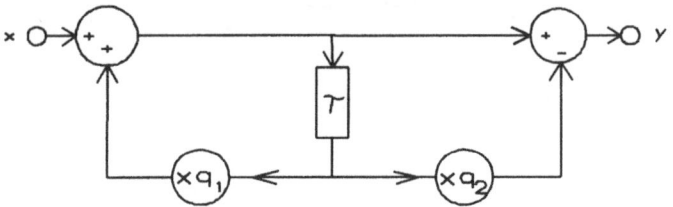

Figure 6.3.4 Series circuit of filters.

When the two q-values are equal the effect of this combination of filters is nil, as the effect of the first is cancelled by the second one. This can be concluded from the impulse responses but even better from the amplitude responses (6.14) and (6.15). The product of these two expressions is equal to 1. Fig.6.3.5 finally depicts the general form of a linear digital filter with poles and zeros (see also Bogner 1975). The design of a filter based on a particular specification (with analog filters usually a particular amplitude characteristic, with digital filters sometimes also a particular impulse response) is in general a difficult task, where moreover large differences occur between the procedures for analog and those for digital filters.

The following remarks refer to analog filters: whether it concerns a mechanical, an acoustical (Helmholtz resonator) or an electrical filter it is typical of an analog filter that it involves a system containing components which, by their physical properties, influence the vibration. With a vibrating string these were elasticity, mass and friction, with the Helmholtz resonator the volume and the size of the opening of the sphere, and the pressure and density of the air, and with the LC-circuit the self-induction, capacity and resistance. The physical quantities can be mathematically described, but a typical complication is that this description is never totally exact and that in practice deviations from the 'ideal' behaviour occur.

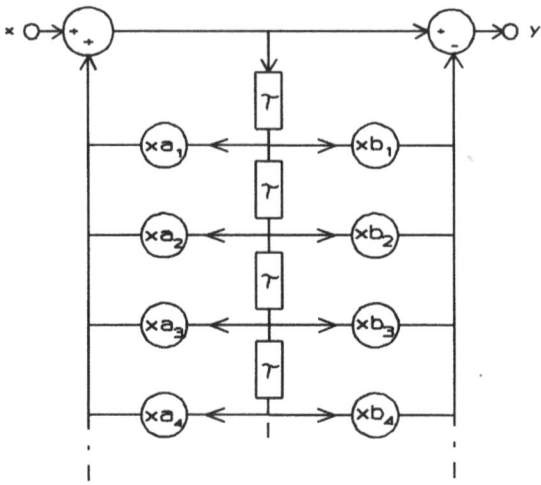

Figure 6.3.5 Digital filter.

The design implies a theoretical analysis of the (idealized) mathematical equation of the system, using a transform technique related to the Fourier transform: the Laplace transform. It involves the use of complex numbers and that is why this method is not dealt with here. For many applications standard designs for both active and passive filters are used, which have optimal specifications as for flank steepness, ripple in the pass band etc. like Butterworth or Tchebychef filters. For active filters frequency dependent feedback is often used, leading to frequency dependent amplification. We shall encounter examples of this.

With digital filters the problem of the deviations from the ideal behaviour of the components does not occur. Other problems arise such as rounding-off errors that are the consequence of the finite word length of the digital samples. Still, digital filters generally offer large advantages. It is possible to design filters that cannot be realized in an analog way. Especially for variable, controllable filters the digital method is advantageous. To vary an analog filter one must usually change the value of a particular component (for example a resistor or capacitor). With digital filters this boils down to changing some factors in the algorithm (algorithm = calculation process).

A very direct method for digital filtering is based upon the Fourier transform. Via an F.F.T. a signal is converted into the corresponding spectrum. Then changes in the amplitude and/or phase spectrum are made (certain frequency components are for example removed, or all phase angles are set to 0, etc.) and then inverse transformation to the time domain takes place. Symbolically:

$$
\begin{array}{ccccc}
\text{FFT} & & \text{changes} & & \text{FFT}^{-1} \\
y(t) \longrightarrow & Y(f) & \longrightarrow & Y'(f) & \longrightarrow y'(t)
\end{array}
$$

A common method in digital filter design is using the impulse response, because there is a technique to derive the filter algorithm from the impulse response. This design procedure is based upon the mathematical technique of the 'Z-transform', which is the time-discrete version of the already mentioned Laplace transform. If the purpose of the filter is to realize a certain frequency response this can be accomplished by first deriving from this the impulse response with the help of the inverse Fourier transform, and then finding the filter algorithm via the Z-transform. Symbolically:

$$(H(f) \xrightarrow{\text{FFT}^{-1}})h(t) \xrightarrow{\quad Z \quad} \text{algorithm}$$

As we have already seen, filter algorithms consist of three operations: time delay, addition and multiplication by a coefficient. All linear digital filters, even the most complicated ones, can be realized by these operations. Of these the multi-plication is the most time-consuming one. If the filter should have an infinite long impulse response (IIR filter) one has to make use of feedback, as for exam-ple in the recursive system of fig.6.3.1. These filters can be relatively compact but because of the feedback there is always the risk of instability. For a FIR filter the principle of the transversal filter of section 5.2 can be applied.

B. *A few linear filter and oscillator circuits*

The fact that filter and oscillator circuits are here named in one breath suggests that there is in principle little difference between them. That is indeed the case. An oscillator may be considered as a system with a particular specified impulse response. We shall encounter examples of this approach.

1. Analog, passive filters.
In this section we return to the simple differentiating and integrating circuits of chapter 2 (fig.2.3.3 in section 2.3.D and fig.2.5.4 in section 2.5.B). We will now study their frequency domain behaviour i.e. their frequency responses. As in the previous section we will find that differentiation corresponds to high pass filter-ing and integration to low pass filtering. In fig.6.3.6 the differentiating network is shown again for which in section 2.3.D the following equation was derived:

$$V_y = RC\frac{dV_x}{dt} - RC\frac{dV_y}{dt} \quad \rightarrow \quad \frac{dV_y}{dt} + \frac{1}{RC}V_y = \frac{dV_x}{dt}$$

We could now ask what the impulse response of the system is. Instead of this we shall have a look at the *step response*: the reaction of the system to an input signal that is 0 for $t < 0$ and has a constant value V_o for $t > 0$, a *step function* (see fig.6.3.7; the impulse response can be derived from the step response, because the derivative of the step function is an infinitely narrow impulse).

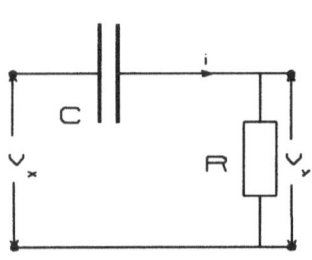

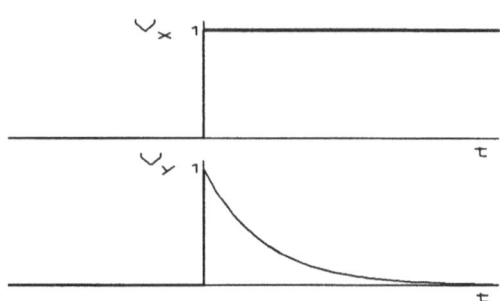

Figure 6.3.6
Differentiating circuit.

Figure 6.3.7 Step function and step response of the differentiating circuit.

For this signal it holds that $dV_x/dt = 0$. The equation thus becomes:

$$\frac{dV_y}{dt} + \frac{1}{RC}V_y = 0 \ ; \qquad \text{Solution:} \quad V_y = V_0 e^{-\frac{t}{RC}}$$

Check via substitution: $\quad \dfrac{dV_y}{dt} = -\dfrac{V_0}{RC} e^{-\frac{t}{RC}} \quad$ and $\quad \dfrac{1}{RC}V_y = \dfrac{V_0}{RC} e^{-\frac{t}{RC}}$

The summation of both expressions yields 0.
To find out what the sine response is we take as an input signal: $V_x = \sin \omega t$ and as trial function: $V_y = B \sin (\omega t - \phi)$. This is a logical choice because we are dealing with a linear system for which thus the sine in/sine out-principle holds. Our task is to determine B and ϕ. Let us substitute V_x and V_y in the equations:

$$\omega B \cos (\omega t - \phi) + \frac{B}{RC} \sin (\omega t - \phi) = \omega \cos \omega t$$

The left expression can be reduced to one cosine function:

$$D \cos (\omega t - \phi + \gamma) = \omega \cos \omega t$$

$$\text{with } D = \sqrt{\omega^2 B^2 + \frac{B^2}{R^2 C^2}} \ \text{ and } \tan \gamma = -\frac{\frac{B}{RC}}{\omega B} = -\frac{1}{\omega RC}$$

or: $D \cos \omega t \cdot \cos (\gamma - \phi) - D \sin \omega t \cdot \sin (\gamma - \phi) = \omega \cos \omega t$

This is the case if $\gamma = \phi$ and $D = \omega$. If in the above expression we replace D for ω we find:

$$\omega = \sqrt{\omega^2 B^2 + \frac{B^2}{R^2 C^2}} \rightarrow \omega^2 = (\omega^2 + \frac{1}{R^2 C^2})B^2 \rightarrow R^2 C^2 \omega^2 = (R^2 C^2 \omega^2 + 1)B^2$$

and thus:

$$H_A(f) = B = \frac{2\pi f R C}{\sqrt{1 + 4\pi f^2 R^2 C^2}}$$

$$\Delta\phi(f) = \phi = \gamma = \tan^{-1} - \frac{1}{2\pi f R C}$$

(6.16)

Graphically:

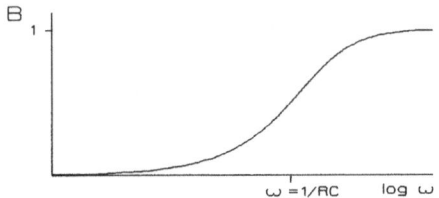

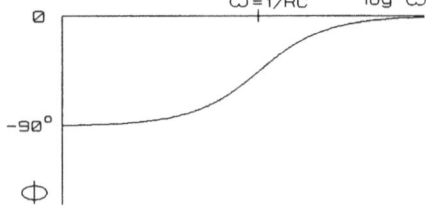

Figure 6.3.8
Amplitude response of differentiating circuit.

Figure 6.3.9
Phase response of differentiating circuit.

For high frequencies $B = 1$ and we thus have a constant output level. With low frequencies B is proportional to ω and we have a 6 dB/octave flank-steepness. The cutoff frequency lies at $\omega_o = 1/RC$ because then $B = 1/\sqrt{2}$ (see fig.6.3.8 and fig.6.3.9). The differentiator is obviously a *highpass filter*.
In the same way we can analyze the integrating RC-network shown in fig.6.3.10.

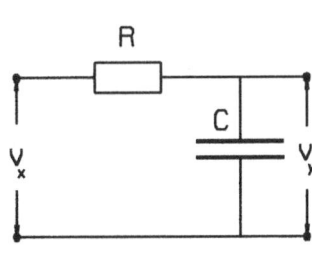

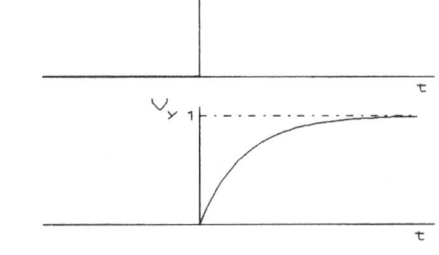

Figure 6.3.10
Integrating circuit.

Figure 6.3.11 Step function and step response of the integrating circuit.

See problem 6.8. It turns out to be a *lowpass filter*. The expression for the step

response is: $V_y = V_0(1 - e^{-\frac{t}{RC}})$ (fig.6.3.11 bottom).

The amplitude and phase response are described by:

$$H_A(f) = B = \frac{1}{\sqrt{1 + 4\pi^2 f^2 R^2 C^2}} \quad \text{and} \quad \Delta\phi(f) = \phi = \frac{1}{2}\pi + \tan^{-1} - \frac{1}{2\pi f R C}$$

The responses are depicted in fig.6.3.12 and 6.3.13.

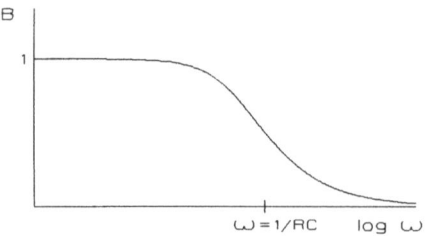
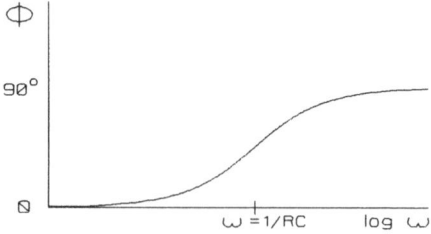

Figure 6.3.12 Amplitude response.   *Figure 6.3.13 Phase response.*

2. Analog, active filters.

Most active filters make use of frequency-dependent feedback. We saw that with an increasing amount of feedback the overall amplification is reduced. If feedback takes place via a filter, the amplification will be large at those frequencies suppressed by the filter and vice versa. The frequency response of the system is therefore the mirrored version of that of the filter. By combining a RC-network with an amplifier a system is realized that functions as an integrator or differentiator with a much more accurate performance than the previous circuits, without amplifiers. Fig.6.3.14 shows a differentiator. Because the current must be equal in both branches it holds:

$$i = \frac{dQ}{dt} = C\frac{dV_x}{dt} = -\frac{V_y}{R} \quad \text{or} \quad V_y = -RC\frac{dV_x}{dt}$$

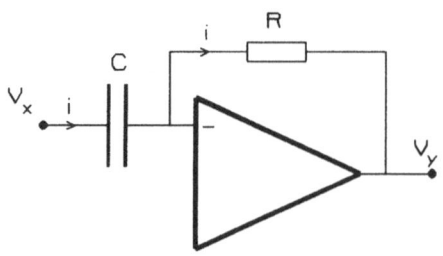

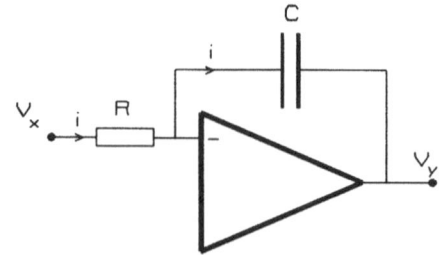

Figure 6.3.14 Differentiator.   *Figure 6.3.15 Integrator.*

In the same way it holds for the integrator of fig.6.3.15:

$$i = \frac{V_x}{R} = \frac{dQ}{dt} = -C\frac{dV_y}{dt} \quad \text{or} \quad V_y = -\frac{1}{RC}\int_0^T V_x dt \quad (+ \text{ a constant})$$

3. Digital filters.
- An 'all-pass' network (Schroeder et al. 1961).
In 1961 Schroeder published a circuit that he considered as a building block for a digital reverberation circuit (see fig.6.3.16). As impulse response the circuit has

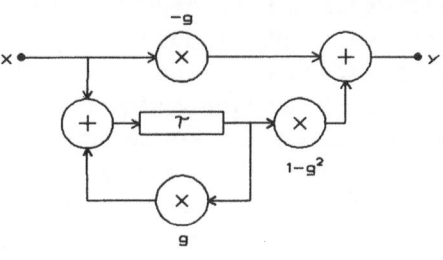

a decreasing series of equidistant pulses (fig.6.3.17), somewhat comparable with the impulse response of the comb filter of fig.6.3.1, but here with a flat amplitude response (fig.6.3.18). Therefore, the characteristic and audible coloration of the comb filter is lacking. We are thus dealing here with a system where conditions are set to both the impulse response and the amplitude response.

Figure 6.3.16 All-pass network.

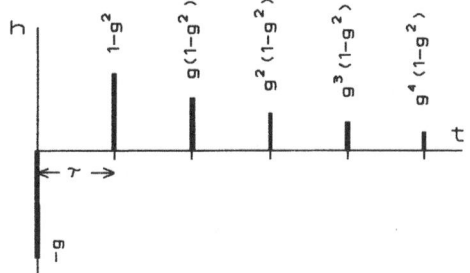

Figure 6.3.17 Impulse response of the all-pass network.

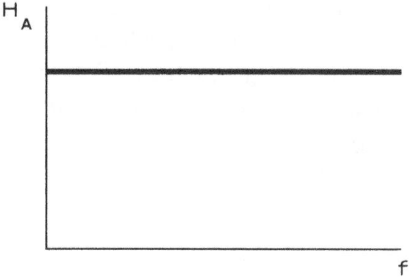

Fig.6.3.18 Amplitude response of the all-pass network.

A complete reverberation circuit can be considered as a similar filter where the impulse response must have such a density of pulses that an exact description of it is impossible (and also unnecessary). Instead of that certain statistical criteria can be applied. Schroeders idea was to build up a digital reverberation circuit from at least five of the above building blocks in a series circuit.

- Linear oscillators (Tempelaars 1982).
We can also set ourselves the task of designing a system with a particular impulse response and by that of keeping spectral considerations out of the discussion in the first instance. Systems that have the task of generating a certain time function are called *oscillators*. A sine wave oscillator produces a sinusoidal

signal. If we should succeed in designing a 'filter' of which the impulse response is this sinusoidal signal then we could have designed just such a sine wave oscillator. We need only supply one single impulse to the system to excite the desired output signal.

For a finite impulse response one could use a transversal filter. The solution is simple: the discrete values of the signal function are used as h-coefficients (thus as multiplication factors) in the scheme of fig.5.2.3. The drawback of a limited duration for the signal can furthermore be avoided by giving the start pulse again at the end of the signal sequence. By this means a periodic signal arises that in principle can be arbitrarily long. If the system is used only in this way (and not also as a 'true' filter with an arbitrary input signal) then it is easier to realize by placing the function values in a memory (analog or digital) and recalling them one after the other. This generator principle (which was earlier mentioned in section 3.2) is known under various names: stored-waveform generator, look-up table generator, sequencer, variable function generator etc. For an infinite impulse response we must make use of a system with feedback. We have already become acquainted with two examples: the digital sine wave generator dealt with in section 3.2 and the generation of a damped sinusoidal signal treated in section 3.4.

Here below are yet three more of such circuits with the corresponding multiplication factors. The circuits are somewhat more complex and that also holds for the signal functions generated: a sinusoidal signal with an envelope with a particular rise and decay time (fig.6.3.19), a VOSIM-signal (fig.6.3.20) and a signal consisting of the product of two sine waves (fig.6.3.21).

- Sinewave with attack and decay; signal function: $y(k) = k \cdot r^k \sin k\gamma$

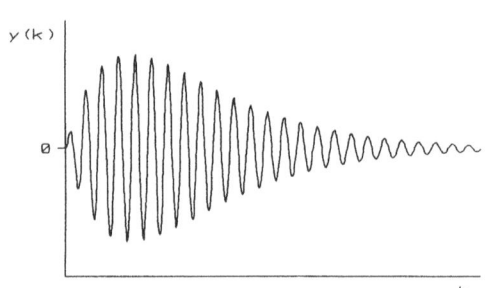

Required coefficients:

$a_1 = -2r \cos \gamma$
$a_2 = 2r^2 (1 + 2 \cos^2\gamma)$
$a_3 = -4r^3 \cos \gamma$
$a_4 = r^4$
$b_1 = r \sin \gamma$
$b_2 = 0$
$b_3 = -r^3 \sin \gamma$
$b_4 = 0$

Figure 6.3.19 Sinewave with attack and decay.

- VOSIM signal; signal function: $y(k) = r^k(\frac{1}{2} - \frac{1}{2}\cos k\gamma)$

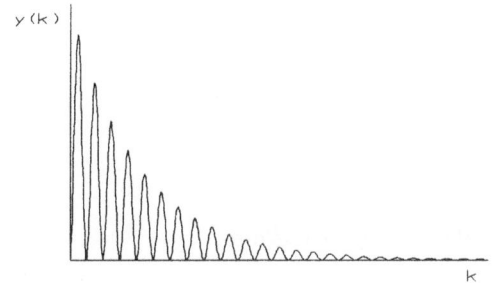

Required coefficients:

$$a_1 = -r(1 + 2\cos\gamma)$$
$$a_2 = r^2(1 + 2\cos\gamma)$$
$$a_3 = -r^3$$
$$b_1 = r(1 - \cos\gamma)$$
$$b_2 = r^2(1 - \cos\gamma)$$
$$b_3 = 0$$

Figure 6.3.20 VOSIM signal.

- modulated sinewave; signal function: $y(k) = \sin k\gamma \cdot \cos k\theta$

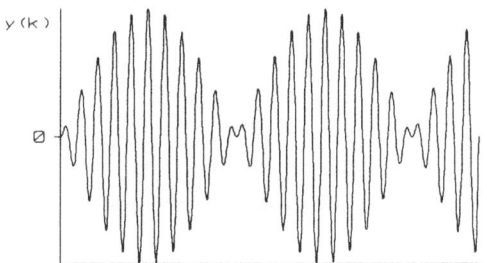

Required coefficients:

$$a_1 = -4\cos\gamma\cos\theta$$
$$a_2 = 2(\cos 2\gamma + \cos 2\theta + 1)$$
$$a_3 = -4\cos\gamma\cos\theta$$
$$a_4 = 1$$
$$b_1 = \sin\gamma\sin\theta$$
$$b_2 = 0$$
$$b_3 = -\sin\gamma\sin\theta$$
$$b_4 = 0$$

Figure 6.3.21 Modulated sinewave.

4. Analog oscillators.

Evidently with the oscillators just dealt with, feedback is made use of. The signal returned is not subtracted from the input signal, but added to it (positive feedback). This can very easily lead to instability because if the input signal becomes larger as a result of feedback, a cumulative process has begun which can no longer be controlled. By means of the accuracy of the calculations in digital systems, things may be kept under control. With analog circuits this is more problematic. For example, let us have a look at the following circuit (fig.6.3.22). The network of two resistors and two capacitors is called a 'Wien bridge' and acts as a band pass filter. This is not surprising as the circuit is a combination of the highpass and lowpass filters of fig. 6.3.10 and 6.3.11. The amplitude and phase response are shown in fig.6.3.23 and fig.6.3.24. The transmission of the network is maximum at the frequency $f_o = 1/2\pi RC$.

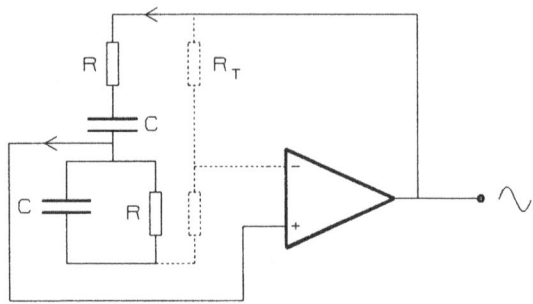

Figure 6.3.22 Wien bridge oscillator.

It amounts to a factor 1/3. The phase shift is then 0°. If the gain factor of the applied amplifier is precisely 3, the total amplitude factor is 1 and the circuit functions as a sine wave generator with frequency f_o.

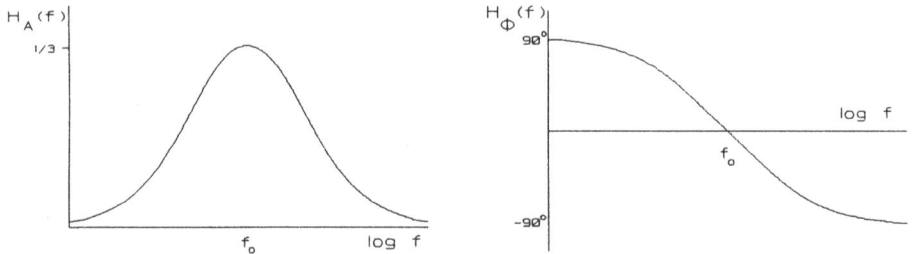

Figure 6.3.23
Amplitude response of Wien bridge.

Figure 6.3.24
Phase response of Wien bridge.

The problem is, however, that an analog amplifier can never exactly be set to a factor 3. One then resorts to a sort of automatic gain control to keep the amplitude of the output signal constant. One method (indicated in fig.6.3.22 with the dashed lines) is to send the output signal to the inverting input of the amplifier as well. The gain factor is determined by the voltage divider. By using a temperature-sensitive type for the uppermost resistor it can be achieved that a larger signal amplitude (thus more heat production) leads to a smaller amplification factor.

Another example is the relaxation-oscillator, by which a sawtooth signal can be generated. This consists of a linear part that generates a voltage increasing proportionally with time, and a nonlinear part that interrupts this process at a particular moment. This can be realized as an analog system in the following way: a current source sends a constant current into a capacitor (fig.6.3.25). Through this the charge of the capacitor increases linearly in time and therefore also the voltage ($Q = C \cdot V$). Via a comparator it is determined that the rising voltage has become equal to a certain reference voltage V_1. The electronic switch S is closed for a short duration, the capacitor discharges itself, the voltage drops to 0, the switch opens again and the process repeats itself. The circuit is also sometimes used in a double version and functions then as a triangular wave generator.

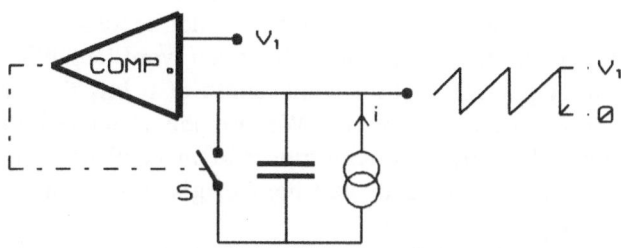

Figure 6.3.25 Relaxation oscillator.

Digitally one uses a counter (a register of which the content is raised by 1 with each clock pulse) and a (numerical) comparator that resets the register to 0 if a certain reference value is reached.

5. Noise generators.
Analog noise generators are particularly easy to construct. Because every electrical current displays statistical fluctuations it is enough to take an amplifier that amplifies the unavoidable noise to such an extent that a sufficiently strong noise signal appears at the output. It is better to place at the input an element in which naturally large statistical fluctuations in electrical current occur. This noise signal has a normal amplitude distribution (random noise) and has a white spectrum. With a 3 dB/octave filter the white noise can be changed into pink noise.
Binary analog noise can be generated by connecting a comparator to this noise generator. As we have seen, a comparator has only two output levels, for example level 1 with a positive input signal and level 2 with a negative one.

   Digital noise is an entirely different problem. As with every digital signal function we are concerned here with a sequence of numbers, here numbers determined by chance. We thus look in fact for a system with a 'random' impulse response. Naturally for this we can use analog noise and an AD-converter. There are however algorithms for generating such random numbers, for example the following procedure: take a number between 0 and 1, add π to it, raise it to the 8th power, take from the result the part after the decimal point. This number is the first one of the sequence to be generated. Repeat the whole calculation with this number to find the next output value. Starting with 0.4 we find the following series:

   0.7786, 0.8883, 0.8843, 0.3077, 0.4699, 0.7680, 0.2090, etc.

We find thus numbers between 0 and 1, and all these numbers are equally probable; the probability density function is flat. If we perform this calculation with a computer then because of the finite supply of numbers it is unavoidable that at a certain moment all numbers have been used and a number appears for the second time. Because the calculation goes on in the same way the whole following sequence is identical with the previous one, in other words, the series repeats itself periodically. We can take care that this will not happen until all the different numbers that can be represented with the given word length have been used (maximum length sequence). With a word length of n bits there are thus 2^n-1 numbers (the zero must not be used). We use the characterization 'pseudo-random' to indicate the difference with true random numbers. The fact that the sequence is reproducible (the same start number gives the same sequence) can sometimes be useful.

A simple and fast, although from a statistical point of view not the most elegant method for generating random numbers, is the following:

Place a binary pattern in a so-called *shift regis-ter*. This is a register where all the bits can be shifted one position to the right (or to the left). Fill the place to the left which has come free because of the shift, in with either a 0 or a 1, depending on the content of two other places in the register. Use the following rule:

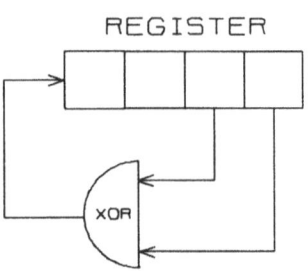

$$0,0 \rightarrow 0; 0,1 \rightarrow 1; 1,0 \rightarrow 1; 1,1 \rightarrow 0$$

(a so-called XOR-gate).

With a 4-bits register (fig.6.3.26), with feedback via a XOR-gate from bit 2 and bit 3 to bit 0, and with the number 1000 as a start number, we get the following pseudo-random sequence of 15 numbers:

Figure 6.3.26
Pseudo-random number
generator.

register	output XOR	decimal		register	output XOR	decimal
1 1000	0	8		9 1010	1	10
2 0100	0	4		10 1101	1	13
3 0010	1	2		11 1110	1	14
4 1001	1	9		12 1111	0	15
5 1100	0	12		13 0111	0	7
6 0110	1	6		14 0011	0	3
7 1011	0	11		15 0001	1	1
8 0101	1	5		(1000)		

The spectrum of such a noise signal is white (see section 4.4.D). Via (digital) filtering it is possible to get other spectral distributions. There are also algorithms that directly give another spectrum or another amplitude distribution, for example algorithms that produce pink noise or brown noise, and spectral slopes in between. To formulate this more exactly: noise with as a spectral distribution proportional to the function $1/f_x$ with for example $0 \le x \le 2$. Similarly it is possible to generate noise with a normal probability distribution instead of a flat one.

6.4 Non-linear and time-variant systems

Linear, time-invariant systems play an important role in signal processing as they are required for the transmission, recording, filtering and reproduction of signals. The essential point here is that no new frequency components arise. The preceding sections all dealt with these LTI-systems. This could cause the impression that non-linear systems are less important. This is not the case. Many phenomena in nature are so complex and even chaotic that the relatively simple description technique of the linear systems is absolutely inadequate. In the meantime mathematics has developed new tools to deal with these non-linear, dynamic and chaotic systems: techniques to solve sets of non-linear differential and difference equations, iterative systems describing chaotic behaviour, fractal geometry etc. These parts of signal processing handling the simulation of existing systems like musical instruments profit most of this development. The knowledge and skill required to be able to use these new mathematical tools exceed the level chosen for this book. Therefore in the next sections just two groups of non-linear systems are discussed that both are closely related to linear systems. For those interested in non-linear and chaotic systems I refer to Schroeder (1989), Truax (1990) and Chua (1993).

A. *Systems with a nonlinear transfer function*

With some systems the input/output relation can be measured by determining which output voltage is produced with a particular constant voltage at the input. The measurement result is then only relevant for the DC-behaviour of the system. With a (DC-) amplifier the relation is represented by $y = A \cdot x$ and is graphically a straight line (fig.6.4.1).

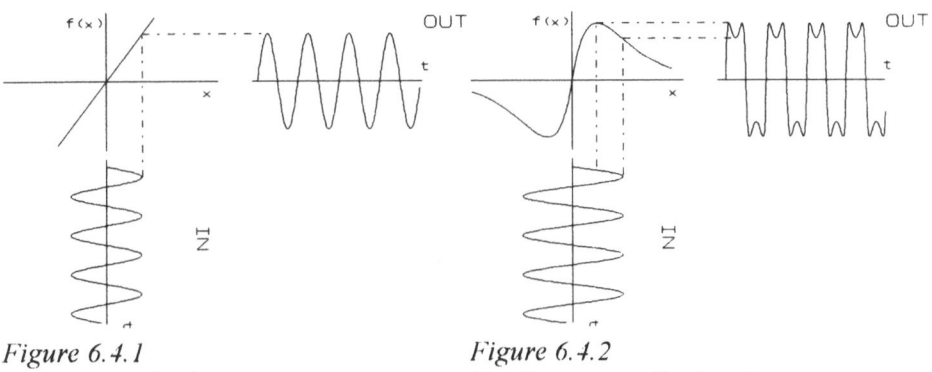

Figure 6.4.1
Linear transfer function.

Figure 6.4.2
Nonlinear transfer function.

This description holds as well for alternating voltages as long as the gain factor is constant (otherwise the steepness of the line changes) and the phase shift is 0, otherwise the line changes into an ellipse or a circle. These limitations show the relative value of this description method. If the input/output relation is non-linear, but has a shape as shown in fig.6.4.2, then it is possible to describe the relation between y and x near a particular x-value with the help of a power series:

$$y = a_1x + a_2x^2 + a_3x^3 + a_4x^4 + \ldots$$

If we use the sum of two sinusoidal signals as input signal for this system, then new frequency components are generated. We have encountered this phenomenon already in chapter 5 (see formula (5.8)) in the discussion of the nonlinear distortion of linear systems. There the rule was stated that the term of the series in which the p-th power occurs causes components with frequencies:

$$|mf_1 \pm nf_2| \quad \text{with} \quad m + n = p$$

This also plays a role in explaining the phenomenon of *combination tones* ('difference tones') which should be considered as distortion products of the hearing organ. Furthermore nonlinear transfer functions are applied in a method for generating complex tones (see section 7.4.D). With digital signals a nonlinear transfer function is simple to realize. This function is then discrete and can be specified as a table or list. By using the discrete values of the input signal as pointer for this (look-up) table every desired corresponding output value can be produced. With analog signals a particular nonlinear transfer function can be approximated with the help of a diode shaping network as shown in fig.6.4.3. A diode can be considered as a sort of valve, a voltage-controlled switch. If the voltage on the side of the triangle is higher than on the other side then the diode functions as a simple connection, and in the other case as an interruption. Symbolically:

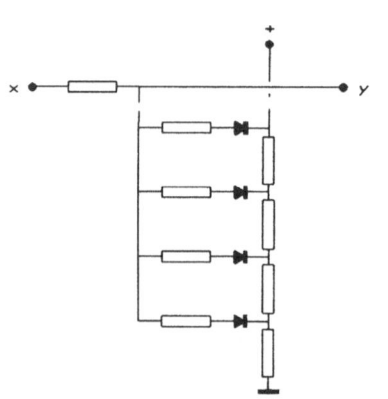

Figure 6.4.3
Diode shaping network.

If in this diode-network the input voltage rises, ever more diodes are conducting and ever more parallel resistors in the lower part of the network will be switched in. We have in fact a voltage divider with an increasingly smaller value of the lower resistance.

In this way the output voltage is increasingly attenuated with regard to the input voltage (see fig.6.4.3) and a 'piecewise linear' approximation of a particular function is possible. This method is applied, for example, in voltage-controlled oscillators to change a triangular wave shape into a sinusoidal one and to provide the oscillator with two signal outputs.

A diode shaping network is an example of an analog system with a nonlinear transfer function, realized with the help of nonlinear components, here diodes. More systems of this type exist, some of them equipped with operational amplifiers to improve their performance. We shall look at some of these systems, where possible in their simplest form without an amplifier, although the use of an operational amplifier is in general to be preferred.

1. Limiter.

With two diodes and two batteries a circuit can be constructed in which the output voltage cannot exceed (in positive or negative sense) the limits determined by the batteries. It is shown in fig.6.4.4. This leads to 'clipping' of the signal peaks. An extreme example of such a limiter is the comparator that we have already become acquainted with. With the comparator the slope of the oblique part of the transfer function is practically vertical.

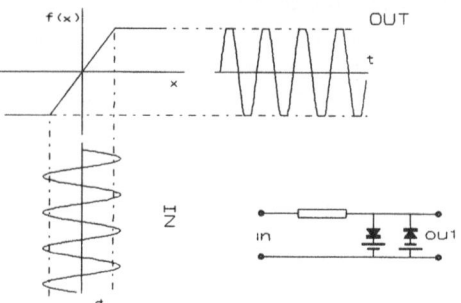

Figure 6.4.4 Limiter.

2. μ-law compression.

This signal-compression technique has been mentioned in section 4.2. It is applied in speech technology to reduce the proportion between the peak value and the RMS-value of a speech signal. Fig.6.4.5 shows the signal compression, fig.6.4.6 the expansion that restores the original signal.

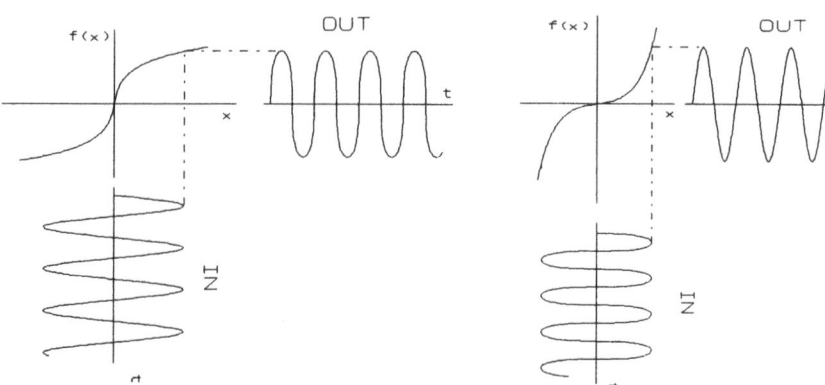

Figure 6.4.5 μ-law compression.   *Figure 6.4.6 μ-law expansion.*

3. Half wave rectification.
With a single diode and a resistor a rectifier can be constructed, a circuit that transmits only the positive (or by reversing the diode only the negative) fragments of the signal. Fig.6.4.7 shows the circuit and the transfer function.

4. Full wave rectification.
With four diodes in the circuit of fig.6.4.8 full wave rectification is possible. With this the negative signal fragments are mirrored into the positive range as shown by the transfer function.

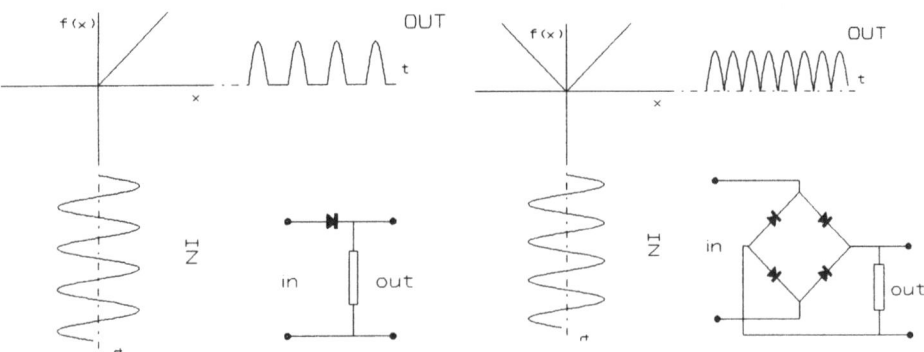

Figure 6.4.7
Half-wave rectification.

Figure 6.4.8
Full wave rectification.

5. Logarithmic and exponential conversion.
The relation between the voltage V across a diode and the current i through it is described by an exponential function: $I = I_0 e^{\frac{V}{V_0}}$ (I_0 and V_0 are constants).

If a diode is placed before the inverting input of an operational amplifier then the output voltage follows the input current (see formula (6.12)) and the input/output relation is thus also exponential. We have constructed an *exponential converter* (fig.6.4.9 and 6.4.10).

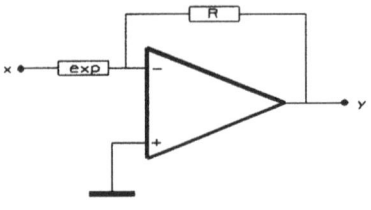

Figure 6.4.9
Exponential converter.

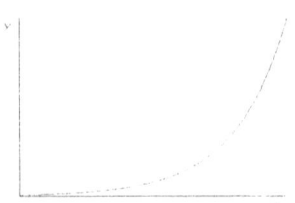

Figure 6.4.10
Exponential transfer characteristic.

By placing the diode in the feedback loop we get the opposite effect and obtain a *logarithmic converter* (figs.6.4.11 and 6.4.12).

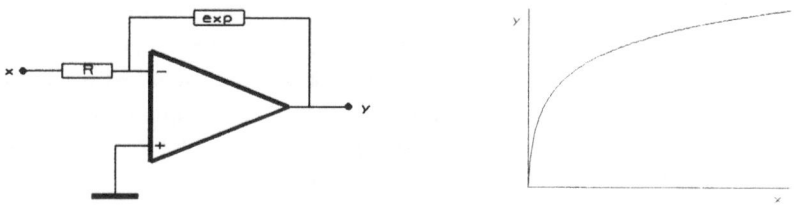

Figure 6.4.11
Logaritmic converter.

Figure 6.4.12
Logaritmic transfer characterstic.

These circuits are used for instance for converting the low-frequency (modulating or control) signals applied in electronic sound synthesis systems ('synthesizers') to achieve, for example, exponentially increasing voltage steps for pitch control, or logarithmic compression of voltage levels to dB-values.

B. *Time-variant systems; amplitude and frequency modulation*

An important category within the systems dealt with in this section is that of linear systems where the performance in time varies. Such systems are in general better suited for mathematical analysis than those discussed under A.
We can think of amplifiers with a non-constant, controllable gain factor, oscillators with a variable frequency and variable filters. In the first case one speaks of *amplitude modulation* (AM), in the second of *frequency modulation* (FM) and in the third case one could speak of spectrum modulation but this term is not used.

Our ear is very sensitive to such variations in the sound signal because as we saw in chapter 1 with acoustical communication the information is conveyed via comparable processes to the signal. AM is perceptible as variation of loudness, FM as variation of pitch, and spectrum modulation as variation of timbre, at least if the maximum modulation speed does not exceed the maximum speed by which we can modulate the sound signal with our muscles. This limits the maximum modulation frequency to about 20 Hz.

Modulation techniques thus play an important role in the synthesis of musical sounds and that of speech sounds. Spectrum modulation is especially important for speech synthesis (LPC). I will return to this in chapter 7. Technically much higher modulation frequencies can be realized. Although such signals are not recognized by our ears as information carriers in acoustical communication, they can still have interesting qualities. This holds in particular for AM and FM. I shall therefore first pay some attention to the theoretical and practical aspects of amplitude and frequency modulation.

1. Amplitude modulation.

If we modify an amplifier so, that the gain factor is no longer constant but can be controlled with the help of a second signal function, the modulating or control signal, in fig.6.4.13 indicated with x_m , we have then realized a system with which AM is possible. The most simple relation between control signal and gain factor is proportion:

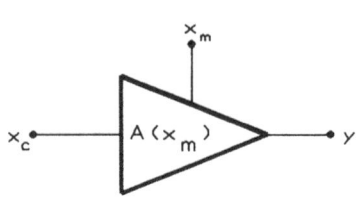

$$\text{gain factor}\quad A(x_m) = G \cdot x_m$$

Figure 6.4.13
Amplitude modulator.

The input/output relation then becomes:

$$y = A \cdot x_c = G \cdot x_m \cdot x_c \qquad (6.18)$$

in which x_c stands for the 'normal' input signal that now functions as a carrier wave. We can thus consider this system a (linear) time-variant system because due to the time dependency of x_m the coefficient A is time-dependent. It is also correct to speak of a nonlinear circuit with two inputs because due to the multiplication of x_m and x_c the superposition principle is no longer in effect.

For this reason there are various names for this system: the names 'voltage controlled amplifier' (VCA), 'envelope modulator' and 'envelope shaper' refer to the first function, and 'product modulator' and 'multiplier' to the second. I shall mostly use the term 'multiplier' as a general indication. Let us have a look at what happens if both x_c and x_m are sinusoidal:

$$\left. \begin{array}{l} x_c(t) = \cos 2\pi f_c t \\[2mm] x_m(t) = \cos 2\pi f_m t \end{array} \right\} \quad y(t) = G \cos 2\pi f_c t \cdot \cos 2\pi f_m t$$

Is this signal periodic? We have discussed this question before, in relation with the time-discrete sinusoidal function (see section 2.6.B). There we saw that this signal is periodic when an integer number of periods of x_c fits in one period of x_m. In other cases there is a fundamental frequency equal to the GCD of f_c and f_m.

Example 1.
Assume f_c = 440 Hz, f_m = 40 Hz (see fig.6.4.14). In one period of f_m fit exactly 11 periods of f_c; the signal is truly periodic, f_0 = 40 Hz.

Example 2.
Imagine f_c = 307 Hz, f_m = 19 Hz. The GCD of f_c and f_m is 1 Hz. Again the signal is periodic, but the period duration is now 1 second. Because with practical systems every frequency value has a limited accuracy there is always a rational quotient to be found. The signal is thus always periodic, but the period duration

can be very long as in the following example. We do not perceive such a signal as periodic.

Example 3.
Imagine $f_c = 307.22541$ and $f_m = 19.02067$; f_0 is now 0.00001 Hz (period duration ca. 28 hours!).

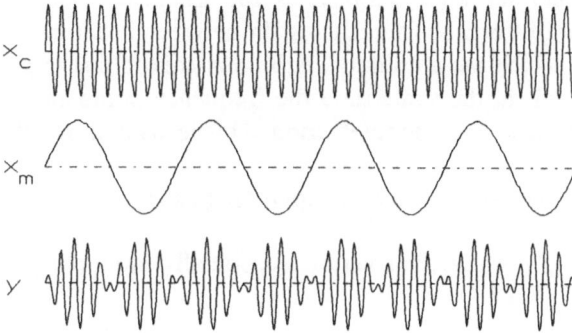

Figure 6.4.14 Amplitude modulation.

2. The spectrum of an AM-signal.

To directly calculate the spectral coefficients we must substitute the function $y(t) = G \cdot \cos 2\pi f_c t \cdot \cos 2\pi f_m t$ in the formulae (4.8) and (4.9). Even without the complications with the periodicity the calculation of the integrals is no sinecure. Fortunately there is a much simpler solution because with the help of rule (2.48) we can write $y(t)$ directly as a series of (co)-sine functions:

$$G \cdot \cos 2\pi f_c t \cdot \cos 2\pi f_m t = \tfrac{1}{2} G \cos 2\pi (f_c + f_m)t + \tfrac{1}{2} G \cos 2\pi (f_c - f_m)t$$

The spectrum consists of only two components, of which the frequencies are equal to the sum and the difference of the frequencies of the two input signals. If f_m is greater than f_c the difference frequency is negative. This is not a problem as $\cos(-\alpha) = \cos \alpha$. If with another choice for the input signals we should have got the term $\sin 2\pi(f_c\text{-}f_m)t$ this would have been no problem either as $\sin(-\alpha) = -\sin \alpha$, so that the component with the negative frequency, is equal to one with the same but positive frequency with a phase shift of $180°$. The result that we have found here is in correspondence with the considerations concerning the periodicity, because the sum and difference frequencies are always multiples of the greatest common divisor of the two original frequencies:

$$f_c + f_m = (k + n)f_0 \, , \quad f_c - f_m = (k - n)f_0$$

In example 2 f_c = 307 Hz and f_m = 19 Hz generate new components with frequencies 307 + 19 = 326 Hz and 307 - 19 = 288 Hz, both multiples of 1 Hz.

Both descriptions of the AM signal, a product function in the time domain and sum and difference frequencies in the frequency domain are mathematically equivalent. As for the perception it holds that with low modulating frequencies the time domain description with the corresponding fluctuating loudness is the most relevant, and for high frequencies the frequency domain description in terms of sum and difference frequencies. In between lies a transition range (f_m = 20 to 40 Hz) in which the amplitude variations are too fast to perceive the loudness variation, and at the same time the frequency components are so close to each other that there is a strong interaction. This leads to a 'rough' sound.

3. Amplitude modulation with a non-suppressed carrier wave.

Let us take for x_m a signal that varies sinusoidally around a particular constant value B:

$$x_m(t) = A \cos 2\pi f_m t + B = B\left(1 + \frac{A}{B} \cos 2\pi f_m t\right)$$

In fig.6.4.15 an example of such a signal function is shown. Observe the difference with fig.6.4.14.

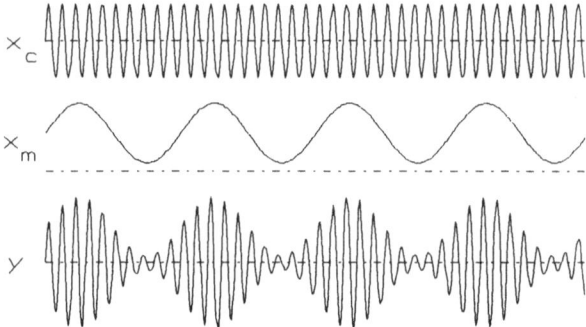

Figure 6.4.15 Amplitude modulation with non-suppressed carrier.

The factor A/B is called the *degree of modulation*. It should be $\leq$ 1 and is usually expressed as a percentage. In this example the degree of modulation is 100%. In the following m is used instead of A/B. The output signal is now:

$$y(t) = G \cdot x_c \cdot x_m = G \cos 2\pi f_c t \cdot B(1 + m \cos 2\pi f_m t)$$
$$= GB(\cos 2\pi f_c t + m \cos 2\pi f_c t \cdot m \cos 2\pi f_m t)$$
$$= GB\{\cos 2\pi f_c t + \tfrac{1}{2}m \cos 2\pi (f_c + f_m)t + \tfrac{1}{2}m \cos 2\pi (f_c - f_m)t\} \quad (6.19)$$

We see that in the spectrum besides the sum and difference frequencies also the original carrier wave frequency f_c appears. In fig.6.4.16 the spectrum of this amplitude modulated signal can be seen. The sum and difference components have equal amplitudes, equal to $\frac{1}{2}m$ times the amplitude of the carrier wave and thus at most equal to half its amplitude.

In the above it is assumed that x_m is a sinusoidal function, eventually with a constant added to it. If that is not the case and x_m is an arbitrary time function, then this function can be split into sinusoidal components as we know, and the above must be applied to each of these components. Imagine, for example, that x_m is a speech signal with a bandwidth of 50 to 8000 Hz. Then when it is amplitude modulated with a carrier wave vibration of 50000 Hz, sum frequencies in the range from 50050 to 58000 Hz and

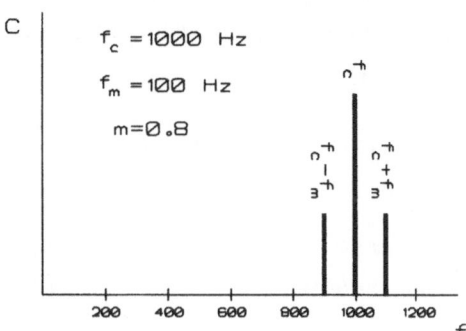

Figure 6.4.16
Spectrum of the signal of fig.6.4.15.

difference frequencies in the range from 42000 to 49950 Hz arise. These two frequency bands are called *sidebands*.

4. Applications of amplitude modulation.

From the above it shall be clear that AM can be used:
- in acoustical communication to give a particular envelope to a signal and in this way to determine the articulation and the loudness fluctuations.
- as sound synthesis method to derive from given input signals vibrations with other frequencies. If the input signals are complex and thus consist of various frequency components then this can lead to interesting results.
- as technical method to enable signal transmission (AM radio).

5. Amplitude demodulation.

Under this heading one understands the reconstruction of x_m from the modulated signal. In 'technical' modulation where a pure sinusoidal carrier is used, this is possible by multiplying the signal $y(t) = G \cdot \cos 2\pi f_c t \cdot \cos 2\pi f_m t$ with the carrier wave $\cos 2\pi f_c t$:

$$y(t) \cdot \cos 2\pi f_c t = G(\cos 2\pi f_c t)^2 \cos 2\pi f_m t =$$

$$= G(\tfrac{1}{2} + \tfrac{1}{2}\cos 2\pi f_c t)\cos 2\pi f_m t =$$

$$= \tfrac{1}{2}G \cos 2\pi f_m t + \tfrac{1}{2}G \cos 2\pi f_c t \cdot \cos 2\pi f_m t =$$

$$= \tfrac{1}{2}G \cos 2\pi f_m t + \tfrac{1}{2}G \cos 2\pi (2f_c + f_m)t + \tfrac{1}{2}G \cos 2\pi (2f_c - f_m)t$$

With the help of a lowpass filter the component with frequency f_m can be iso-lated. One calls this method 'synchronous detection' because an extra signal synchronous with the carrier is required.

For demodulation of analog AM-signals with a non-suppressed carrier wave (and a degree of modulation ≤ 100%) and also in the case of 'naturally' modulated signals where the carrier wave can be very complex, a simple envelope follower can be used, consisting of a diode, a capacitor and a resistor (fig.6.4.17).

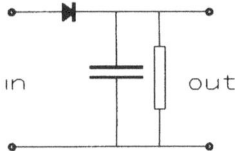
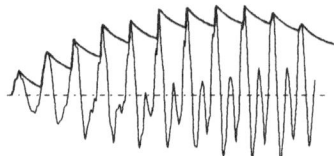

Figure 6.4.17   *Figure 6.4.18*
Envelope follower.   *Principle of envelope detection.*

Via the diode the capacitor is charged until the voltage across it equals the peak value of the input signal. If the voltage drops, the diode is blocked and the capac-itor discharges (slowly) via the resistor. As can be seen in fig.6.4.18 the capacitor voltage has a ripple that can be removed by means of a lowpass filter. The prog-ress of the peak values of a digital signal is generally quite simply determined.

Fig.6.4.19 shows the relation between a particular input signal and the corre-sponding output signal of an amplitude demodulator.

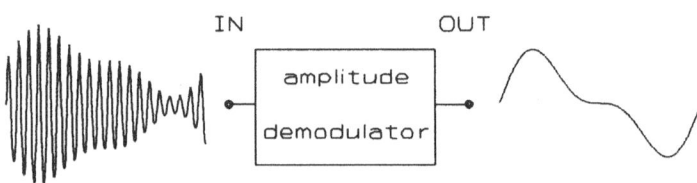

Figure 6.4.19 Amplitude demodulation.

6. Frequency modulation.
As the term says here the *frequency* of a sinusoidal signal $y(t) = \sin 2\pi ft$ varies. Let us assume this variation to be sinusoidal as well, around a value f_c, as shown in fig.6.4.20. Here f_i is the instantaneous frequency, T_m the modulation period (= 1/modulation frequency f_m) and Δf the maximum frequency deviation. The func-tion rule for f_i is:

$$f_i = f_c + \Delta f \cdot \cos 2\pi f_m t \qquad (6.20)$$

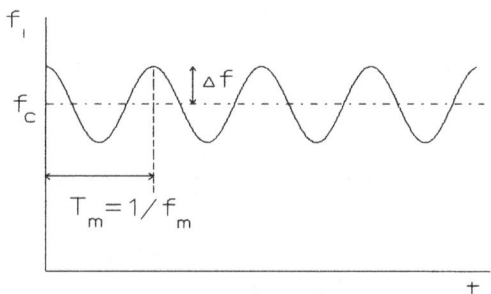

Figure 6.4.20 The instantaneous frequency.

It now seems easy to derive the function rule of the FM-signal, just by replacing frequency f in $y(t)$ by expression (6.20):

$$y(t) = \sin (2\pi f_c\, t + 2\pi\, t \cdot \Delta f \cdot \cos 2\pi f_m\, t)$$

This result cannot be correct because the product $t \cdot \Delta f$ implies that the frequency deviation grows proportional to time and thus becomes infinitely large. The problem is that the concept 'frequency' here loses its meaning. With a stationary sinusoidal signal the frequency is the number of periods per second. In the vector model (the rotating arrow in fig.6.4.21, introduced in chapter 2 as the 'bicycle-step' model) the frequency is the number of rotations of the arrow per second. When we vary the frequency, the rotation speed will vary, and the arrow will alternately move faster and slower. With large fluctuations this could lead to the vector coming to a standstill in the slow phase or even reversing the direction of the rotation. It can also happen that no complete rotation takes place and with this the concept of 'number of rotations per second' (=frequency) becomes meaningless. To solve this problem we introduce therefore a new definition of the concept of frequency, via the relation with the phase angle $\alpha(t)$. This new definition will turn out to be identical with the old definition when the vibration is stationary, but it will enable us to derive the correct expression for $\alpha(t)$ with a FM-signal, and then the function rule itself follows immediately from $y(t) = \sin \alpha(t)$. The relation between the frequency and α can be found via the rotation *speed*.

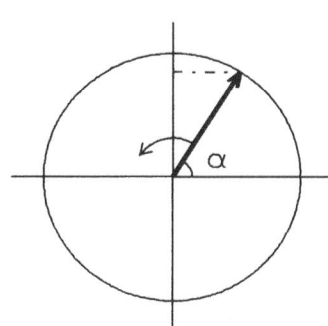

Figure 6.4.21
Angle modulation.

For 'frequency' we have: frequency   =   number of rotations/sec
                                          =   number of 2π rad/sec
and for 'rotation speed':   rotation speed   =   number of radians/second

From this we can conclude: frequency = rotation speed/2π

The rotation speed follows from the rotation angle by differentiating this angle

with regard to time, or $\qquad$ rotation speed $= \dfrac{d\alpha(t)}{dt}$

Combining these two results yields: $\quad$ frequency $= \dfrac{1}{2\pi}\dfrac{d\alpha(t)}{dt}$ $\qquad$ (6.21)

For a stationary vibration with $\alpha(t) = 2\pi f t + \phi$ this gives the expected result:
$$\text{frequency} = f.$$
Let us now replace 'frequency' by the expression (6.20) for f_i:

$$2\pi f_i \;=\; 2\pi(f_c + \Delta f \cdot \cos 2\pi f_m t) \;=\; \dfrac{d\alpha(t)}{dt}$$

or $\quad \alpha(t) = 2\pi f_c t + 2\pi\Delta f \dfrac{1}{2\pi f_m}\sin 2\pi f_m t + \alpha_0 = 2\pi f_c t + \dfrac{\Delta f}{f_m}\sin 2\pi f_c t + \alpha_0$

The proportion $\Delta f/f_m$ is called the *modulation index* and will be abbreviated from now on to m. The function rule for the FM-signal now is:

$$y(t) \;=\; A\sin\alpha(t) \;=\; A\sin(2\pi f_c t + m\sin 2\pi f_m t) \qquad (6.22)$$

The relation between the modulating signal, the carrier and the modulated signal is shown in fig.6.4.22.

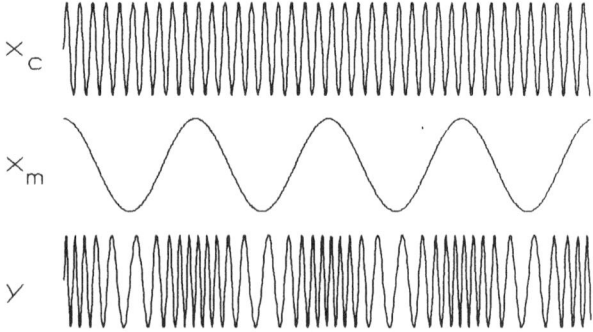

Figure 6.4.22 Frequency Modulation.

7. The periodicity of the FM-signal.
Regarding the periodicity of the FM-signal the same applies as with the AM-signal. There is a fundamental frequency that is equal to the GCD of the frequencies f_c and f_m.

8. The spectrum of the FM-signal.

For the same reason as with the AM-signal we must attempt to determine the spectral coefficients by working out $y(t)$ with trigonometric means to a sum of sinusoidal vibrations. For this we must make use of a special mathematical function, the *Bessel function*.

In 1824 while working on a certain mathematical problem, the German astronomer Friedrich Bessel was confronted with the following differential equation:

$$\frac{d^2y}{dt^2} + \frac{1}{z}\frac{dy}{dt} + (1 - \frac{p^2}{z^2})y = 0$$

The solution to this equation is a function of the complex variable z and the parameter p. Bessel proved several properties of these function, which are usually given with the letter J and were later named after him. For us only the case that z takes real values x is important. We shall thus work with the functions $J_p(x)$, 'Bessel functions of the first kind and of order p'. The reason Bessel functions are important for our problem is that the following relations can be proven:

$$\cos(m \sin b) = J_0 + 2\{J_2(m) \cos 2b + J_4(m) \cos 4b + \ldots\}$$

$$\sin(m \sin b) = 2\{J_1(m) \sin b + J_3(m) \sin 3b + J_5(m) \sin 5b \ldots\} \qquad (6.23)$$

We shall see then while working out the function rule of the FM-signal we will encounter expressions of this type. Rule (6.23), however, is only useful if it is possible to determine the values of the Besselfunctions J. This is indeed feasible with the following formula:

$$J_p(x) = (\frac{x}{2})^p \sum_{n=0}^{\infty} \frac{\left(-\frac{x^2}{4}\right)^n}{n!(p+n)!} \qquad (6.24)$$

Sometimes it works faster to use the following recursion formula to calculate higher order Bessel functions from lower order ones:

$$J_{n+1}(x) = \frac{2n}{x} J_n(x) - J_{n-1}(x)$$

The time function of the FM-signal (rule 6.22) can be reduced to the form:

$$\sin(a + m \sin b) = \sin a \cdot \cos(m \sin b) + \cos a \cdot \sin(m \sin b)$$

$$= \sin a [J_0(m) + 2\{J_2(m) \cos 2b + J_4(m) \cos 4b + \ldots\}] +$$

$$+ \cos a \cdot 2\{J_1(m) \sin b + J_3(m) \sin 3b + J_5(m) \sin 5b + \ldots\}$$

Bringing respectively $\sin x$ and $\cos x$ within brackets leads to products of the type:

$$\sin a \cdot \cos nb = \frac{1}{2}\sin(a + nb) + \frac{1}{2}\sin(a - nb) \quad (n \ \textit{even})$$

$$\cos a \cdot \sin nb = \frac{1}{2}\sin(a + nb) - \frac{1}{2}\sin(a - nb) \quad (n \ \textit{odd})$$

If we replace a by $2\pi f_c t$ and b by $2\pi f_m t$ and if we further assume that the amplitude factor A is equal to 1 and the initial phase angle to 0 (these two constants play no further role of importance), we see that we find for $y(t)$ the following expression.

$$y(t) = \sin(2\pi f_c t + m \sin 2\pi f_m t) =$$

$$= J_0(m) \sin 2\pi f_c t +$$

$$+ J_1(m) \sin 2\pi (f_c + f_m)t - J_1(m) \sin 2\pi (f_c - f_m)t +$$

$$+ J_2(m) \sin 2\pi (f_c + 2f_m)t + J_2(m) \sin 2\pi (f_c - 2f_m)t +$$

$$+ J_3(m) \sin 2\pi (f_c + 3f_m)t - J_3(m) \sin 2\pi (f_c - 3f_m)t +$$

$$+ \ldots$$

Or shorter:

$$y(t) = J_0(m) \sin 2\pi f_c t +$$

$$+ \sum_{k=1}^{\infty} J_k(m)\{\sin 2\pi (f_c + kf_m)t + (-1)^k \sin 2\pi (f_c - kf_m)t\} \qquad (6.25)$$

The spectrum thus contains the frequency components $f_c \pm k \cdot f_m$ (with $k = 0$, 1, 2, 3,...). The amplitude factor of each component is found by substituting the modulation-index m in the corresponding Bessel function, e.g. with rule (6.24) or by using the graphs of fig.6.4.23.

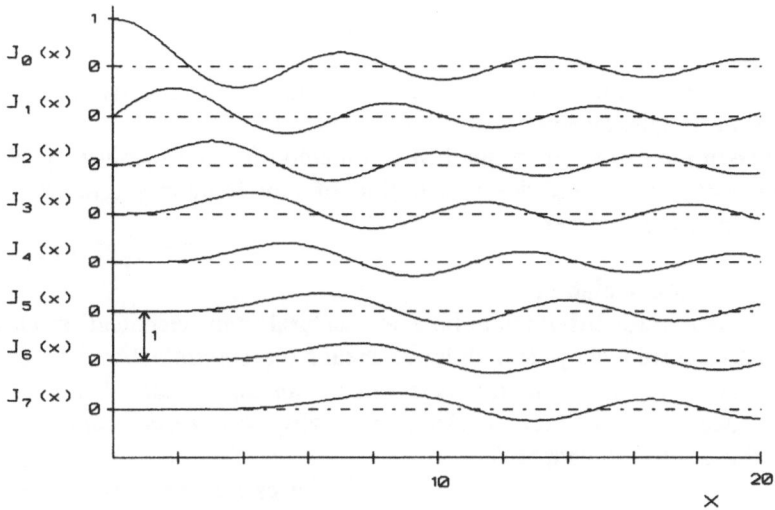

Figure 6.4.23 Bessel functions.

Example: $f_m = 300$ Hz, $f_c = 3400$ Hz, $\Delta f = 1500$ Hz
The frequency components are 300 Hz apart. The modulation-index is equal to
1500/300 = 5. This gives the following amplitude values:

frequency component	Bessel function	amplitude
f_c	$J_0(5)$	- 0.1776
$f_c \pm f_m$	$J_1(5)$	- 0.3276
$f_c \pm 2f_m$	$J_2(5)$	0.0466
$f_c \pm 3f_m$	$J_3(5)$	0.3648
$f_c \pm 4f_m$	$J_4(5)$	0.3912
$f_c \pm 5f_m$	$J_5(5)$	0.2611
$f_c \pm 6f_m$	$J_6(5)$	0.1310
$f_c \pm 7f_m$	$J_7(5)$	0.0534

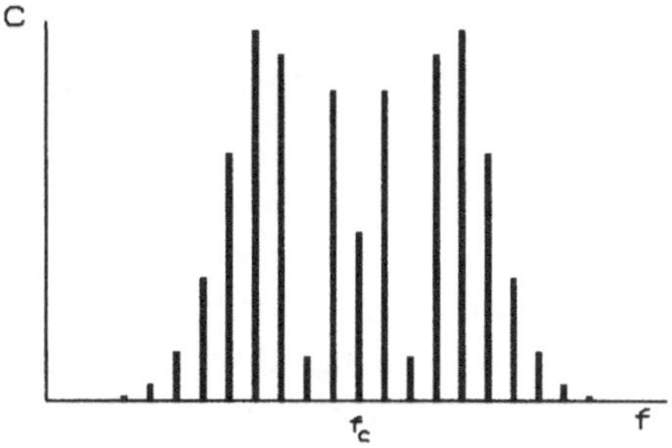

Figure 6.4.24 Spectrum of a FM-signal.

This spectrum is shown in fig.6.4.24 (the minus signs are neglected). It is clear
(and visible in fig.6.4.23) that the value of J_p decreases rapidly with increasing
value of p. As a rule of thumb we can assume that for m > 4 the coefficient $J_m(m)$
gives the last 'large' amplitude value (in our example $J_5(5)$).
This corresponds with the frequency components:

$$f_c \pm m f_m = f_c \pm \frac{\Delta f}{f_m} f_m = f_c \pm \Delta f$$

The 'practical' bandwidth of the FM signal is thus approximately $2\Delta f$ Hz.

9. Applications of frequency modulation.
- in acoustical communication FM is the method to convey information to the signal via control of the pitch (intonation).
- FM offers interesting possibilities for the synthesis of sounds, consisting of equidistant frequency components. See chapter 7.
- In the communication technique FM is a method for high-quality signal transmission (FM radio) and for registration of low frequency and DC signals (instrumentation recorder).

10. Frequency demodulation.
Here again one must differentiate between 'natural' and 'technical' modulation. In the first case frequency demodulation means the determination of the fundamental frequency of the signal that is in general non-sinusoidal. This is an important and sometimes difficult problem, especially with speech signals. I shall return to this subject in chapter 7.

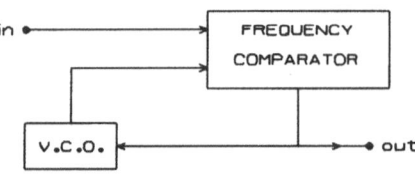

Figure 6.4.25 Frequency demodulation with Phase Locked Loop.

If the carrier wave is sinusoidal (or can be made sinusoidal by removing higher frequency components with the help of a filter) one can with analog systems make use of a so-called PLL-circuit (PLL = Phase Locked Loop) (fig.6.4.25). The frequency is compared with that of a voltage controlled oscillator. The frequency comparator gives a (DC-)voltage that is proportional to the value of the frequency difference. With this voltage the oscillator frequency is adjusted until it is equal to that of the input signal. The control voltage reflects the frequency variations and is thus the demodulated signal.

In a digital FM-signal with sinusoidal carrier wave the positions of the zero-crossings (either ascending or descending) are easy to determine. The distance in between is the period duration; the reciprocal of this is the frequency value. In fig.6.4.27 a frequency demodulator with input and output signal is shown.

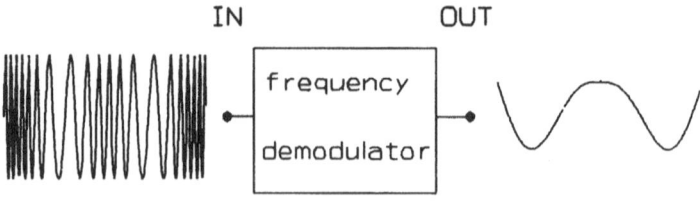

Figure 6.4.26   Frequency demodulation.

6.5   Problems

6.1   The output voltage of a pre-amplifier without load resistance and with a R_{out} of 4 kΩ is 1 volt. Now two inputs of other devices are connected (in parallel). The two input resistances are 1 and 9 kΩ respectively. Which voltage is found on these inputs? How much energy does each of them withdraw from the pre-amplifier?

6.2   Now a transformer is used for an optimal match between the pre-amplifier of problem 6.1 and the load. What should be the proportion of the number of windings N_s/N_p? Which voltage is now measured at the input and how much power is taken?

6.3   A resistor of 1 kΩ is connected to the output of some apparatus. As a result the output voltage decreases from 2 volts to 1.6 volts. What is the output resistance of this apparatus?

6.4   The output of device A (with $R_{out} = R_A$) is connected, via a potentiometer, to the input of device B (with $R_{in} = R_B$).

a.   Consider the potentiometer as a part of device B. Which load 'sees' device A?

b.   Consider the potentiometer as a part of apparatus A. Calculate the output resistance.

6.5   Calculate the open-loop and closed-loop gain (in dB) of the amplifier of fig.6.2.4. (The dB is here -incorrectly - used as a measure to compare amplitudes!) Use $A = 10000$, $R_1 = 10$ kΩ, $R_2 = 40$ kΩ.
        Assume that the amplifier (without feedback) has an output resistance of 1 kΩ. Calculate the output resistance of the circuit with feedback.

6.6   A sound source is located at 1.75 m above a hard, reflecting floor. Let us check the effect of the interaction between the direct sound and the reflected sound at a distance of 5 metres from the source at the same level. Calculate the frequencies of the maxima and minima of the comb-filter characteristic and the peak/valley proportion in dB. Assume an attenuation factor due to absorption, of 0.9. Speed of sound: 340 m/s.

6.7   Derive formula (6.15) by giving the filter of fig.6.3.1b an input signal $x(t) = \sin \omega t$ and assuming an output signal $y(t) = A \sin(\omega t + \phi)$

6.8   Derive the formulae (6.17) in the same way as (6.16).

6.9   In electronic circuits the RC-network of fig.6.3.6 is often used to separate the DC-voltage, which is required for a proper functioning of the transistors, from the AC-voltage of the signal. Calculate the value of C required for a cutoff frequency of 20 Hz when R is 5 kΩ. The input resistance of the following circuit is much larger than 5 kΩ.

6.10   Amplitude demodulation with the circuit of fig.6.4.17 causes a sawtooth-like ripple. This can be removed with the RC-filter of fig.6.3.10. Give reasonable values for C and R, when the carrier frequency is 5 kHz, and the highest modulation frequency is 200 Hz.

6.11   Construct a list of 100 random numbers r ($0 < r \le 1$). Use a calculator that has this possibility, or by applying the π-algorithm of section 6.3.D.5. Subdivide the 100 numbers in 20 groups of 5 numbers and calculate the average of each of these groups. Calculate the global frequency distribution of these averages by counting how often the value is between 0.2 and 0.4, between 0.4 and 0.6, between 0.6 and 0.8 and so on. Sketch the distribution curve. Does it look like a normal distribution?

6.12
   a.   Calculate the frequencies and amplitudes of the components of the AM-signal
$$y(t) = \cos 2\pi f_c t (1 + m \cdot \cos 2\pi f_m t)$$
when $m = 0.2$, $f_c = 1000$ Hz and $f_m = 200$ Hz.

   b.   Work out the function rule of the following FM-signal:
$$y(t) = \sin(2\pi f_c t + m \cdot \cos 2\pi f_m t)$$

   c.   Calculate the frequencies and amplitudes of the three central components of this FM-signal (higher and lower order components can be neglected) when $f_c = 1000$ Hz, $f_m = 200$ Hz and $\Delta f = 40$ Hz.

   d.   Notice the correspondence between the spectra of the AM- and the FM-signal. What is the most conspicuous difference?

6.13   Calculate the frequencies and amplitudes of the significant spectral components of a FM-signal with $f_c = 940$ Hz, $f_m = 200$ Hz and $\Delta f = 2000$ Hz.

CHAPTER 7

Analysis and Synthesis Techniques

To analyse a signal function is to determine the values of quantities related to the signal function in question, as for example the average energy of the signal. With such analyses one is mainly concerned with the following two areas of application:

1. The realization of optimal efficiency and quality in the (technical) process of signal transmission, registration and reproduction.
2. The study of the process of acoustical communication and the systems that play a role in this.

Examples of the first area of application are:

a. the determination of the peak value of a signal to prevent overload in a registration or reproduction process;
b. making the transmission of speech more efficient and thus less expensive by not transmitting the complete signal but only the modulation parameters (Vocoder, LPC).

In relation to the study of speech and musical signals I now return to the subject discussed in the introductory chapter, namely the fact that during acoustical communication the acoustical signal is an intermediate phase, presenting two points of contact:

- we can derive from it properties of the sound source,
- we can attempt to predict what the effect of the signal on the receiver will be.

In the last case the analysis becomes a simulation of the demodulation process that takes place in the perception of acoustical signals.

Pitch, loudness and *timbre* are the principal attributes derived by the ear from the modulated carrier wave. These attributes are connected with the physical characteristics *periodicity (frequency), amplitude* and *spectrum*. If we wish to study this process, we must be able to measure these physical quantities and furthermore we must know the relation with perceptive qualities. For this last subject I refer to the specialized literature in this field such as the book "Aspects of Tone Sensation" by R. Plomp (1975). In this book it is explained how amplitude and spectrum measurements can be performed so that the results are relevant for the perception of loudness and timbre.

Measurement of amplitude of both analog signals (with a voltmeter) and digital signals (using the sample values) is easily performed and does not need to be dealt with here.

The techniques for spectral analysis are treated in chapter 4. In the present chapter therefore only the analysis of periodicity will be discussed and also in which way detailed information concerning the sound source may be derived from the signal.

In the second part of this chapter synthesis techniques will be discussed. Here we can differentiate between technical and scientific applications, e.g. the synthesis of speech signals as a part of a system for speech transmission (technical) and synthesis of musical signals as a method to check the accuracy of a preceding analysis (scientific). See Rabiner et al. (1985) and Witten (1982).

There is yet a third area of application, the artistic-creative use of synthesis techniques in electronic and computer music and the use of synthesizers. In all three cases the same question arises, how in the most efficient and effective way a modulated vibration can be produced. See Bateman (1980), Dodge (1985) and Roads et al.(1985).

7.1 The analysis of periodicity, autocorrelation

We are facing two problems here: it must be decided if an arbitrarily given signal is or is not periodic, and if so, the period duration must be determined. In the case of an approximately periodic signal the situation is not too difficult. One can look, for example, at the spectrum to determine the fundamental frequency of the harmonic series of frequency components that occurs there. Problems arise when the signal is quasi-periodic (which is always the case with signals from speech and music), because then the question is how large the deviation from a purely periodic signal can be (and how to measure this deviation) before the signal must be considered non-periodic. Clearly the properties of the hearing organ play an important role here. Actually this is a pattern recognition problem. Thanks to our well-developed ability to recognize visual patterns we immediately recognize the period in the registration of a harpsichord signal shown in fig.7.1.1. It is quite difficult, however, to develop measuring equipment or computer programs that can take over this task.

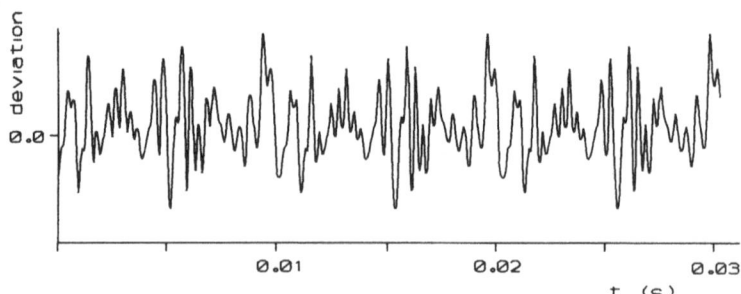

Figure 7.1.1 Registration of a harpsichord tone.

A periodic sound vibration is perceived as a tone with a certain pitch, corresponding to the period T. This is also the case when the spectrum is 'incomplete', i.e. when one or more harmonics are 'missing'. It is especially important to know that for this pitch impression the presence of the first harmonic is not essential. A vibration without a first harmonic still evokes a pitch that corresponds to this 'missing fundamental'. The capability to hear this 'low' or 'virtual' pitch is a remarkable property of our perception system. A condition is that the composition of the spectrum is not too extreme. A vibration with frequency components of 300, 400 and 500 Hz (thus the third, fourth and fifth harmonics of 100 Hz) will be periodic with a period duration T of 10 ms and will have a pitch corresponding to 100 Hz. See fig.7.1.2.

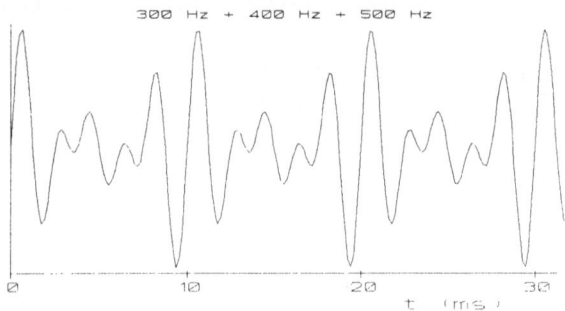

Figure 7.1.2 Periodic vibration with missing fundamental.

A vibration consisting of the 34-th and 35-th harmonics (3400 Hz and 3500 Hz), although still exhibiting a period of 10 ms, no longer provokes this pitch impression. The frequency of 100 Hz in these examples may be interpreted as a subharmonic of the actual frequency components. Just as the n-th harmonic of frequency f_0 has the frequency nf_0, the n-th subharmonic has as frequency f_0/n and 100 Hz is thus the third subharmonic of 300 Hz, the fourth of 400 Hz etc. Certain theories on pitch perception assume that such subharmonics play a role in this mechanism.

There are time domain and frequency domain methods for determining the periodicity of a signal function with a computer. The use of the autocorrelation function, to be discussed later in this section, is an example of the former approach. There are several other methods, with less computational complexity but often based on rather heuristic principles with which conclusions about the periodicity or quasi-periodicity are derived from the shape and location of the peaks of the signal function or from the pattern of zero-crossings. It seems therefore also attractive to consider a frequency domain method. The spectrum consists of lines at certain frequencies and the repetition frequency is equal to the frequency difference between two subsequent components. Several complications however occur in practice:

- Some components may be missing.
- With quasi-periodic signals other components may occur that are not harmonically related to the fundamental.
- The spectrum consists of peaks instead of lines, so an algorithms for estimating the frequency of the relevant component is required. As we have seen in section 4.3.G, the frequency resolution is N/f_s.

An effective method to detect the fundamental frequency f_0 is spectral compression (Hermes, 1988). From the given spectrum new spectra with subharmonics are derived via multiplication by 1/2, 1/3, 1/4, ... respectively. Then all spectra are added. It is very likely that in each of them f_0 occurs. If the fundamental is for example 100 Hz, it will occur in the second spectrum, because 200/2 = 100, in the third spectrum, because 300/3 = 100 etc. This component will therefore be strongly represented in the sum spectrum. See fig.7.1.3. Due to the limited frequency resolution the maximal compression factor is ca. 5. This can be improved by using a logarithmic subdivision of the frequency axis.

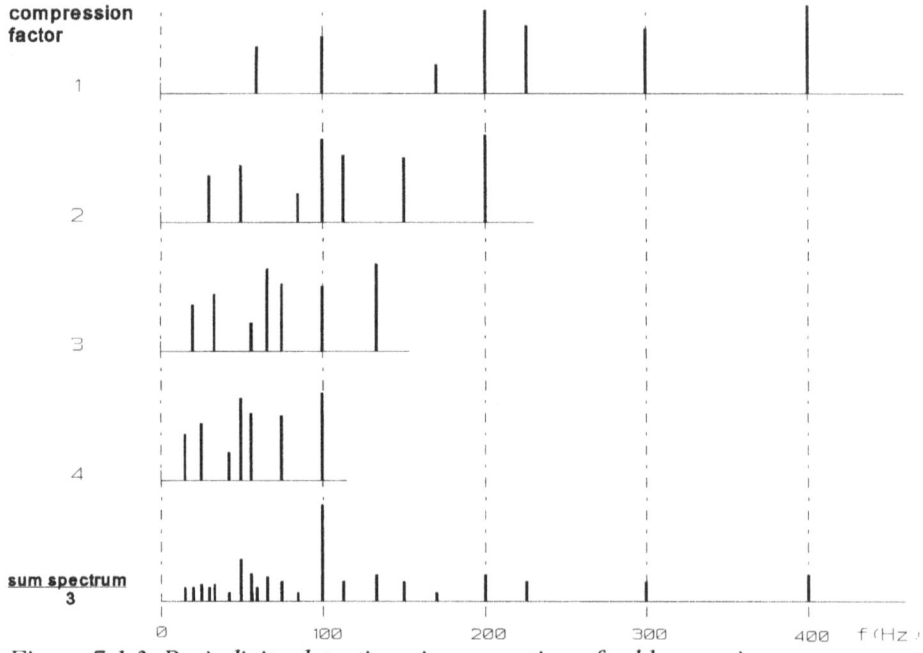

Figure 7.1.3 Periodicity detection via summation of subharmonics.

Let us now return to the time domain. Here there are two methods with a strong theoretical basis that produce reasonably reliable results. They both are computationally quite complex. The first is autocorrelation to be discussed now, and the second is cepstrum analysis to be discussed in section 7.2.

autocorrelation

This method is based on the determination of the *autocorrelation function R.*
What kind of function this is, can be made clear by explaining the calculation of
R for a time-discrete signal. The starting point is the finite, time-discrete signal
function $y(k)$, shown in fig.7.1.4 ($y(k) = 0$ if $k < 1$ and if $k > N$).

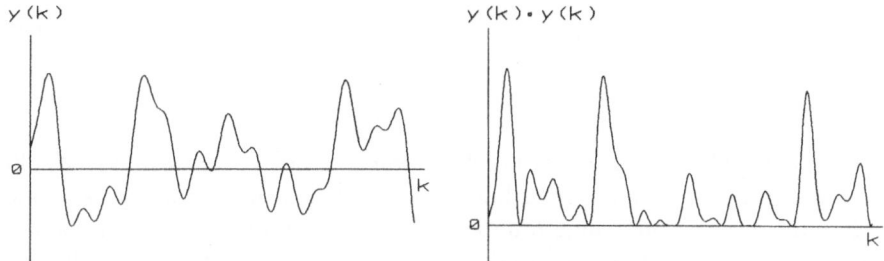

Figure 7.1.4 A signal function. y(k).   *Figure 7.1.5*
The product function y(k) · y(k).

First we multiply this function by itself. The product $y(k)\cdot y(k)$ is shown in
fig.7.1.5. Next we add up all values of this product function. This gives one
particular sum value that we (for reasons yet to be clarified) designate as $R(0)$.

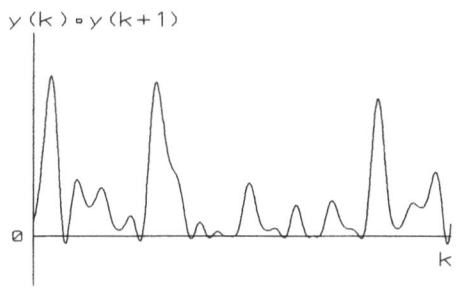

Figure 7.1.6
The product function y(k) · y(k+1).

Thus: $$R(0) = \sum_{k=1}^{N} y(k)\cdot y(k)$$

Next we repeat the whole process
with one modification: we multiply
the signal by a version of itself shifted
by one sample. This product function
is shown in fig.7.1.6. The sum value
is determined again and designated
with $R(1)$:

$$R(1) = \sum_{k=1}^{N} y(k)\cdot y(k+1)$$

The value of $R(1)$ will probably be slightly less than $R(0)$ because the products
$y(k)\cdot y(k)$ are always positive while it may occur with the products $y(k)\cdot y(k+1)$
that two y-values have different signs that cause the product to be negative and
thus gives a negative contribution to the sum value. It will meanwhile be clear
that the index of R gives the number of samples to which the shift has occurred.
We go on in the same way and calculate $R(2)$, $R(3)$ etc. The general expression
for the autocorrelation function R is:

$$R(m) = \sum_{k=1}^{N} y(k) \cdot y(k+m)$$

We may set the summation limits at $+\infty$ and $-\infty$ because outside the given range the function y has the value 0:

$$R(m) = \sum_{k=-\infty}^{\infty} y(k) \cdot y(k+m) \tag{7.1}$$

If the function $y(k)$ is irregular in structure, and thus noisy, then with values of m > 0 the positive and negative contributions to the sum will lay close to each other which will cause the sum value R to be in the neighbourhood of 0 (after the peak at $m = 0$). If the function has a periodic structure, something totally different happens. At first the value of R decreases with increasing m, but as the shift becomes larger the moment draws near that the signal is multiplied by a version of itself shifted over one period. Then the situation is exactly the same as with $m = 0$. With increasing m the fluctuation of R around the zero-line changes into an increase of R, and at $m = T$ a maximum is reached. After the exact coincidence the process repeats itself because with a shift of two periods there is again a maximum. Fig.7.1.7 shows the autocorrelation function for a noise signal, and fig.7.1.8 that for a periodic signal.

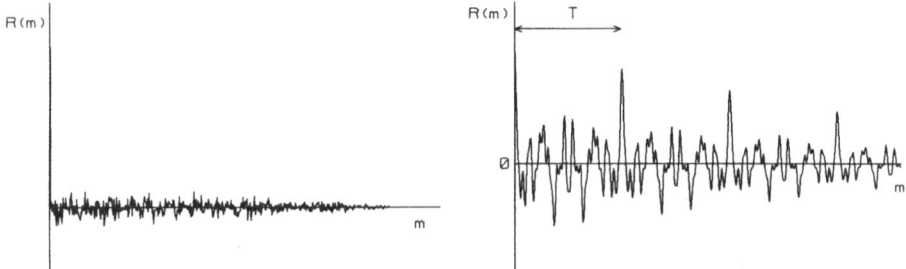

Figure 7.1.7 Autocorrelation function of a noise signal.   *Figure 7.1.8 Autocorrelation function of a (quasi-)periodic signal.*

Conclusion: the autocorrelation function of a periodic signal is itself also periodic with the same value of the period. The question is now whether the periodicity of R is easier to detect than that of y. The answer to this question is negative if the signal is truly periodic and affirmative if the signal function contains random elements (this generally happens with quasi-periodic signals), because these are averaged with the calculation of R and reduced in that way. However, these are the signals we are interested in.

It is possible to say more about the relation between the wave shape of y and that of R via a frequency domain approach. Beforehand we must have a look at a few derived definitions of R.

The original definition of R for resp. time-discrete and time-continuous functions is:

$$R(m) = \sum_{k=-\infty}^{\infty} y(k) \cdot y(k+m) \text{ and } R(\tau) = \int_{-\infty}^{\infty} y(t) \cdot y(t+\tau) dt \qquad (7.2)$$

For the analysis of quasi-periodic signals where the (quasi-)period can gradually change one restricts oneself to a rather short signal fragment in the calculation of R. The signal is given the desired finite duration by multiplying it by a (rectangular) window $w(k)$:

$w(k) = 1$ for $k = 1, 2, ..., N$ and $w(k) = 0$ for other k-values.

This allows keeping the number of calculations to a reasonable amount. In this way the *short time autocorrelation function* is formed:

$$R_s(m) = \sum_{k=-\infty}^{\infty} y(k) \cdot w(k) \cdot y(k+m) \cdot w(k+m) \qquad (7.3)$$

For theoretical applications with time-continuous signals one must in principle proceed from the general definition given above. With a non-finite signal this leads to problems. In this case the *average autocorrelation function* should be used:

$$\overline{R}(\tau) = \lim_{T \to \infty} \frac{1}{T} \int_{-\frac{1}{2}T}^{\frac{1}{2}T} y(t) \cdot y(t+\tau) dt \qquad (7.4)$$

With periodic signals this average is equal to the average over one period. Then it holds:

$$\overline{R}(\tau) = \frac{1}{T} \int_{-\frac{1}{2}T}^{\frac{1}{2}T} y(t) \cdot y(t+\tau) dt \quad (T = period\ duration)$$

We shall go on from this latter definition to derive the relation between y and R via the frequency domain. This is done by replacing y by the corresponding Fourier series:

$$\bar{R}(\tau) = \frac{1}{T} \int_{-\frac{1}{2}T}^{\frac{1}{2}T} \left\{ \sum_{n=1}^{\infty} C_n \cos(2\pi n ft + \phi_n) \cdot \sum_{m=1}^{\infty} C_m \cos(2\pi m f(t+\tau) + \phi_m) \right\} dt$$

The second Fourier series can be written as

$$\sum_m \{ C_m \cos(2\pi m ft + \phi_m) \cos 2\pi m f\tau - C_m \sin(2\pi m ft + \phi_m) \sin 2\pi m f\tau \}$$

If both series are multiplied by each other and are integrated term for term, integrals result of the type:

$$\cos 2\pi m f\tau \int C_n \cos(2\pi n ft + \phi_n) C_m \cos(2\pi m ft + \phi_m) dt$$

and

$$\sin 2\pi m f\tau \int C_n \cos(2\pi n ft + \phi_n) C_m \sin(2\pi m ft + \phi_m) dt$$

In the same way as in section 4.3 it can be shown that all these integrals result in 0 except the cosine product with $n = m$. For this we find:

$$\int C_n^2 \cos^2(2\pi n ft + \phi_n) dt = \frac{1}{2} T C_n^2$$

The expression for R finally becomes:

$$\bar{R}(\tau) = \frac{1}{2} \sum_{n=1}^{\infty} C_N^2 \cos 2\pi n f\tau \qquad (7.5)$$

This is a Fourier series! Constructing a signal function with the help of the spectral components is called Fourier synthesis or inverse Fourier transform. (symbol: F^{-1}, see section 4.3.G). This takes place here with the squared amplitude coefficients (the power spectrum coefficients) of the function $y(k)$ as spectral coefficients. Therefore, we may also write for R:

$$\bar{R}(\tau) = F^{-1}\{|Y(f)|\}^2 = F^{-1}\{|F[y(t)]|\}^2 \qquad (7.6)$$

or

$$|F[y(t)]|^2 = F[\bar{R}(\tau)] \qquad (7.7)$$

There are thus two possibilities for the calculation of R: directly from the time function according to the definitions given on the previous page, or via the inverse Fourier transform of the power spectrum. For the second possibility use can be made of the FFT and this is usually the fastest method. It appears further from this that the wave shape of R contains the same harmonic components as the signal function with squared amplitude coefficients.

For the time domain calculation time can be saved by working with the *pseudo* autocorrelation function R_p. Here the signal function y is not multiplied by itself but by a function p that is defined as follows:

$$p(k) = \begin{cases} +1 \ \ when \ \ y \geq d \\ 0 \ \ when \ \ -d < y < d \\ -1 \ \ when \ \ y \leq -d \end{cases}$$

in which d represents a particular limit-value that must be chosen in connection with the average signal amplitude. The pseudo autocorrelation function is now:

$$R_p(m) = \sum_k y(k) \cdot p(k+m)$$

In fact no real multiplications are performed because the terms of the series are either equal to $y(k)$ itself, or equal to 0 or equal to $-y(k)$. This naturally saves much time and for the detection of periodicity, the result is practically the same as that of the normal method because R_p shows peaks if the signal function is (quasi-)periodic. Periodicity detection with the help of the (pseudo) auto-correlation function boils down to the detection of peaks and the determination of their position.

With a similar method it is possible to test two different signal functions for correspondence in the wave shape. To that end one calculates the *cross correlation function* $C(\tau)$:

$$C(\tau) = \int_{-\infty}^{\infty} y_1(t) \cdot y_2(t+\tau) \, dt \tag{7.8}$$

which also shows peaks if y_1 and y_2 are similar (when for example one function is a delayed or mirrored version of the other). The cross correlation function plays an important role in the research on directional hearing and in stereophonic recording technique.

With rule (7.7) we can show that the noise signal produced by a random number generator has an almost white spectrum. The argument is the following:

The autocorrelation function of a noise signal that comes from an ideal pseudo-random number generator is a periodic pulse (fig.7.1.9):

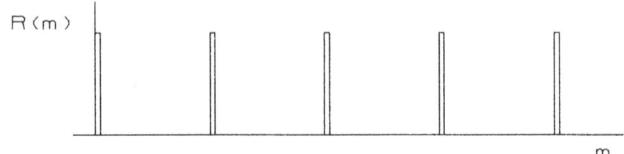

R(m)

m

Figure 7.1.9 Autocorrelation function of a pseudo-random signal.

Explanation:
- pulse shape, because due to the lack of any systematic relation between consecutive samples, the R-value is already (ca.) 0 with a shift of one sample;
- periodic, because the number series is periodic (see section 6.3.B.5).

The Fourier transform of such a function has the well-known sin x/x shape with zero points at multiples of the clock frequency. Over the frequency range 0 - ½f_s the power spectrum of this noise signal is almost flat and so the spectrum is white.

7.2 Cepstrum analysis

A. *Deconvolution*

This analysis method can be applied for determining the periodicity (via the frequency domain) of a signal function and also for the analysis of properties of the signal source. In the discussion of the general model of a linear system (shown in fig.7.2.1, see also fig.5.2.8):

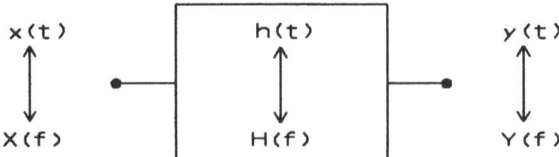

x(t)   h(t)   y(t)

X(f)   H(f)   Y(f)

Figure 7.2.1 Linear system.

in which an input signal $x(t)$ (with spectrum $X(f)$) is connected to a system with an impulse response $h(t)$ (or frequency characteristic $H(f)$) and in this way produces an output signal $y(t)$ (with spectrum $Y(f)$), we have seen that:
1. It does not matter with which function of each of the three function-pairs we work, because we can always switch over from the time domain to the frequency domain via the Fourier transform F and from the frequency domain to

the time domain via the inverse Fourier transform F^{-1}. In practice the calculations are performed with the help of a FFT-program.

2. We must know at least two of the three function-pairs to be able to calculate the third. This calculation takes place as follows
   - in the time domain with $y(t) = x(t) * h(t)$ (convolution)
   - in the frequency domain with $Y(f) = X(f) \times H(f)$.

Cepstrum analysis is a technique for deriving information about the input signal and/or the system from the output signal only. This is only possible - and then merely to a certain extent - if we have some global information about x or X and h or H. That is for example the case with speech (and with certain musical instruments) if we base ourselves on the linear source/filter model. Then $x(t)$ is the excitation signal produced by the vocal chords, and $H(f)$ is the frequency response of the vocal tract conceived as a linear filter. This frequency response is also called the 'formant characteristic'. It is known that in this response several resonance peaks (formants) occur. If we wish to derive x or X and h or H from y or Y then we should work in the frequency domain, because obviously it is more difficult to separate two function which are convoluted with each other (this process is called 'deconvolution') than to separate two functions which are multiplied by each other. The starting point is thus the speech signal $y(t)$ (see fig.7.2.2), of which we calculate first the spectrum $Y(f)$ (fig.7.2.3) with a FFT-routine:

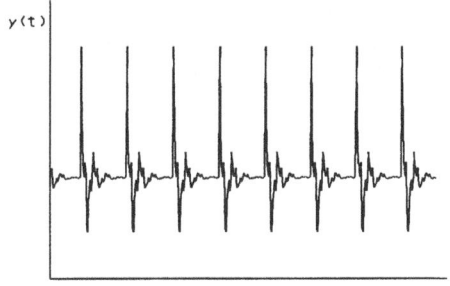

Figure 7.2.2 The speech signal y(t).   *Figure 7.2.3 The spectrum Y(f).*

We cannot separate X and H from each other by means of a filter as subtractive operation filtering cannot be applied to a product. This problem can be solved by switching over from Y to log Y (fig.7.2.4) because if $Y(f) = X(f) \cdot H(f)$ then it holds that log $Y(f) = $ log $X(f) + $ log $H(f)$

Let us first look at what we know about X and H:

1. If we restrict ourselves to a vowel signal, we know that the signal is quasi-periodic, and that the spectrum is a line spectrum. If we take the logarithm, the graph is 'compressed' in a vertical direction, and the calibration of the vertical

axis is changed, but the 'shape' of the function does not change substantially.
2. *H* is the formant characteristic of the vocal tract. The shape of this characteristic also does not alter greatly either when swicthing over to the logarithm of the function *H*. We know that the function log *Y* is the sum of a fast and periodically fluctuating function (log *X*, the logarithm of the line spectrum of the excitation signal) and a slow undulating function (log *H*, the logarithm of the formant characteristic).
We can separate the 'rapid' and the 'slow' components by means of 'filtering' the spectrum, by treating the spectrum so to speak as a time function. This may seem a strange step to take but a computer cannot differentiate between these two sorts of functions. Every function is simply a collection of numbers. For the filter to be used with log *Y* we apply the principle described in section 6.3.A:
- calculate by means of the FFT the spectrum of the function (log *Y*) to be filtered;
- remove from this spectrum the components to be filtered out;
- then reconstruct the function with the inverse FFT.
The reconstructed function will as a consequence of the filtering have a shape that differs from the original shape.

If we calculate the spectrum of log *Y*, we in fact calculate the spectrum of the logarithm of the spectrum of *y*(*t*). We must pay attention here not to allow any confusion to arise. To prevent this the authors of the first paper devoted to this method (Bogert et al.1963) introduced a number of new terms: the spectrum-of-the-(logarithmic)-spectrum is called the *cepstrum*, the variable along the horizontal axis is designated by the word *quefrency*, possible peaks as *rahmonics*, while the word filter is replaced by *lifter*, magnitude by *gamnitude* and phase by *saphe*. If we calculate the spectrum of log *Y* with a FFT, we get the cepstrum, see fig.7.2.4.

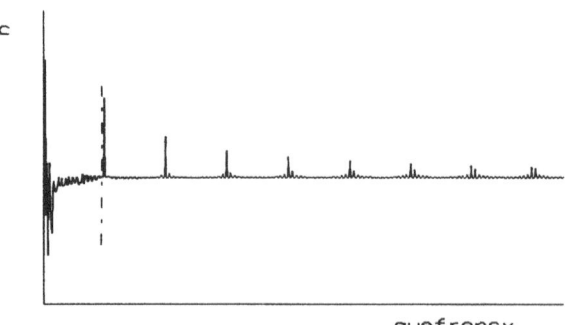

Figure 7.2.4 The cepstrum.

The slowly varying component of log *Y* (= log *H*) is to be found in the low-quefrency part of the cepstrum and the rapid periodic fluctuations (log *X*) in the high-quefrency part.

Proceeding from the cepstrum various operations can be performed:

1. From the position of the rahmonics the periodicity in the original spectrum can be calculated, and from there again the repetition frequency in the original time function. Applied in this way cepstrum analysis is one possibility for periodicity detection.

2. We can leave the cepstrum unmodified and return backwards along the same path:

   a. with FFT^{-1} reconstruct log Y;

   b. via exponentiation reconstruct Y (Note e$^{\ln Y}$ = Y);

   c. with FFT^{-1} reconstruct $y(t)$.

   With step b we get the original spectrum, with step c the original time function. Therefore after all nothing has happened, but we can in this way test the correct operation of the computer programs.

3. We follow the original plan and filter log Y by modifying the cepstrum. Obviously one must either remove the high-quefrency part (right of the dashed line in fig.7.2.4) or the low-quefrency part, to the left of the dashed line. 'Remove' means 'make equal to 0'. We speak of 'short-pass liftering' and 'long-pass liftering' respectively.

short-pass liftering

From the cepstrum of fig.7.2.4 a part is left over, as shown in fig.7.2.5. When we follow the same reconstruction procedure as described in point 2, we reconstruct first log Y with FFT^{-1}. But log Y is now actually equal to log H, because log X is removed by the filter process. See fig.7.2.6.

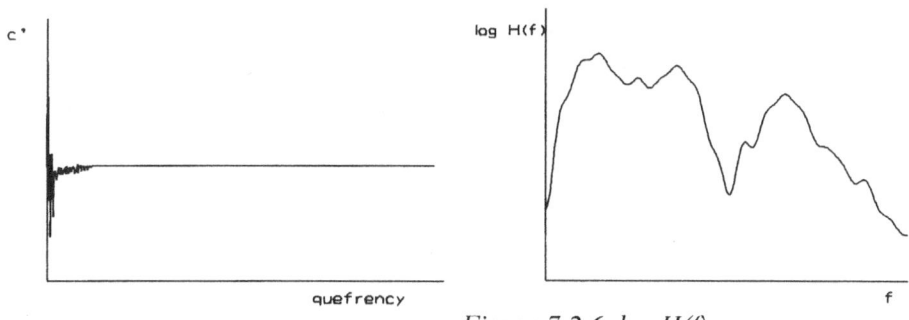

Figure 7.2.5
Cepstrum after short-pass liftering.

Figure 7.2.6 log H(f).

- from log H we calculate H via exponentiation: e$^{\ln H}$ = H.
  In this way we find the formant characteristic of the vocal tract.
- from H we can with FFT^{-1} subsequently calculate the impulse response $h(t)$ of the vocal tract.

long-pass liftering

The modified cepstrum now looks as follows (fig.7.2.7): The reconstruction procedure leads consecutively to

- the reconstruction of log Y via FFT^{-1}. But log Y is now equal to log X, because log H has been filtered away.
- From log X we calculate $X(f)$ via $e^{\ln X} = X$.
  We find here the spectrum of the vocal chord signal.
- From $X(f)$ we calculate with FFT^{-1} the time function $x(t)$ of the vocal chord signal (fig.7.2.8).

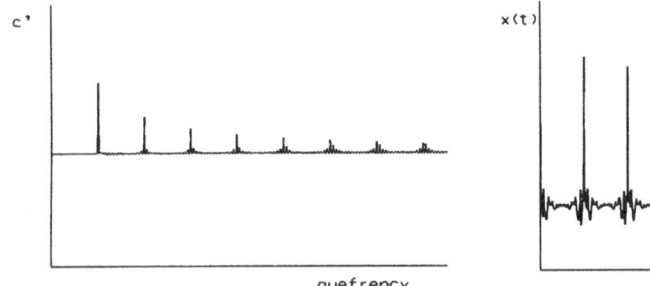

Figure 7.2.7   *Figure 7.2.8*
Cepstrum after long-pass liftering.   *The reconstructed source signal.*

The complete procedure looks schematically like this:

```
                                    (FFT⁻¹)
          FFT           log          FFT
y(t) ———→ Y(f) ———→ log Y ———→ cepstrum
                                       ↓
                         (FFT)   modification of cepstrum
        FFT⁻¹        expon       FFT⁻¹     ↓
y'(t) ←—— Y'(f) ←—— log Y' ←—— cepstrum'
```

If the excitation signal and the filter response are determined then based on these the speech signal can be re-synthesized. In the diagram it is indicated that instead of the second FFT the inverse transform can also be used with the result that on the return path the first FFT^{-1} should be replaced by a forward transform. This is done because of the parallel with the autocorrelation functions. Cepstrum analysis is also performed with the power spectrum. One speaks of the 'power cepstrum' C_p, and with this exchange of forward and inverse transforms it holds:

$$C_p(\tau) = F^{-1} \log |F\{y(t)\}|^2$$

The correspondence with the autocorrelation function

$$R(\tau) = F^{-1} | F \{ y(t) \} |^2$$

is obvious.

Although it may appear that we are dealing here with a drastic operation, it can be shown that in this case it makes no difference whether we work with the forward or with the inverse transform. The power spectrum is a pure amplitude spectrum, because the phase angles are not relevant for the energy (see section 4.3.F). We may thus set all phase angles equal to 0 and this means that all *b*-coefficients are 0 (this follows from tan $\phi = -b/a$). Let us also assume $y(t) = 0$ for $t < 0$.

The forward transform is then

$$a(f) = 2 \int_0^\infty y(t) \cos 2\pi f t \, dt$$

and the inverse transform is

$$y(t) = \int_0^\infty a(f) \cos 2\pi f t \, df$$

If we wish to calculate the forward transform of a particular function we must substitute the function in the first formula in the place of $y(t)$ and for the inverse transform in the second formula in the place where $a(f)$ occurs. It will be clear that (except for a constant factor) the result in both cases is the same.

B. *Formant determination*

Even after the periodic component via cepstrum analysis has been removed from the spectrum and we have isolated the formant characteristic it is still not easy to localize the formant peaks, because not every maximum in the spectrum is a formant. If we restrict ourselves to the first three formant peaks, we can use the fact that the formants must lie in the frequency range from 100 to 3500 Hz and that their bandwidth must be less than 500 Hz.

All peaks that do not satisfy this can be neglected. If there are exactly three peaks with these properties the situation is clear. If there are more or less than this number then comparison with analysis results from preceding fragments may yield results.

C. *The excitation signal in speech synthesis*

If with the help of cepstrum analysis (or another method) the formant characteristic has been determined, the result can be used to synthesize the speech signal, by sending the vocal chord signal through this formant filter. The question is which wave shape the excitation signal should have. It is elegant to use the x-signal likewise determined via cepstrum analysis for this but it is not necessary (and with other analysis methods not yielding $x(t)$ it is not possible). It is known that the vocal chord or glottis signal is rather constant in shape and that this shape is roughly a triangle (see fig.7.2.9).

t

Figure 7.2.9 The glottis-excitation signal.

In a triangular signal the amplitudes of the spectral coefficients decrease by 12 dB/octave (see section 4.3.F). If this signal is sent through a formant filter (with an average spectral slope of 0 dB/oct) and afterwards emitted through the mouth that, as a consequence of its relatively small size (see section 1.1), has the effect of a high-pass filter with a steepness of +6 dB/octave, then the final spectral slope is:

$$-12 \text{ dB/oct} + 0 \text{ dB/oct} + 6 \text{ dB/oct} = -6 \text{ dB/oct}$$

We could thus proceed from a triangular wave shape in speech synthesis. It is however actually simpler to work with a pulse-shaped signal, but then spectral correction must take place.

With a narrow pulse an approximately flat spectrum corresponds. After the formant filter the spectrum is on the average still flat and now to get the desired 6 dB/oct decrease in the spectrum a separate lowpass filter with this characteristic must be used.

With LPC-synthesis dealt with in the following section a pulse-like excitation signal is indeed used and so this extra correction must be applied.

7.3 LPC analysis and synthesis

A. *Coding the speech signal*

The abbreviation LPC means Linear Predictive Coding. In chapter 4 we learnt about the principle of predictive coding, being a method to code a signal efficiently. The principle is once again shown in fig.7.3.1.a. The output signal that is normally written as y, is indicated here with x, because it is identical with the input signal. The increased efficiency follows from the fact that not the signal itself is coded, but only the difference between the predicted and the true value. When this difference signal e has a smaller dynamic range than the signal itself, fewer bits are required for coding it.

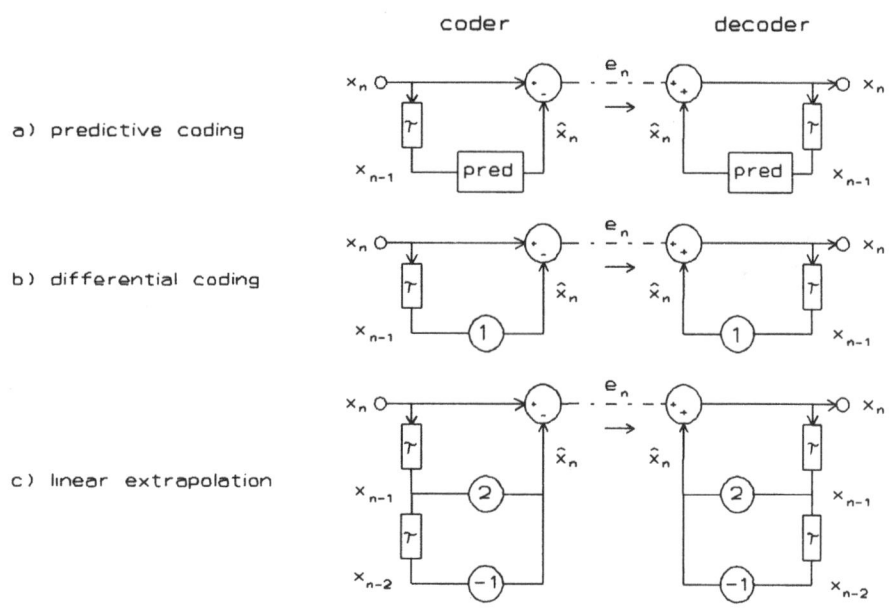

Figure 7.3.1 Predictive Coding.

The simplest practical version of this system is that of differential coding, already discussed in section 4.2.B.2 and shown again in fig.7.3.1.b. Here the previous signal value x_{n-1} is used as prediction $\hat{x}_n$ of the next signal value, thus $\hat{x}_n = x_{n-1}$.

We have analysed this coding/decoding circuit in more detail in section 6.3.A. There we found that the amplitude responses of both circuits compensate each other exactly and we will encounter this property with other versions of the system, still to be discussed. The transmission process can be written symbolically as follows: $x \rightarrow e \rightarrow x$

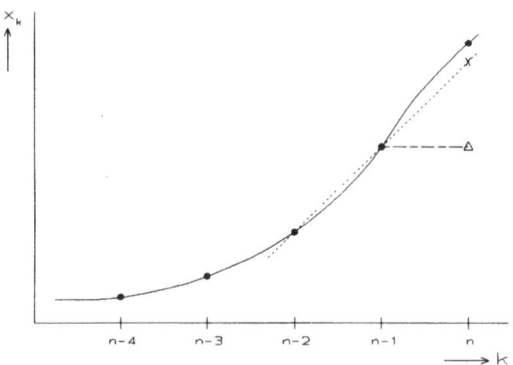

Figure 7.3.2   Some different predictions.

In fig.7.3.2 some signal samples are plotted. The value predicted via differential coding is indicated with a triangle.Obviously, more intelligent predictions are possible. Linear extrapolation based on the last two samples for example yields the value marked with a cross. As can be checked easily the following relationship now exists:

$$\hat{x}_n = x_{n-1} + (x_{n-1} - x_{n-2}) = 2x_{n-1} - x_{n-2}$$

The circuit for this calculation is shown in fig.7.3.1.c. The decoder here is identical with the resonator from chapter 3 and again the signal chain is: $x \rightarrow e \rightarrow x$. By using even more samples for the prediction a still better result can be achieved. With LPC use is made of a so-called 'linear combination' of the N last samples and the expression for $\hat{x}_n$ then becomes:

$$\hat{x}_n = a_1 \cdot x_{n-1} + a_2 \cdot x_{n-2} + \ldots + a_N \cdot x_{n-N} = \sum_{k=1}^{N} a_k \cdot x_{n-k}$$

A difference with the previous systems is also that the a-coefficients are not constant. They are regularly updated to achieve an optimal prediction of the signal. We call such a system an *adaptive* system.

The number of samples N can be chosen freely. In the following examples I shall use the quite normal value $N = 10$. Clearly we can create such a linear combination by extending the diagram of fig.7.3.1.c. That leads to the diagram of fig.7.3.3.

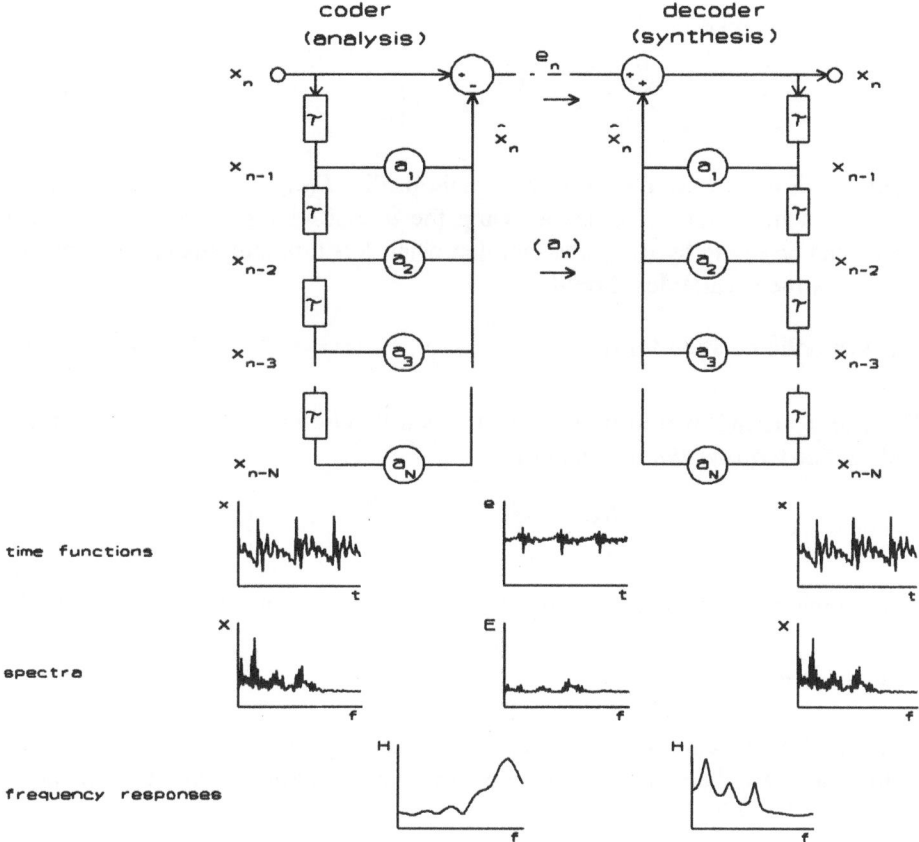

Figure 7.3.3   Complete LPC system.

Left below the circuit diagram an input signal x is shown, a fragment from a speech signal with its spectrum X. The signal chain here is: $x \rightarrow e, (a_n) \rightarrow x$. The brackets indicate that the coefficients a_n are transmitted only once per signal fragment, while the e-samples are clock-synchronous.

To further understand how the system works it is necessary to know how to find the optimal values of the a-coefficients. We know that they should be chosen in such a way that the difference signal e is as small as possible. For e we have:

$$e_n = x_n - \hat{x}_n = x_n - \sum_{k=1}^{N} a_k x_{n-k}$$

A useful criterion for e being minimal is the condition that the energy $E = \sum e_n^2$ is minimal. This implies that when N is large enough, the spectrum of the e-

signal is almost white. Let us thus make E minimal for $N = 10$ and for a signal fragment consisting of M samples.

$$E = \sum_{n=1}^{M} e_n^2 = \sum_{n=1}^{M} \left\{ x_n - \sum_{k=1}^{10} a_k x_{n-k} \right\}^2 \qquad (7.9)$$

Evidently E is a function of a_1 to a_{10}; symbolically: $E(a_1, a_2, \ldots a_{10})$.

We can thus make E minimal using the a-coefficients. The optimal coefficients can be found by using the fact that when E is minimal, the derivatives of E to each of the a-variables should be 0.

The symbolic notation for this is: $\dfrac{\delta E}{\delta a_i} = 0 \qquad (i = 1, 2, \ldots 10)$

We use δ instead of d to indicate that E is a function of more that one variable and use the term *partial differentiation*.

Let us work this out: $\qquad \dfrac{\delta E}{\delta a_i} = \sum_{n=1}^{M} 2 \left\{ x_n - \sum_{k=1}^{10} a_k x_{n-k} \right\} x_{n-i} = 0$

After division by 2 and bringing the first summation inside the brackets, this

changes into: $\qquad \sum_{n=1}^{M} x_n x_{n-i} = \sum_{n=1}^{M} \sum_{k=1}^{10} a_k x_{n-k} x_{n-i} = \sum_{k=1}^{10} a_k \sum_{n=1}^{M} x_{n-k} x_{n-i}$

We are already familiar with sums of products of mutually displaced function values: they are the values of the autocorrelation function R. Therefore the above result may be written in a shorter form as:

$$R(i) = \sum_{k=1}^{10} a_k R(k - i)$$

By writing this out for $i = 1$ to $i = 10$ we get:

$$R(1) = a_1 R(0) + a_2 R(1) + a_3 R(2) + \ldots + a_{10} R(9)$$

$$R(2) = a_1 R(1) + a_2 R(0) + a_3 R(1) + \ldots + a_{10} R(8)$$

$$\cdot \quad \cdot \quad \cdot \quad \cdot \quad \cdot \quad \cdot \quad \cdot \quad \cdot \quad \cdot \quad \cdot \quad \cdot \quad \cdot \quad \cdot \quad \cdot \quad \cdot$$

$$R(10) = a_1 R(9) + a_2 R(8) + a_3 R(7) + \ldots + a_{10} R(0)$$

These are 10 equations with 10 unknowns, the coefficients a_1 to q_0 we were looking for. Due to the symmetrical structure of this set of equations it can be solved rapidly and efficiently by a computer, using the Durbin-Levinson algorithm. Then we know the filter coefficients.

Let us study an example. In fig.7.3.4 the input signal x from fig.7.3.3 is shown again, with its autocorrelation function R.

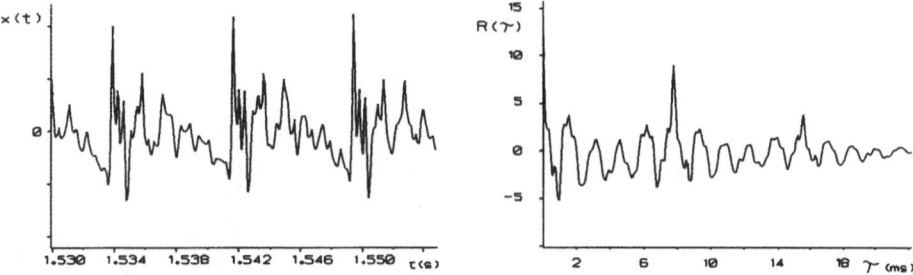

Figure 7.3.4   Input signal and Autocorrelation function.

Eventually from the autocorrelation function the period duration of the signal may be determined, but here we are interested in the first 11 R-values, given in the following list:

$R(0) = 14.366$	$R(4) = 2.336$	$R(8) = -1.842$
$R(1) = 9.697$	$R(5) = -0.627$	$R(9) = -4.647$
$R(2) = 3.193$	$R(6) = -2.624$	$R(10) = -5.200$
$R(3) = 2.240$	$R(7) = -1.556$	

When we substitute these values in the equations and solve them, we find the following a-values:

$a(1) = 0.998$	$a(5) = -1.178$	$a(9) = -0.958$
$a(2) = -0.332$	$a(6) = 0.945$	$a(10) = 0.244$
$a(3) = -0.389$	$a(7) = -0.669$	
$a(4) = 0.962$	$a(8) = 0.528$	

How accurate the prediction based on these coefficients is can be seen in the figure below, which gives a part of the signal function (drawn line) and of the predicted values (dashed line).

With this the transmission problem is solved. In the decoder the transmitted e-signal is combined with the prediction leading to an exact reconstruction of the input signal x. We know that the changes in a speech signal caused by modulation are so slow that a signal fragment of some tens of milliseconds may be considered (quasi-)stationary. The fragment shown in fig.7.3.5 for example has a duration of 25 ms. Updating the a-coefficients should thus be done within this time interval, which means that once per 25 ms new a-values should be transmitted.

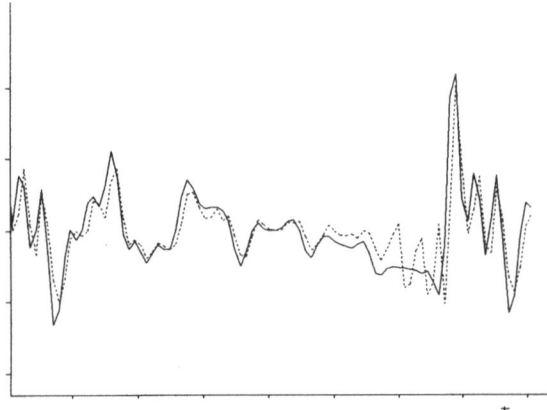

t

Figure 7.3.5 Compar.son of the true and the predicted values.

LPC has more possibilities than efficient signal transmission; it can be used for the analysis of speech and for producing synthetic speech.

B. *Speech analysis*

Let us look at fig.7.3.3 again. When the spectrum E of the difference signal is almost 'white', then the coder or analysis stage of the circuit must have smoothed the spectrum X of the input signal. The amplitude response of this analysis filter must thus have a dip for each peak in the spectrum. It is therefore the mirror image of the formant spectrum. How could we determine the filter response? By calculating the Fourier transform of the impulse response (the frequency response of a linear system = the spectrum of its impulse response). How could we find the impulse response? By using a single sample (for example a '1') as input signal for the system and finding what output signal is produced. From fig.7.3.3 it follows that then the following 11 numbers appear at the output:

$$1, \quad -a_1, \quad -a_2, \ldots, \quad -a_{10}$$

This sequence is thus the impulse response. See fig.7.3.6, on the left. We can calculate the spectrum of this signal. By adding 0-samples we can increase the number of samples until it is a power of 2, e.g. 512. Then we can use a FFT program for the calculation of the spectrum. Adding zero-samples is allowed as we have seen in section 4.3.G. With 512 samples a good resolution is achieved. The spectrum that we find is thus also the frequency response of the filter. In fig.7.3.6 we see from left to right: the impulse response, the amplitude response H and its reciprocal version $1/H$, which reflects the formant structure of the signal. It is easy to recognise the formant peaks. Their position may be determined in the same way as with cepstrum analysis. These two characteristics are also shown in fig.7.3.3.

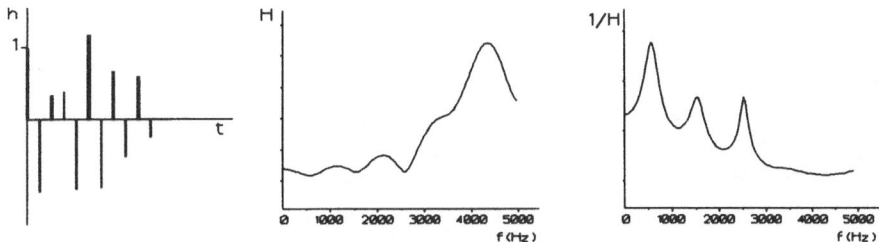

Figure 7.3.6 Impulse response and amplitude responses.

The procedure can be summarized as follows: $x \rightarrow a_n \rightarrow H$ (n = 1, ... N).
There is yet another possibility of locating the formant peaks. Let us look at the
synthesis part of the circuit. This filter is the counterpart of the analysis filter,
because from the flat *E*-spectrum it reconstructs the original *X*-spectrum. The *H*-
characteristic of the synthesizer is thus identical with the 1/*H*-characteristic of the
analyser and thus with the formant characteristic. The coder and the decoder have
the same a_k-coefficients. There is a procedure to convert the synthesis filter with
its 10 feedback loops to a series circuit of 5 filter sections, each with 2 feedback
loops. The technique to do this cannot be treated here, because it requires the *Z*-
transform. It however boils down to writing a polynomial of order 10 as the
product of 5 second order-polynomials:

$$z^{10} - a_1 z^9 - a_2 z^8 - a_3 z^7 - a_4 z^6 - a_5 z^5 - a_6 z^4 - a_7 z^3 - a_8 z^2 - a_9 z^1 - a_{10} =$$

$$(z^2 + p_1 z + q_1)(z^2 + p_2 z + q_2)(z^2 + p_3 z + q_3)(z^2 + p_4 z + q_4)(z^2 + p_5 z + q_5)$$

The first expression corresponds to the synthesizer part of fig.7.3.3; the second
expression to the diagram below (fig.7.3.7):

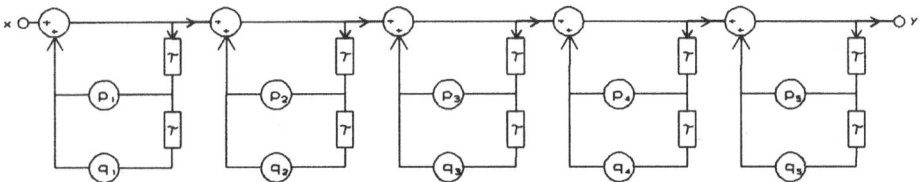

Figure 7.3.7 Equivalent series circuit of five second-order filters.

This conversion is also called *factorisation*. For its calculation a computer is
required. Several algorithms for this exist. Starting from the values of the a-
coefficients on the previous page and using the so-called Bairstow algorithms,
we find the following *p*- and *q*-coefficients for the five filter sections:

k	p	q	f_0	B	Q
1	1.683	0.812	581.3	332.2	1.75
2	-0.603	-0.690	3091.1	590.6	5.23
3	-0.993	0.607	3599.2	793.5	4.54
4	-0.085	0.888	2571.7	188.4	13.65
5	0.996	0.807	1565.0	340.8	4.59

The filter sections of fig.7.3.7, however, are the well-known second-order band filters. Each section accounts for one formant peak and as we know the coefficients of each section we can calculate the resonance frequency f_0 and the bandwidth B (and thus also the Q-factor) of each resonance peak. In this direct and straightforward way the values of f_0, B and Q in the table above were calculated. Symbolically: $x \rightarrow a_n \rightarrow p_k q_k \rightarrow f_{0,k}, B_k, Q_k$ $(n = 1, .., N; k = 1, .., \frac{1}{2}N)$

The formulae we need for this are the following ones:

$$\left. \begin{array}{l} p_k = c \cdot r \\ \\ q_k = -r^2 \end{array} \right\} \quad \text{thus} \quad r = \sqrt{|q_k|}, \ c = \frac{p_k}{\sqrt{|q_k|}}$$

With formula 3.23 we can now calculate f_0:

$$f_0 = \frac{1}{2\pi} f_s \cdot \cos^{-1} \frac{1}{2} c = \frac{1}{2\pi} f_s \cdot \cos^{-1} \frac{\frac{1}{2} p_k}{\sqrt{|q_k|}}$$

and with formula 3.24 we find B:

$$B = -\frac{1}{\pi} f_s \ln r = -\frac{1}{2\pi} f_s \ln(|q_k|)$$

A characteristic consequence of this technique is that always $\frac{1}{2} N$ values are found. Whether all of them are significant remains to be seen. In our example peaks no. 1, 4 and 5 are certainly significant. The f_0-values correspond well with the position of the peaks in fig.7.3.6.

Due to the factorisation this method is more time-consuming, but also more attractive than the former one, as the formant data are produced directly and not via some peak detection algorithm analysing the H-characteristic. If there is time to calculate the p- and q-coefficients, they may be used during transmission instead of the a-coefficients. As synthesizer the circuit of fig.7.3.7. should then be used. This brings us to the next subject:

C. *Speech Synthesis*

Although the method described under A. already considerably reduces the amount of information to be transmitted, substantially larger reductions are still possible. When we realise that the *e*-signal has a white spectrum and that the synthesizer of fig.7.3.3 acts as a formant filter, which is even more obvious after the conversion to the diagram of fig.7.3.7, we see that we are using a synthesis system according to the source-filter model, well-known from phonetics (see fig.7.3.8):

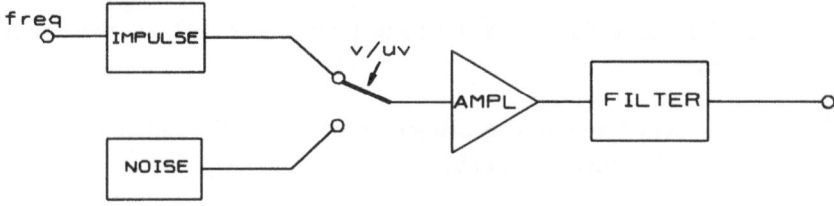

Figure 7.3.8   The source-filter model.

This model works with a source signal that is either periodical (with voiced speech sounds) or a noise signal (with unvoiced sounds); the filter is the formant filter. Instead of using the 'true' *e*-signal in our LPC-system, we could generate an artificial source signal (a pulse for voiced sounds and noise for unvoiced sounds) at the input of the synthesizer. Both signals have white spectra! Then there is no need anymore to transmit the *e*-signal; it is sufficient to transmit the v/uv-data, the gain factor and with voiced sounds, the repetition frequency. The values can be transmitted simultaneously with the filter coefficients, thus once per 10 to 25 ms. The amount of information to be transmitted is reduced drastically as we shall see in an example. The whole process thus runs as follows:

Analysis:

1. The signal is divided into frames of 10 to 25 ms;

2. The autocorrelation function R is calculated per frame;

3. From R are derived:

    a.   the v/uv-indication (presence of peaks in R, or from the proportion $R(0)/R(1)$),

    b.   with 'v': the period duration (from the position of the peak or from another pitch tracking method),

    c.   the filter coefficients a_k (and from them eventually the p, q-coefficients,

d.   the amplitude factor G $\left(G^2 = R(0) - \sum_{k=1}^{10} a_r R(k) \right)$ (without proof).

These values can be transmitted or be stored for later use.

Synthesis:

4. An excitation signal is generated with the proper amplitude and in the case of a pulse signal, with the proper frequency;

5. This signal is sent through a formant filter controlled by the calculated a_k-coefficients;

6. The signal is sent through a -6 dB/octave filter (for the radiation correction, see section 7.2.C), and made audible.

The example that follows and that represents an existing LPC-scheme, shows how much data reduction can be achieved

- v/uv + pitch: 8 bits = 256 numbers.
  By working in steps of 2 Hz, frequencies between 0 and 510 Hz can be coded. The number 0 can be used as uv-indication;
- amplitude factor: 5 bits = 32 amplitude levels;
- filter coefficients. For

$a_1 - a_4$	: 5 bits (32 coefficients)	=	20 bits,
$a_5 - a_8$	: 4 bits (16 coefficients)	=	16 bits,
a_9	: 3 bits (8 coefficients)	=	3 bits,
a_{10}	: 2 bits (4 coefficients)	=	2 bits,

(The filter coefficients are themselves stored in a look-up table, so that the parameter memory only needs to contain the table addresses.)

The total number of bits amounts to 8 + 5 + 20 + 16 + 3 + 2 = 54. If for the frame duration a value of 22.5 ms is used, it means that for coding the speech signal 1000/22.5 · 54 = 2400 bits per second must be stored. This is an enormous (and still not the maximum) data reduction in comparison with linear PCM coding of the original speech signal, for which with 8 bits word length and with a clock frequency of 8 kHz 8 · 8000 = 64000 bits per second must be stored. It is furthermore not necessary that the parameters are derived from a real speech signal; they can also be chosen on another basis. This leads to true synthetic speech.
Besides the autocorrelation method described above for the calculation of the predictor coefficients there is a second method that has been given the name

covariance method. The calculation method here is different and has in certain circumstances some advantages over the autocorrelation method.

There is also the possibility of making use of another filter model where the vocal tract is considered to consist of a series of coupled tubes with varying diameters (fig.7.3.9):

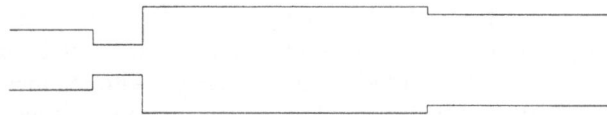

Figure 7.3.9 The coupled-tubes model.

At each junction a part of the sound energy is reflected. The whole system is described by a series of reflection and transmission coefficients. The digital (synthesis) filter can be designed based on these coefficients. It is called a *lattice* filter and looks as follows (fig.7.3.10):

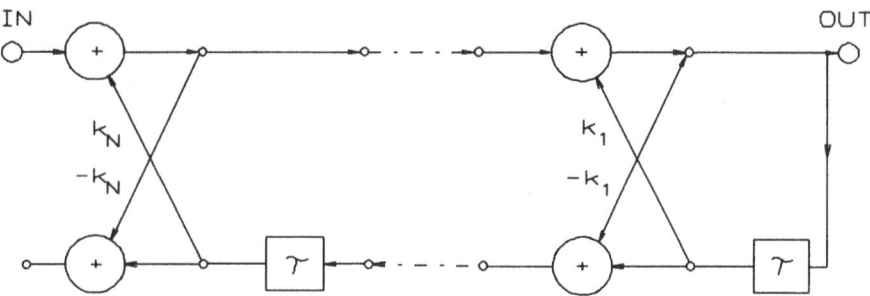

Figure 7.3.10 Lattice filter for LPC synthesis.

The advantage of such a filter is that the absolute value of all coefficients must be < 1, so that eventual incorrect coefficients that can lead to instability can be detected and corrected. With the autocorrelation and covariance methods stability problems are less easy to prevent. From the reflection coefficients the a_k-coefficients can be calculated and from them the formant characteristic.

7.4 Sound synthesis

As was stated in chapter 1, in speech and musical signals we are dealing with the simultaneous modulation of frequency, amplitude and spectrum. We saw how in LPC-synthesis such modulations are realized. One of the reasons why it has taken so long for electronic sound generation to find its place in musical practice is in my view the fact that the importance of modulation aspects has been underestimated. With electronic musical instruments from before the second world war there was hardly any form of microstructure modulation; with pitch this remained limited to vibrato (with fixed amplitude and frequency), with loudness to a more or less stereotype envelope consisting of an attack, sustain and an exponential decay, and timbre modulation was not applied at all. The electronic sound remained therefore far behind instrumental sounds in expressive possibilities. Furthermore, the technical possibilities required of achieving refined modulations did not yet exist either.

After the second world war a method was discovered how to give electronic sounds the necessary complexity: the splicing and manipulation (in the broadest sense) of tape on which sounds are recorded. This was the start of the development of what in a restricted sense is called 'electronic music'. The procedure for making tape music consists of several subsequent phases and therefore does not allow real time music production. The equipment of the studios developed for this kind of electronic music soon turned out to be so flexible, especially after the introduction of the technique of *voltage control*, that the line leading to real time sound production could be taken up again with the construction of the *synthesizer*. See Weiland et al. 1982.

The new techniques and insights formed moreover an excellent starting point for the approximately simultaneously developed technique for digital sound synthesis. I shall now deal with a few sound synthesis techniques and I shall in particular devote attention to those techniques that are interesting from the viewpoint of signal theory.

A. *Formant synthesis via the wave shape*

A special form of timbre modulation arises when signals are generated with spectra with fixed formant peaks. On the one hand this means that with a varying fundamental frequency the proportions between the harmonics also change, so that in a certain sense we are dealing with a dynamic spectrum, and on the other hand it is known from perception research that such a formant structure is perceived as characteristic for a particular signal or particular class of signals. The recognition of vowels is for example based upon this, but also musical sounds are sometimes identified by means of a particular formant structure.

Formant peaks can be generated with the help of filters, as in LPC, but one can also work with signal functions that in themselves already have this quality.

As we saw in chapter 4 this is the case with a pulse signal, and even more pronounced with the so-called VOSIM signal (see section 4.3.F) (Kaegi et al. 1978). This signal function has several free variables; besides the period duration that determines the pitch and the pulse width that in its turn determines the position of the formant peak, one can vary the amplitude, the number and the damping of the $\sin^2$-pulses, and also the duty cycle. This is the proportion between the time duration of the pulse series and that of the 0-interval between the last pulse and the beginning of the next period. Thanks to this 0-interval it is furthermore possible to modulate the period duration in a simple way. By adjusting all these variables fast enough it is possible to realize the necessary variation (= modulation) in time. By combining several VOSIM functions with identical period durations but different pulse widths a signal arises with as many different formant peaks. The VOSIM sound synthesis system is, as is the LPC-system, a direct implementation of the modulation model of acoustical communication dealt with in chapter 1, but avoiding the nasty problems that can be introduced by filters.

A comparable approach has been used by Rodet (Rodet et al. 1984) in the CHANT project. The waveform is a sinewave of finite duration, with attack and exponential decay, a so-called FOF (Fonction d'Onde Formantique) like the signal function shown in chapter 6, fig.6.3.19. Several of these signal functions are generated with different parameters and are then added.

B. *The Karplus-Strong Algorithm*

This algorithm (Karplus et al. 1983) is a synthesis technique for producing plucked string and drum sounds. It is based on the principle of the look up table or wavetable generator as described in section 3.2.B in which numbers that are stored in a table are fetched one by one and transported to the output. Schematically:

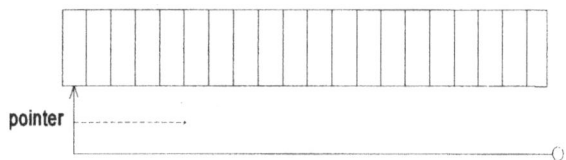

Figure 7.4.1 A look-up or wave table generator.

Another and in this context more suitable way to describe this principle is to consider the table a shift register with feedback from the last to the first cell. In this way the numbers circulate and the pointer is fixed for example at the last cell (fig.7.4.2):

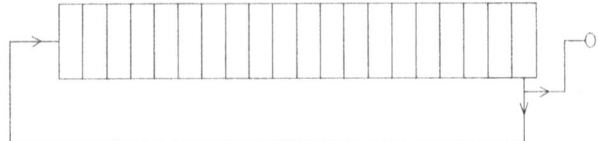

Figure 7.4.2 A look up table generator with shift register.

The look up table generator produces only periodic signals without any modulation. Such signals are useless for musical purposes, but they can be applied in synthesis systems such as FM or waveshaping. Another possibility of getting rid of the stationary character of the signals is to continuously change the contents of the table. This is how the Karplus-Strong algorithm works. The circuit of fig.7.4.2 is modified as follows (fig.7.4.3):

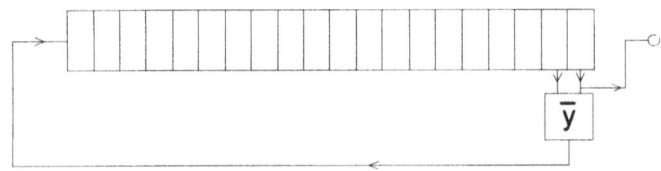

Figure 7.4.3 A look-up table generator with shift register and averager.

Not the numbers themselves are written back, but each number is replaced by the mean value of two subsequent table values. This averaging process is a form of smoothing and thus equivalent with filtering. It mainly affects the higher frequency components, as large differences between subsequent sample values are rapidly reduced. The final result is a decaying tone of which the higher frequency components die out faster than the lower ones. This property is also characteristic for the tone of a plucked string. The initial waveform is not very important. It should only contain sufficient high frequency components. A simple and effective procedure therefore is to fill the table with random numbers.

C. Synthesis with the help of orthogonal functions

As was discussed in section 4.5, every arbitrary signal function can be composed from elementary functions, if these functions form an orthogonal system. The sinusoidal Fourier functions and the squarewave-like Walsh functions satisfy this condition. For the synthesis of an arbitrary function we must thus have at our disposal a (large) number of elementary functions ϕ_k ; we should give each of these functions its own amplitude factor C_k , and we must add all functions together.

1. Fourier synthesis.

We have here the choice between elementary functions of the type $a_n\cos n\omega t$ and $b_n\sin n\omega t$, or of the type $C_n\cos(n\omega t + \phi_n)$. In the last case we only need to generate only half as many elementary functions as in the first case, and therefore preference is given to this method.

For an analog synthesizer we must generate several sinusoidal vibrations with given phase and amplitude relations. The latter is not difficult, but the former is indeed, and therefore one has come no further in the construction of analog Fourier synthesizers than systems with only ca. 6 harmonics.

Digitally this problem is much easier to solve: from a look-up table with one sine function one can derive sinusoidal vibrations with harmonically related frequencies by fetching a sample from the table and then skipping 0, 1, 2, 3, ... samples before fetching the next one.

The phase relation can be fixed by choosing for every harmonic a particular (different) starting address in the table. In this way it is for example possible to build a Fourier synthesizer with 31 independent harmonics. The sounds that can be generated by such a synthesizer only become interesting when they are non-stationary and modulated. This means that in principle all amplitudes and phase angles must be continuously adjusted and this leads very quickly to a large amount of control data. A substantial reduction of this is possible by approximating the control functions that determine the course of values of the parameters with straight line segments. Then only the endpoints of these line segments need be specified.

2. Walsh synthesis.

It is much easier to generate Walsh functions, thanks to their binary structure. Fig.7.4.4 shows the diagram of a circuit that can produce the first 15 Walsh functions. It contains only two types of components: binary scalers (indicated with :2) and XOR-gates (indicated with **X**). The input signal to a binary scaler or flip-flop is a squarewave signal. The output signal is also a squarewave but with half the frequency. This is because the scaler only reacts to a voltage jump from low to high and not to a voltage jump in the opposite direction. A XOR-gate (see also section 6.3.B.5, fig. 6.3.26) produces a low output voltage when the (binary) input signals are equal (both high or both low), and a high output voltage level when the input signals are not equal.

Check yourself whether by combining both functions according to the configuration shown below indeed the correct Walsh functions are produced. For the amplitude control the same considerations hold as for the corresponding case of the Fourier synthesis.

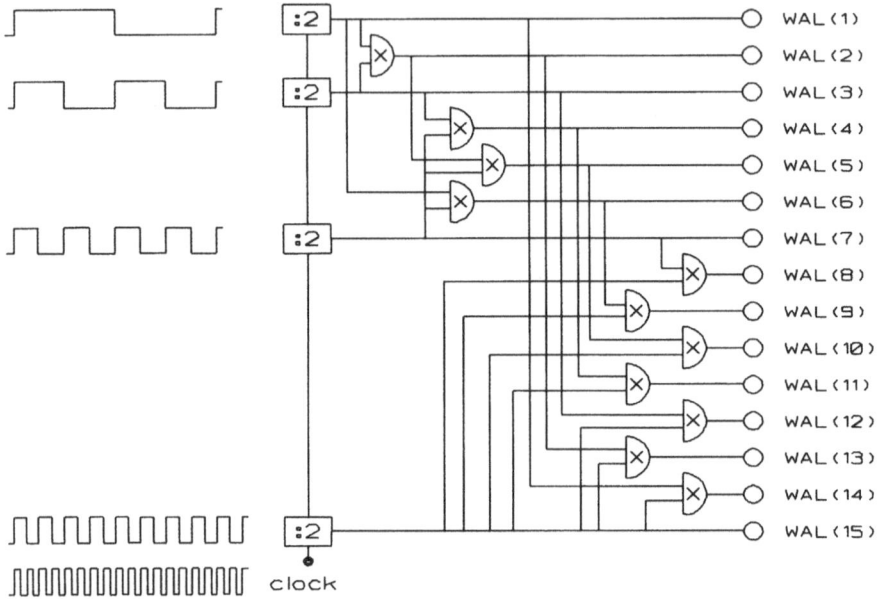

Figure 7.4.4 Walsh synthesizer.

D. *Non-linear synthesis via modulation*

1. Amplitude modulation.

There are various sound synthesis systems based upon nonlinear operations. The advantage is that by the sine in/sine out-principle not being in effect, a complex output signal with many harmonics can be produced with a simple input signal. We shall go first into the matter of the application of amplitude and frequency modulation, which we became acquainted with in chapter 6.

Amplitude modulation (AM) (Dashow, 1978) is described (in a simplified way) by the following relation (rule 6.19):

$$\cos 2\pi f_c t (1 + m \cos 2\pi f_m t) =$$
$$= \cos 2\pi f_c t + \frac{1}{2} m \cos 2\pi (f_c + f_m) t + \frac{1}{2} m \cos 2\pi (f_c - f_m) t$$

and frequency modulation (FM) by rule (6.24):

$$\sin (2\pi f_c t + m \sin 2\pi f_m t) =$$
$$= J_0(m) \sin 2\pi f_c t + \sum_{k=1}^{\infty} J_k(m) \left\{ \sin 2\pi (f_c + f_m) t + (-1)^k \sin 2\pi (f_c - f_m) t \right\}$$

In both cases we see on the right-hand side of the equal-sign a summation of sinusoidal vibrations that is equivalent to the 'closed' expression on the left-hand side. The right-handed expression can thus be conceived as the spectrum of the signal function of the left-handed expression. This means that if we generate a signal function according to the closed expression, we get the corresponding sine series as spectrum.

The generation of these signal functions is comparatively simple. We can directly implement the two formulae digitally for example with look-up tables, and for analog generation we need in the first case a voltage-controlled amplifier and in the second case a generator with an externally controlled frequency, for example a voltage-controlled oscillator (VCO). When using two sinusoidal input signals with frequencies f_c and f_m, we get back output signals with three or with many sinusoidal components respectively. Both techniques can therefore be applied for generating complex sounds. In auditory perception research this technique has been used for years to produce stationary, complex sounds, but the possibility to generate non-stationary sounds in this way is also attractive for musical applications.

With AM we must usually repeat the operation several times (cascade circuit) to achieve the necessary complexity, with FM this is not necessary. This latter method has become very popular and is applied in synthesizers. Therefore I shall limit myself to the FM method for producing time-variant spectra, first described by Chowning (1973).

2. Frequency modulation.

Three aspects are important here: the frequency composition of the spectrum, the energy distribution over the spectrum and the time-dependence.

a. The frequency components.
First this remark: in the above expression the value of $f_c - k \cdot f_m$ can become negative. In the vector model of the sinusoidal vibration a negative frequency means that the vector rotates in the opposite direction. This mirroring can be interpreted as a phase shift of 180°, as we have seen before. We can thus neglect the minus sign (and the phase effect) and treat the frequency of the concerned component as a 'normal' positive frequency. If we choose carrier and modulating frequencies so that their GCD is a frequency within the auditory range, thus

$$f_c = k \cdot f_0 \; , f_m = n \cdot f_0 \; , f_0 > \text{ca. 100 Hz} \; , \; k \text{ and } n \text{ integer}$$

then a normal complex tone results with f_0 as fundamental frequency, for example:

$$f_c = 900 \text{ Hz}, f_m = 400 \text{ Hz} \; \rightarrow \; f_0 = 100 \text{ Hz}$$

Components of the FM signal:

$$f_c\text{-}f_m \quad f_c \quad f_c\text{+}f_m$$
$$900$$

500	1300
100	1700
(-)300	2100
(-)700	2500

etc.

More interesting sounds come about if we give f_c (or f_m) a small frequency deviation f_d in relation to the integer multiple of f_0: $f_c = k \cdot f_0 + f_d$, $f_m = n \cdot f_0$
For example: $f_c = 910$ Hz, $f_m = 300$ Hz, $f_0 = 300$ Hz, $f_d = 10$ Hz, (GCD: 10 Hz.)

$$f_c\text{-}f_m \quad f_c \quad f_c\text{+}f_m$$
$$910$$

610	1210
310	1510
10	1810
(-)290	2110

etc.

As long as f_d is not larger than ca. 30 Hz a non-harmonic complex sound will be heard with a pitch corresponding here to a frequency slightly above 300 Hz. Positive and negative frequencies no longer coincide exactly but become close neighbours (for example the 310 and the -290 component). This causes beats and/or roughness. Due to these interactions and the non-harmonic structure of the spectrum these tones are much more lively and interesting. If the frequency deviation f_d increases then the impression of a 'tone' becomes weaker and the sound changes gradually into a mixture of independent frequency components that could even get noise-like characteristics.

b. The energy distribution.
As we saw in section 6.4 the amplitudes of the frequency components (and thus the distribution of the energy over the spectrum) are determined by Bessel functions, with the modulation index m (= $\Delta f / f_m$) as parameter. With a given modulation frequency the energy distribution thus depends only upon the frequency deviation Δf. The frequency components of the FM-signal with relatively large amplitude coefficients lie in the range from $f_c - \Delta f$ to $f_c + \Delta f$. To get in the above example an audible contribution of the negative frequencies it is necessary that $\Delta f > f_c$ (or $m > 1$). This means that the instantaneous frequency f may become less than 0, see fig.6.4.20. The larger m, the broader the spectrum and the more important the contribution of the negative frequencies. Technically this demand can bring complications with it. It is not simple to design an analog voltage-controlled oscillator with a control range that crosses the zero-line. Digitally this

is no problem. There a negative frequency means that the look-up table is passed through in the opposite direction.

c. The time-dependence.

By using for Δf not a constant, but a time-dependent value we can let the spectrum of the generated sound vary in time, with which we introduce a property that as we know is characteristic for many instrumental sounds. For musical sounds this is perhaps the most important form of micro-modulation. In particular this aspect, the ability to produce easily these dynamic or time-variant spectra is to one to which FM-technique owes its large popularity. Certain instrumental timbres (bell sounds, brass sounds) can be very naturally simulated, and also non-instrumental sounds with the same degree of complexity can be generated.

E. *Discrete summation*

There are yet other possibilities of generating sine complexes via an equivalent (closed) function rule (Moorer, 1976), e.g. by making use of:

$$\sin x + \sin 2x + \ldots + \sin Nx = \sum_{k=1}^{N} \sin nx = \frac{\sin x - \sin (N+1)x + \sin Nx}{2 - 2 \cos x}$$

The correctness of this relation can be proven by showing that the expressions on both sides of the equal-sign are equal to a third expression. We first prove:

$$\sin x + \sin 2x + \ldots + \sin Nx = \frac{\sin \frac{1}{2}(N + 1)x \cdot \sin \frac{1}{2}Nx}{\sin \frac{1}{2}x}$$

or

$$\sin x \cdot \sin \frac{1}{2}x + \sin 2x \cdot \sin \frac{1}{2}x + \ldots + \sin Nx \cdot \sin \frac{1}{2}x = \sin \frac{1}{2}(N + 1)x \cdot \sin \frac{1}{2}Nx$$

With the help of rule (2.47) the left-hand side can be worked out to:

$$\left(-\frac{1}{2}\cos \frac{3}{2} + \frac{1}{2}\cos \frac{1}{2}x\right) + \left(-\frac{1}{2}\cos \frac{5}{2}x + \frac{1}{2}\cos \frac{3}{2}x\right) + \ldots$$

$$+ \left(-\frac{1}{2}\cos (N + \frac{1}{2})x + \frac{1}{2}\cos (N - \frac{1}{2})\right)x$$

All terms in this series appear twice, once with a plus and once with a minus sign, and thus cancel each other, with the exception of the first and the last term.

Conclusion:

$$(\sin x + \sin 2x + \ldots + \sin Nx)\, \sin \tfrac{1}{2}x = \tfrac{1}{2}\cos \tfrac{1}{2}x \, \tfrac{1}{2}\cos (N + \tfrac{1}{2})x$$

$$= -\tfrac{1}{2}\{\cos (N + \tfrac{1}{2})x - \cos \tfrac{1}{2}x\}$$

$$= -\tfrac{1}{2}\{-2 \sin \tfrac{1}{2}(N + 1)x \cdot \sin \tfrac{1}{2}Nx\}$$

Dividing left and right by sin ½x gives the desired result. Now we prove that the right expression is equal to the closed expression from which we proceeded, by multiplying the numerator and denominator by 4sin ½x:

$$\frac{\sin \tfrac{1}{2}(N + 1)x \cdot \sin \tfrac{1}{2}Nx \cdot 4 \sin \tfrac{1}{2}x}{4 \sin \tfrac{1}{2}x} =$$

$$= \frac{-2 \sin \tfrac{1}{2}(N + 1)x \cdot (-2 \sin \tfrac{1}{2}Nx \cdot \sin x)}{2 - 2 \cos x}$$

$$= \frac{-2 \sin \tfrac{1}{2}(N + 1)x \{\cos (N + 1)\tfrac{1}{2}x - \cos (N - 1)\tfrac{1}{2}x}{2 - 2 \cos x}$$

$$= \frac{-2 \sin \tfrac{1}{2}(N + 1)x \cdot \cos \tfrac{1}{2}(N + 1)x + 2 \sin \tfrac{1}{2}(N + 1)x \cdot \cos \tfrac{1}{2}(N - 1)x}{2 - 2 \cos x}$$

$$= \frac{- \sin (N + 1)x + \sin Nx + \sin x}{2 - 2 \cos x}$$

With this the proof is completed. Moorer (1976) describes a number of variants of this relation that can be changed into a time-domain version by replacing every x by $2\pi ft$.

F. *Synthesis via nonlinear distortion (wave-shaping)*

With the help of a system with a nonlinear transfer function (as e.g. shown in fig.6.4.2) a sinusoidal input signal can be altered into a non-sinusoidal output signal with the same frequency. This signal thus has the same pitch, but contains, in contrast to the input signal, harmonics and has therefore another timbre. We could ask ourselves if it is possible to give the transfer function such a shape that a certain pre-specified amplitude spectrum can be achieved. This is indeed possible (Schaefer, 1970 and Le Brun, 1979).

We look first at a simpler problem. Is it possible to find a transfer function T_k

so that the output signal is equal to the k-th harmonic of the sinusoidal input signal? T_k must thus have the following property:

$$T_k(\cos 2\pi f t) = \cos 2\pi \, k f t$$

For $k = 0$ and $k = 1$ T_k is easy to find:

$$k = 0 : \quad T_0(\cos 2\pi f t) = \cos 2\pi \, 0 f t = \cos 0 = 1$$
$$k = 1 : \quad T_1(\cos 2\pi f t) = \cos 2\pi \, 1 f t = \cos 2\pi f t$$

If we for the sake of brevity $\cos 2\pi f t$ replace by x, we can write for T_0 and T_1:

$$T_0(x) = 1$$
$$T_1(x) = x$$

For higher k-values we can derive a recursive relation as follows:

$$\cos 2\pi \, k f t \cdot \cos 2\pi f t = \tfrac{1}{2}\cos 2\pi(k + 1)f t + \tfrac{1}{2}\cos 2\pi(k - 1)f t$$

or: $\quad T_k(\cos 2\pi f t) \cdot \cos 2\pi f t = \tfrac{1}{2}T_{k+1}(\cos 2\pi f)t + \tfrac{1}{2}T_{k-1}(\cos 2\pi f t)$

and with x instead of $\cos 2\pi f t$: $\qquad T_k(x) \cdot x = \tfrac{1}{2}T_{k+1}(x) + \tfrac{1}{2}T_{k-1}(x)$

We can now express T_{k+1} in T_k and T_{k-1}: $\quad T_{k+1}(x) = 2xT_k(x) - T_{k-1}(x)$

Because we already know T_0 and T_1, we can now calculate the other T-functions. These functions are called *Tchebycheff-polynomials*. We find the following expressions for the first seven T_k-functions:

$$T_0(x) = 1$$
$$T_1(x) = x$$
$$T_2(x) = 2x^2 - 1$$
$$T_3(x) = 4x^3 - 3x$$
$$T_4(x) = 8x^4 - 8x^2 + 1$$
$$T_5(x) = 16x^5 - 20x^3 + 5x$$
$$T_6(x) = 32x^6 - 48x^4 + 18x^2 - 1$$
$$T_7(x) = 64x^7 - 112x^5 + 56x^3 - 7x$$

With a given input signal $x = \cos 2\pi f t$ every Tchebycheff-polynomial generates one harmonic. We can now generate a series of harmonics by using a combination of these polynomials. Imagine that we wish to generate a signal with a first, fourth and fifth harmonic. We take then as transfer function $f(x)$:

$$f(x) = T_1(x) + T_4(x) + T_5(x)$$

With $x = \cos 2\pi ft$ the output signal $y(t)$ now becomes:

$$y(t) = f(\cos 2\pi ft)$$

$$= T_1(\cos 2\pi ft) + T_4(\cos 2\pi ft) + T_5(\cos 2\pi ft)$$

$$= \cos 2\pi ft + \cos 2\pi 4ft + \cos 2\pi 5ft$$

The function rule for this transfer function is thus:

$$f(x) = x + 8x^4 - 8x^2 + 1 + 16x^5 - 20x^3 + 5x$$

$$= 16x^5 + 8x^4 - 20x^3 - 8x^2 + 6x + 1$$

The general expression for a transfer function with which an arbitrarily long Fourier series with amplitude coefficients C_k can be generated is:

$$f(x) = \frac{1}{2}a_0 + \sum_{k=1}^{\infty} C_k T_k(x)$$

Dynamic spectra can also be generated in this way, because if the amplitude of the input signal is made less than 1, also a smaller part of the transfer function is used, with consequently a smaller amount of nonlinear distortion. The amplitudes of the distortion products, the harmonics are thus smaller as well. With a very small input signal the transfer function is approximately linear and no harmonics are generated. The amplitude coefficient of the input signal thus controls the strength of the spectral components.

G. *Granular synthesis*

The two elementary signals we use for describing linear systems, the impulse and the sinewave, are antipoles as regards their time and frequency domain behaviour. The impulse is exactly defined in the time domain but has a spectrum that stretches over the whole of the frequency axis (see fig. 5.2.6) whereas the sine wave as time function goes from -∞ to +∞, but with just a single spectral line it is sharply defined in the frequency domain.

We could ask ourselves which signal function is the optimal compromise between these two extremes; 'optimal' meaning here a combination of reasonably sharp definitions in both the time and the frequency domain. This turns out to be a sinewave with a Gauss curve (see fig.4.4.8) as an envelope. The spectrum of this signal has as spectral envelope a Gauss curve as well. For this reason this signal function is often applied as test signal in psycho-acoustical experiments. The signal function is shown in fig.7.4.5.

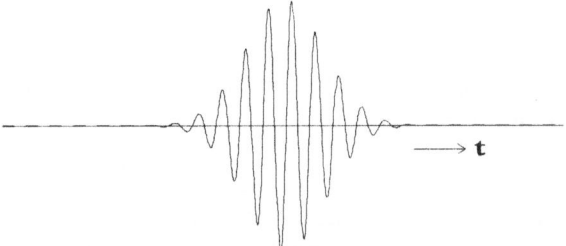

Figure 7.4.5 A sound 'grain'.

A few years after the second world war the British scientist Dennis Gabor formulated the hypothesis that any acoustical signal can be constructed by adding (a large number of) these elementary signal functions, although they do not form an orthogonal set. Some years later this hypothesis was proven, but even without bothering too much about the formal and theoretical aspects, it can be shown that on these 'grains' (as these elementary functions are usually called) a very flexible and interesting sound synthesis system can be based (Roads, 1978; Truax 1988). We have here in fact a very straightforward and effective method to control the 'micro structure modulation' (see chapter 1).

The flexibility stems from the large number of variables: the shape of the envelope (Gaussian, Hamming, triangular etc.), the waveform of the carrier (sinusoidal, frequency-modulated sinewave, recorded sounds etc.) and the way in which the grains are distributed along the time and frequency axis with variables as density and spread.

With the previously described methods the arsenal of possibilities for the generation of sounds is by no means exhausted. An interesting development in sound synthesis is the computer simulation of vibrating physical structures like musical instruments, for example by solving numerically the differential equations of that system. This 'physical modelling' technique is a very sophisticated version of the elementary approach used in chapter 3 to analyse the primitive model of a vibrating string.

I direct those who wish to go deeper into this to the relevant literature on these subjects, in particular The Journal of The Audio Engineering Society (Ed. Audio Engineering Society, New York), The Computer Music Journal (ed. M.I.T. Boston) and the book "Representations of Musical Signals" (De Poli et al. 1991).

7.5 Problems

7.1
a.   What is the average autocorrelation function of a sinusoidal signal?
b.   And that of a square wave?

7.2
a.   Consider a simple LPC-system with only 2 coefficients. Analysis of the input signal gives for the first three coefficients of the autocorrelation function the following values: $R(0) = 1$, $R(1) = 0.918$, $R(2) = 0.690$. Calculate the filter coefficients a_1 and a_2.
b.   What does the impulse response of the analysis stage look like?
c.   Calculate from a_1 and a_2:
            the resonance frequency f_0,
            the bandwidth B,
            the quality factor Q.
The clock frequency is 10000 Hz. Calculate first the values of the coefficients c and r of the harmonic oscillator.

7.3   Speech sounds are radiated more efficiently when the frequency is higher. This effect can be described as a filter process. The signal is filtered by a high-pass filter with a steepness of 6 dB/octave. In a LPC-system the circuit of fig.6.3.1 can be applied for this purpose, with $q = 1$. Calculate the amplitude response of this filter, and check the flank steepness.

7.4   Verify that the third Tchebycheff polynomial generates the third harmonic.

7.5   The peak value of the crosscorrelation function of two signals is called the *degree of coherence* of the two signals.
a.   Presume that in a concert hall two microphones are placed at a short distance from each other. What will be the shape of the 'short time' degree of coherence?
b.   It is possible to convert a monophonic signal into a 'pseudo'-stereo signal by deriving from the original signal two new signals with a low degree of coherence. Find a way to do this.

CHAPTER 8

Acoustical Communication Revisited

It is clear that the techniques, discussed in the previous chapters are extremely useful for the description of all kinds of acoustical, electrical and other systems that can be found between the sound signal source and the hearing organ. We could however ask ourselves whether they could also be used for the description of the source and hearing organ themselves. A partial answer to that question (that was also asked at the end of chapter one) has already been given: in section 7.3 we have seen that by considering the vocal organ a linear system a good and effective method for efficiently coding natural speech and generating synthetic speech could be developed: Linear Predictive Coding. In this final chapter we will check to which extent a similar approach is possible to the hearing organ.

8.1 The hearing organ, a linear system?

A. *Filtering and critical bandwidth*

Determining the properties of a system is possible by comparing the output signal with the input signal. By now it will be clear why it is a good idea to use a sinusoidal input signal for that purpose. With the hearing organ however the output signal is not available and a direct measurement is impossible. The only alternative we have is an indirect measurement via hearing. Our only escape is to accept as a working hypothesis the assumption that the presence of a sinusoidal component can be confirmed via hearing and that perhaps its level (but certainly not its phase angle!) may be estimated in that way. This is absolutely not a trivial supposition, but it is our only option. It is especially important to note that often implicitly the system is assumed to be linear, on the analogy of our artificial technical transmission systems.

Perhaps the most remarkable property of hearing is our capability to distinguish the individual components in a mixture of simultaneously arriving signals. A possible explanation is the existence of some filter system that separates the individual components via band filtering.

There are several indications, both from psycho-acoustical and from (neuro)-physiological and anatomical research that the inner ear indeed contains a filter mechanism. The first question that then immediately comes to one's mind is, what the bandwidth of this filter is. Once again, a straightforward answer to this question can not be given due to the impossibility of measuring the filter output. Filtering manifests itself through the presence or absence of certain forms of interaction between a primary sine tone and one or more secondary sine tones. Therefore it is assumed that the filter is characterized by that frequency distance at which those interactions just disappear. This limit value is called the *critical bandwidth*. Fig.8.1.1 shows how this bandwidth depends upon the frequency of the primary tone.

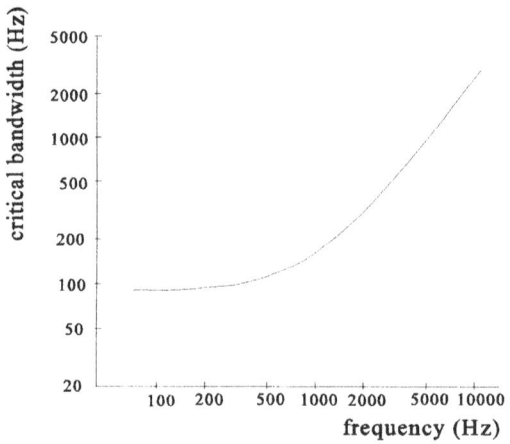

Figure 8.1.1 Critical bandwidth.

Its shape can be described concisely by saying that above 500 Hz the critical bandwidth is proportional to frequency and that below 500 Hz it gradually changes into a constant value of ca. 100 Hz. In the upper frequency range the factor of proportionality is ca. 0.20 and there the critical bandwidth corresponds to an interval between a major and a minor third. In many important aspects of auditory perception the critical bandwidth plays a central role.

To simulate our hearing organ in this respect, a filter bank, a set of third octave bandpass filters is often used (which means that for convenience' sake the critical bandwidth is interpreted as a 'normal' filter bandwidth). At lower frequencies this simulation is of course not very realistic. There are filter banks that simulate the critical bandwidth more accurately. The bandwidth of these filters is sometimes indicated with the name 'bark' (after the German physicist Barkhausen). One imperfection of the filter bank simulation is that they do not do justice to the 'gliding', continuously tunable filters of the inner ear.

To distinguish between several simultaneous signals it would be useful to have very sharp auditory filters. This would, however, destroy the temporal structure (and thus the modulation) of the signals. This is demonstrated in the figures 8.1.2 to 8.1.4. In fig.8.1.2 a and b two signal functions are shown that are prototypes of the signals that play a role in acoustical communication. Each signal consists of a sinusoidal carrier that is also sinusoidally amplitude modulated. When we listen to both signals simultaneously our ears receive the sum signal depicted in fig.8.1.2.c. It is hardly possible to visually recognize the two components.

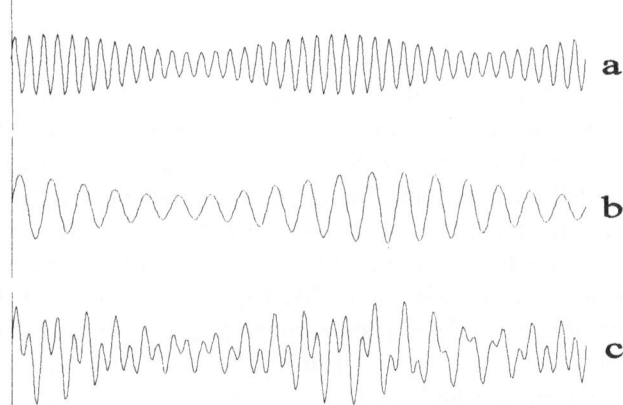

Figure 8.1.2 Two amplitude modulated signals and their sum.

When the inner ear filtering would be very sharp, this sum signal would be split into six stationary sinusoidal components: the two carrier waves each with its two sidebands. Any temporal variation has disappeared here (see fig.8.1.3). When the filtering is done in such a way that the one modulation spectrum is separated from the other whereas the sidebands are not isolated from the carrier then the sum signal is split into the two original amplitude modulated sine waves as shown in fig.8.1.4.

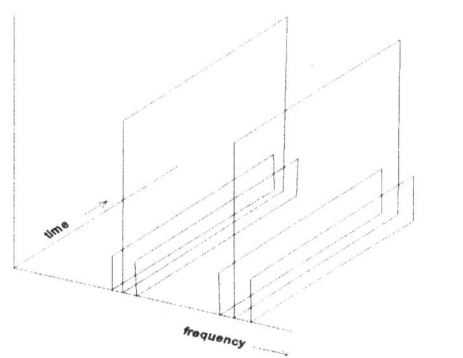

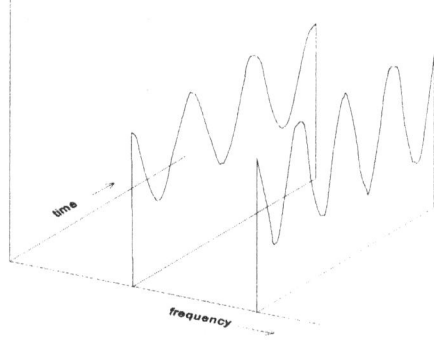

Figure 8.1.3
Spectrum after sharp filtering.

Figure 8.1.4
Spectrum after less sharp filtering.

The filtering should thus not be too sharp. The optimal bandwidth is a compromise between a reasonable resolving power to separate different signals from each other and a sufficient width to keep the temporal structure intact. The bandwidth of the auditory filters indeed satisfies these two conditions.

B. *Masking*

Still accepting the hypothesis that perceiving or hearing a tone may be interpreted as a measurement, it seems that one essential condition for linearity, superposition, is not always satisfied. A sine tone that can be heard as such can under certain conditions, become inaudible (although physically still present) when at the same time a second sine tone is presented. Clearly, this is a form of mutual interaction which is not allowed by the superposition principle.

To find out under what conditions this happens we could do the following experiment. First the lowest level is determined at which the first tone (from now on named 'the masked tone') can be heard while the second tone ('the masker') is absent. Then the second tone (or better a narrow band of noise to avoid other interactions like beats) is switched on and the measurement is repeated. The minimally required level is higher now and the level difference in dB is plotted in a graph with the frequency of the masked tone as independent variable. The measurement is repeated for other frequencies of the masked tone, while we keep the level and the frequency of the masker constant at 80 dB and 400 Hz respectively. The result is shown in fig.8.1.5.

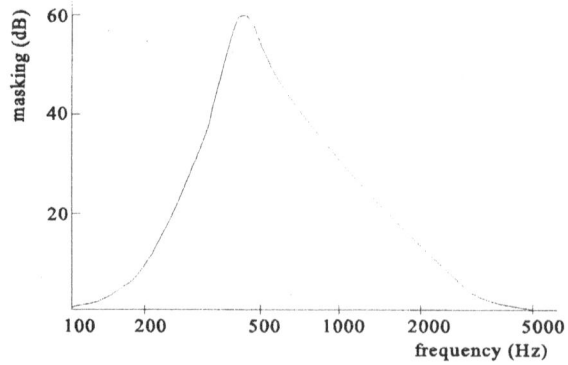

Figure 8.1.5 The masking curve.

The asymmetry of the curve is striking. Below 100 Hz no longer any masking occurs. On the side of the high frequencies this point is not reached until 5000 Hz. Masking is clearly an important phenomenon with regard to loudness perception because when frequency components are hardly audible or not audible at all, they do not contribute to the perceived loudness either. Masking can also be used with digital coding to realize a substantial data reduction. This topic is discussed in section 8.3.

C. *Non-linear distortion*

In one respect the hearing organ seems to behave like a conventional linear system: it exhibits non-linear distortion. In the 18th century Tartini described the phenomenon that, with two simultaneously sounding tones, sometimes a third, extra tone is audible with a pitch that corresponds qua frequency with the difference in frequency between the two primary tones. It was therefore called the 'difference tone'. In the next century Helmholtz found the correct explanation by assuming a non-linear transfer function (see also section 5.1.C) somewhere in the hearing organ. He could not explain why we hear only a difference tone and not a sum tone as predicted by theory. We now know that explanation: an eventual sum tone is masked by the primary tones.

A lot of research has been done to discover the properties of the 'combination tones' as these distortion products are usually called. The power series model of section 5.1.C allows the prediction of the frequencies and levels of the combination tones. The prediction proved to be correct for the normal difference tone but not for the higher order combination tones that are caused by the third, fourth and higher terms of the power series. We now also know that even when the combination tones are not audible as individual tones, they play an important role in all kinds of psycho-acoustical phenomena. For more information on this subject I refer to Plomp (1976), Moore (1991) and other relevant literature.

8.2 The transfer of modulation

In chapter 1 we have seen that with speech and music the acoustical signal is provided with information via a modulation mechanism. For this information to reach its destination it is therefore important that the modulation patterns are not affected during transmission through the communication channel. This channel usually consists of a series of (more or less) linear systems, for example the space in which the sound is transmitted, a microphone, amplifier, recorder, loudspeaker etc. Each subsystem that satisfies the two criteria for distortion-less transmission (see section 5.4), a straight horizontal amplitude response and a linear phase response, keeps the waveform intact and that of course also goes for all amplitude and frequency fluctuations in the signal, in other words for the modulation and the information it contains. There are however subsystems that do not satisfy these two conditions, for example the electronic filters in amplifiers and so on, and also the earshell that, due to all the resonances it causes, also acts as a filter. These filters are however sufficiently smooth to avoid a drastic disturbance of the spectrum of carrier and sidebands. The modulation is thus hardly affected.

In one of the subsystems a serious distortion of the modulation can occur: the (acoustical) space. The numerous reflections that occur in such a space cause a dense pattern of interference maxima and minima leading to whimsical amplitude and phase responses with large deviations. With the technique described in chapter 5, it is of course possible to use these responses (or equivalently, the impulse response) to predict the resulting wave form of the output signal with its modulation and then to assess the 'damage' done to the modulation. A more general approach is to concentrate directly on the modulation. For example via amplitude demodulation of the acoustical signal the modulating signal can be obtained and then the spectrum of this slowly fluctuating signal function could be found. By doing so both at the sound source and at the receiver (the listener) the frequency response (for the modulation frequencies!) can be derived. Plomp (1983) has done a measurement of this type with bandpass filtered speech signals. Modulation spectra for several centre frequencies are shown in fig.8.2.1.

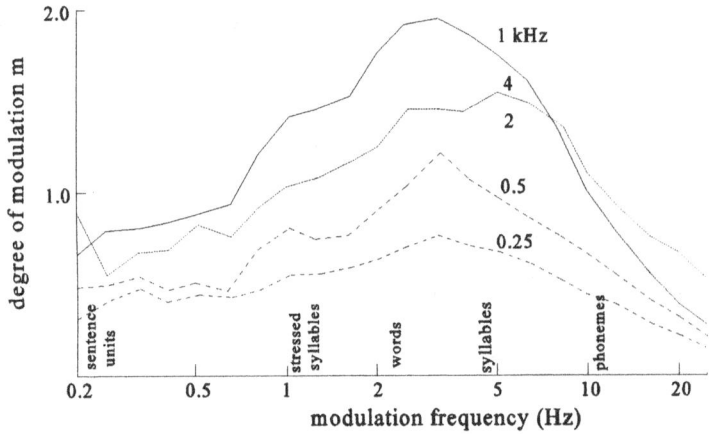

Figure 8.2.1   Modulation spectra. From Plomp (1983) by permission of the author.

Along the vertical axis now the degree of modulation m (see section 6.4.B.3) is plotted. They contain roughly frequencies between 0.3 Hz and 15 Hz (confirming the explanation of the lower limit of the auditory range, given in chapter 1), and have their maximum at 4 Hz. Peaks corresponding to linguistic units can be recognized. In fig.8.2.2 the frequency response of a space for modulation frequencies (here called the 'temporal modulation transfer function') is depicted. By combining these measurements with data from experiments to determine the intelligibility of the speech, a 'speech transmission index' is acquired, that is a good indication of the 'quality' of that space with regard to speech intelligibility.

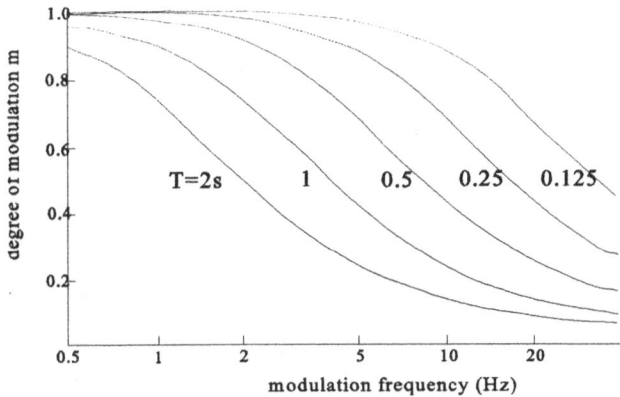

Figure 8.2.2 Temporal modulation transfer functions for some different values of the reverberation time. From Plomp (1983) by permission of the author.

8.3 Perceptual coding

In section 4.2.B several possibilities were discussed of reducing the amount of information during digital coding to a lower value than for example the 768 kbit/s required for linear PCM with a sample frequency of 48 kHz and a word length of 16 bits. One possibility not mentioned there is to use compression techniques based on the statistical properties of the bit patterns. This is a common technique for decreasing the size of computer files (programs and data). The degree to which this is possible varies from almost nil with program files to substantial degree with text and image files. The advantage of this method is that it leads to an exact reconstruction of the original files. Nothing is lost and, in the case of sound files, no noise is added to the signal. Sound files however turn out to be so complex that only a very modest amount of compression is possible, varying from 1 (= no reduction) to 2.

Very substantial data reduction is possible by making use of the masking phenomenon. We will now delve into this so-called *perceptual coding*. As we will see, here the audio signal is split into a number of frequency bands by a filter bank. It is important to understand that with a narrow-band signal a proportional lower sampling rate can be used. To explain this the following example: suppose that we have a digital signal with frequencies between 7 kHz and 8 kHz, sampled at 20 kHz. As we have seen in section 4.4.A we then have the following spectrum (fig.8.3.1.a):

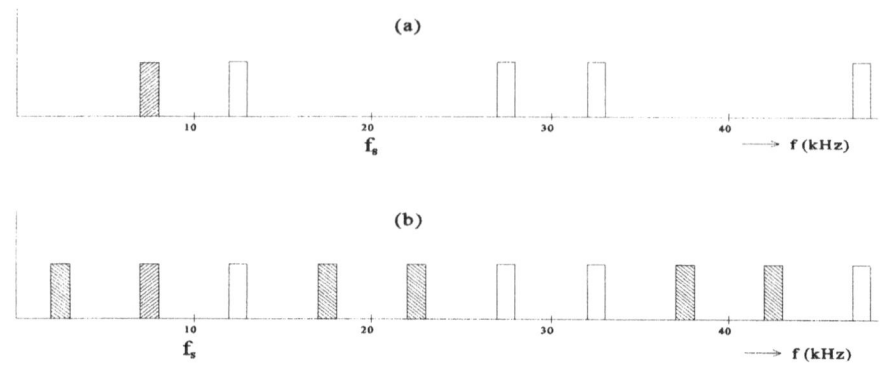

Figure 8.3.1 Subsampling.

When we now lower the clock frequency to 10 kHz, the expected aliasing occurs, but, thanks to the restricted bandwidth of the signal, there is <u>no</u> overlap (see fig.8.3.1.b). This means that it is still in principle possible to reconstruct the original signal! Reducing the signal's bandwidth permits a proportional lowering of the clock rate. Using the method for changing the sampling rate, discussed in section 4.4.C, we can now work with a narrow-band signal according to the following procedure. First with a filter, we reduce the bandwidth of the signal by a factor L. Then the sampling rate is lowered by the same factor by means of omitting L-1 samples from each group of L samples. To reconstruct the signal we first increase the sampling rate to the original value by inserting L-1 zero-samples between each pair of samples. Then using the same filter as before we admit only that portion of the spectrum we want and suppress everything else. In this way the necessary interpolation is achieved and the original signal is reconstructed. The complete diagram (Stoll 1993) is shown in fig.8.3.2:

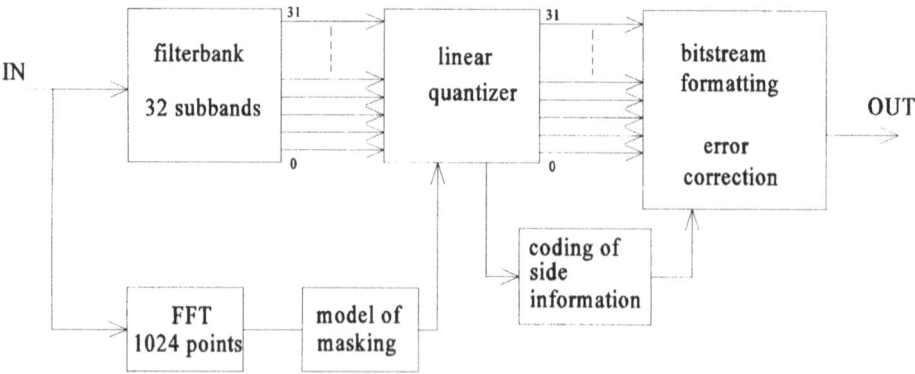

Figure 8.3.2 Perceptual Coding System.

The original signal (for example a 16-bit digital signal, sampled at 48 kHz, thus corresponding with a bit stream of 768 kbits/s) is split with a filter bank into 32 equally wide bands. The sampling rate is lowered by the same factor using the method just described. The bitstream per band is thus reduced to 24 kbits/s, but as we have 32 channels, the total bit stream is still $32 \times 24 = 768$ kbits/s. At the same time a 1024-point FFT is made to get an accurate representation of the spectrum of the signal being processed. This information is combined with a psycho-acoustical model of masking. When in a certain band the signal contains a strong frequency component, the masking threshold in the neighbouring bands can be calculated from the model. Then the word length of the samples from these bands is reduced. This increases the level of the quantization noise, but that noise remains inaudible as long as its level stays below the masking threshold. It is with this step that the data reduction is achieved. Finally the signal is reformatted to form one single stream of bits that contains the samples plus a number of bits with information about the data structure, in total 192 kbits/s, a reduction with a factor 4! In the decoder this stream of bits is subdivided again into the 32 original channels, the word length and sample frequency are reset to their original values and the 32 output signals are added together to form again one single audio signal. The circuit of the decoder is much simpler than that of the encoder, because here no psycho-acoustical modelling is required anymore.

From the discussion in this chapter it will be clear that the possibilities of applying the theory of linear systems to the hearing organ are very limited. This organ should not be considered a simple linear transmission system, a kind of 'living microphone'. Together with the brains it forms a complex information procession system of a much higher order. We still have to learn a lot about this system.

References

Ahmed, N. & Rao, K. R. (1975). *Orthogonal Transforms for Digital Signal Processing*. New York:Springer Verlag

Bateman, W. (1980). *Introduction to Computer Music*. New York: J.Wiley & Sons.

Batschelet, E. (1975). *Introduction to Mathematics for Life Scientists*. New York, Heidelberg, Berlin: Springer-Verlag.

Blesser, B. (1978). Digitization of Audio: a Comprehensive Examination of Theory, Implementation, and Current Practice. *J. Audio Eng. Soc.*, 26, 739-771.

Bogert, B.P., Healy, M.J.R. & Tukey, J.W. (1963). Quefrency analysis of time series for echoes: cepstrum, pseudo-autocovariance, cross-cepstrum and saphe cracking. In *Proceedings of the Symposium on Time Series Analysis*, (pp. 209-243). New York: Wiley.

Bogner, R.E., Constantinides, A. (1975). *Introduction to Digital Filtering*. New York: Wiley.

Burgess, J. (1975). On digital spectrum analysis of periodic signals. *J. Acoust. Soc. Am* 58, 556-567.

Chowning, J.M. (1973). The Synthesis of Complex Audio Spectra by Means of Frequency Modulation. *J. Audio Eng. Soc.*21, 526-534.

Chua, L.O., Hu, C.W., Huang, A. & Zhong, G.Q. (1993). A universal circuit for studying and generating chaos, part I: Route to chaos. *IEEE Transactions on Circuits and Systems*, E746-A, 704-734.

Cooley, J. W. and Tukey, J. W. (1965). An Algorithm for the Machine Calculation of Complex Fourier Series. *Math. Computation* 19, .297-301.

Corliss, E. (1990). The Ear as a Mechanism of Communication. *J. Audio Eng. Soc.* 38, 640 - 651.

Dashow, J. (1978). Three Methods for the Digital Synthesis of Chordal Structures with Non-Harmonic Partials. *Interface* 7, 70-94.

De Poli, G., Piccialli, A. & Roads. C. (eds.). (1991). *Representations of Musical Signals*. Cambridge, MA: MIT Press.

Dodge, Ch., Jerse, Th.A. (1985). *Computer Music; Synthesis, Composition and Performance*. New York, London: Schirmer Books.

Gabel, R., Roberts, R.A. (1973). *Signals and Linear Systems*. New York: Wiley.

Goldman, S. (1948). *Frequency Analysis, Modulation and Noise*. New York: Dover.

Graeme J., Tobey, G. & Huelsmann, L. (Eds.). (1971). *Operational Amplifiers; Design and Applications*. New York: McGrawHill.

Hintze, G. (1966). *Fundamentals of Digital Machine Computing*. New York, Heidelberg, Berlin: Springer Verlag.

Hsu, H.P. (1970). *Fourier Analysis*. New York: Simon and Schuster.

Kaegi, W. , Tempelaars, S. (1978). VOSIM, a New Sound Synthesis System. *J. Audio Eng. Soc.*, 26, 418-425.

Karplus, K., Strong, A. (1983). Digital Synthesis of Plucked-String and Drum Timbres. *Computer Music Journal*, 7(2), 43-55.

Lax, P., Burstein, S. & Lax, A. (1976). *Calculus with Applications and Computing*. New York, Heidelberg, Berlin: Springer-Verlag.

Le Brun, M. (1979). Digital Wave-shaping Synthesis. *J. Audio Eng. Soc.*, 27, 250-265.

Mayr, H. (1980). Acoustical Communication Theory. *J. Audio Eng. Soc.*, 28, 331-336.

Moore, B. (1991). *An Introduction to the Psychology of Hearing*. London: Academic Press.

Moorer, J.A. (1976). The Synthesis of Complex Audio Spectra by Means of Discrete Summation. *J. Audio Eng. Soc.*, 24, 717-727.

Morse, P. (1948). *Vibration and Sound*. New York: McGraw-Hill.

Oppenheim, A.V., Schafer, R. W. (1975). *Digital Signal Processing*. Englewood Cliffs: Prentice-Hall.

Pierce, J. R. (1983). *The Science of Musical Sound*. New York: Scientific American Library.

Plomp, R. (1976). *Aspects of Tone Sensation*. London: Academic Press.

Plomp, R. (1984). Perception of Speech as a Modulated Signal. In *Proceedings of the 10-th International Congress of Phonetic Sciences* (pp. 29-40). Dordrecht: Foris Publications.

Poularikas, A. D., Seely, S. (1985). *Signals and Systems*. Boston: PWS Engineering.

Preis, D. (1982). Phase Distortion and Phase Equalization in Audio Signal Processing - A Tutorial Review, *J. Audio Eng. Soc.*, 30, 774-794.

Rabiner, L. R. (1982). Digital Techniques for Changing the Sampling Rate of Signals. In *Digital Audio, papers from the 1982 A.E.S.Conference*.

Rabiner, L. R., Schafer, R. W. (1985). *Digital Processing of Speech Signals*. Englewood Cliffs, NJ: Prentice-Hall.

Roads, C. (1978). Automated Granular Synthesis of Sound. *Computer Music Journal*, 2(2), 61-62.

Roads, C. , Strawn, J. (eds.). (1985). *Foundations of Computer Music*. Cambridge, MA: MIT Press.

Rodet X., Potard, Y. & Barrière, J. (1984). The CHANT project: From the Synthesis of the Singing Voice to Synthesis in General. *Computer Music Journal*.

Schaefer, R.A. (1970). Electronic Musical Tone Production by Nonlinear Waveshaping. *J. Audio Eng. Soc.*, 18, 413-417.

Schroeder, M.R., Logan, B. F. (1961). 'Colorless' Artificial Reverberation. *J. Audio Eng. Soc.*, 9, 192-197.

Schroeder, M. R. (1989). Self-Similarity and Fractals in Science and Art. *J. Audio Eng. Soc.*, 37, 795-808.

Stoll, G. (1993). Status and Future Activities in Standardization of Low Bit-Rate Audio Codecs. In *Proceedings 12-th A.E.S. International Conference on The Perception of Reproduced Sound*. Copenhagen.

Szabo, I., Wellnitz, K. & Zander, W. (1974). *Mathematik*. New York, Heidelberg, Berlin: Springer-Verlag

Tempelaars, S. (1977). The VOSIM Signal Spectrum. *Interface*, 6, 81-96.

Tempelaars, S. (1982). Linear Digital Oscillators. *Interface*, 11, 109-130.

Truax, B. (1988). Real-Time Granular Synthesis with a Digital Signal Processor. *Computer Music Journal*, 12(2), 14-26.

Truax, B. (1990). Chaotic Non-linear Systems and Digital Synthesis: an Exploratory Study. In *Proceedings of the ICMC* (pp. 100-103). San Francisco.

Weiland, F. , Tempelaars, S. (1982). *Elektronische Muziek*. Utrecht: Bohn, Scheltema and Holkema.

Witten, I. (1982). *Principles of Computer Speech*. London: Academic Press.

Appendix
solutions to problems

Chapter 2

2.1 Fig.2.1.2 T changes from 3.9 ms to 4.4 ms, $f \approx 238$ Hz ('c')
Fig.2.1.3 $T = 4.56$ ms, $f = 219.12$ Hz ('a')
Fig.2.1.4 $T = 4.53$ ms, $f = 220.8$ Hz ('a')
Fig.2.1.9 $T = 5$ ms, $f = 200$ Hz ('g')

2.2

a. $y(t) = 1$ if $0 \le t \le \frac{1}{2}T$, $y(t) = -1$ if $\frac{1}{2}T < t \le T$

b. $y(t) = \dfrac{4t}{T} + 1$ if $-\frac{1}{2}T < t \le 0$, $y(t) = -\dfrac{4t}{T} + 1$ if $0 \le t \le \frac{1}{2}T$

c. $y(t) = \dfrac{2t}{T}$ if $-\frac{1}{2}T \le t \le T$

d. $y(t) = \left| \dfrac{\sin 2\pi t}{T} \right|$

For all functions: $y(t + T) = y(t)$

2.3 (a) 1 (b) $1/\sqrt{3}$ (c) $1/\sqrt{3}$ (d) $1/\sqrt{2}$
2.4 (a) 1.3386 (b) $1.0520 \cdot 10^{-8}$ (c) not defined (d) 1.3973 (e) 3.2175
2.5 (a) 1.3348 ($\leftrightarrow$ 1.3333) (b) 1.6818 ($\leftrightarrow$ 1.6667) (c) 1.8877 ($\leftrightarrow$ 1.8750)
2.6 1200 cent; $2^{1/1200} = 1.00057779$
2.7 6 dB
2.8 8.7 dB
2.9 80.4 dB
2.10 54.1 dB

2.11 470 μs $R = \dfrac{V}{i}$, $C = \dfrac{Q}{V}$,   thus $RC = \dfrac{Q}{i} = \dfrac{Q}{\frac{Q}{t}} = t$

2.12

$$f(x) = \frac{1}{\sqrt{x^2 a^2 + (b - cx^2)^2}} = \frac{1}{\sqrt{c^2 x^4 - 2bcx^2 + a^2 x^2 + b^2}}$$

$$f'(x) = -\frac{1}{2}\frac{4c^2 x^3 + 2(a^2 - 2bc)x}{\sqrt{(c^2 x^4 - 2bcx^2 + a^2 x^2 + b^2)^3}} = 0$$

$f'(x) = 0$ when the numerator of this expression equals 0, thus:

$$2c^2 x^3 + (a^2 - 2bc)x = 0 \quad ; \quad \text{divide left and right by } x:$$

$$2c^2 x^2 = 2bc - a^2 \rightarrow x = \sqrt{\frac{2bc - a^2}{2c^2}} = \sqrt{\frac{b}{c} - \frac{a^2}{2c^2}}$$

2.13

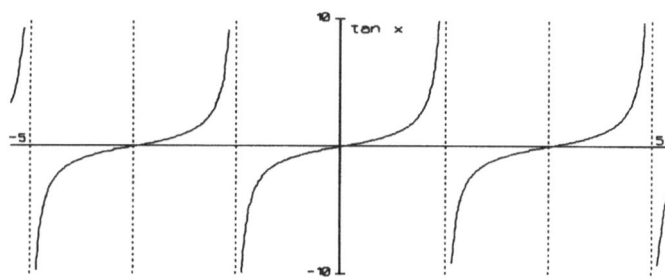

2.14

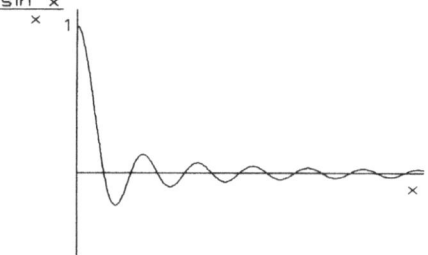

$$\lim_{x \to 0}\frac{\sin x}{x} = \lim_{x \to 0}\frac{1}{x}(x - \frac{x^3}{3!} + \frac{x^5}{5!} - \frac{x^7}{7!} + \frac{x^9}{9!} - \ldots) = \lim_{x \to 0} 1 - \frac{x^2}{3!} + \frac{x^4}{?}$$

zero crossing at $x = k \cdot \pi$; maxima (after numerical solution of $\tan x = x$) at

$$\begin{array}{rl}
x = & 4.493409 \\
& 7.725252 \\
& 10.904122 \\
& 14.066194 \\
& 17.220755 \\
& 20.371303 \\
& \text{etc.}
\end{array}$$

2.15

a. $\dfrac{1}{T} \int\limits_0^T \sin 2\pi f t \cdot \sin 2\pi f m\, t \;=$

$$= \frac{1}{T} \int\limits_0^T \left\{ \frac{1}{2} \cos 2\pi (n-m)ft - \frac{1}{2}\cos 2\pi(n+m)ft \right\} dt \;=$$

$$= \frac{1}{2T} \left\{ \frac{T}{2\pi(n-m)} \sin \frac{2\pi(n-m)}{T} t \,\Big|_0^T - \frac{T}{2\pi(n+m)} \sin \frac{2\pi(n+m)}{T} t \,\Big|_0^T \right\} =$$

b. $\dfrac{1}{T} \int\limits_0^T \cos 2\pi f t \cdot \cos 2\pi f m\, t \;=$

$$= \frac{1}{T} \int\limits_0^T \left\{ \frac{1}{2} \cos 2\pi (n+m)ft + \frac{1}{2}\cos 2\pi(n-m)ft \right\} dt \;=$$

$$= \frac{1}{2T} \left\{ \frac{T}{2\pi(n+m)} \sin \frac{2\pi(n+m)}{T} t \,\Big|_0^T - \frac{T}{2\pi(n-m)} \sin \frac{2\pi(n-m)}{T} t \,\Big|_0^T \right\} = C$$

c. $\dfrac{1}{T} \int\limits_0^T \sin 2\pi f t \cdot \cos 2\pi f m\, t = \dfrac{1}{T} \int\limits_0^T \left\{ \dfrac{1}{2} \sin 2\pi(n+m)ft + \dfrac{1}{2}\sin 2\pi(n-m)ft \right\} \ell$

$$= \frac{1}{2T} \left\{ \frac{T}{2\pi(n+m)} \cdot -\cos \frac{2\pi(n+m)}{T} t \,\Big|_0^T - \frac{T}{2\pi(n-m)} \cdot -\cos \frac{2\pi(n-m)}{T} t \,\Big|_0^T \right\} =$$

2.16 $C = 1.39$, $\tan \phi = -0.58$, $\phi = -30.26°$

2.17 $y_k = 2.4 \cdot \sin(0.097k + 0.559)$

2.18 (a) 0.716 (b) 240.6423 (c) 3.386 (d) -3.7

2.19 (a) 97.403° (b) 240.6423° (c) 900°

2.20 (a) 71.43 Hz (b) 11627.9 Hz (c) 0.893 Hz (d) $5.787 \cdot 10^{-6}$ Hz

2.21 $f = 162.34$ Hz

2.22 (a) $\sin b - \sin a$ (b) $\frac{1}{2}(b-a) + \frac{1}{4}(\sin 2b - \sin 2a)$

2.23 127.32 Hz, 143.239 Hz, 111.408 Hz

2.24

$$\lim_{\Delta x \to 0} \frac{\displaystyle\int_a^{x+\Delta x} y(t)\,dt \; - \; \int_a^{x} y(t)\,dt}{\Delta x} = \lim_{\Delta x \to 0} \frac{\displaystyle\int_x^{x+\Delta x} y(t)\,dt}{\Delta x}$$

$$= \lim_{\Delta x \to 0} \frac{Y(x+\Delta x) \; - \; Y(x)}{\Delta x} = Y'(x) = y(x)$$

Chapter 3

3.1 (a) $f = 100$ Hz (b) $f_o = 600$ Hz (c) $Q = 5$ (d) $B = 120$ Hz

3.2 See fig.3.4.1; $c = 2 \cos \dfrac{2\pi 600}{20000} = 1.9646, \quad r = e^{\frac{-\pi 120}{20000}} = 0.9813$

the coefficients are thus: $c \cdot r = 1.9279$ and $-r^2 = -0.9630$.

3.3

a.

$$y(t) = e^{-Pt}(A_1 \cos \omega_d t + A_2 \sin \omega_d t)$$

$$v(t) = \frac{dy}{dt} = -p e^{-Pt} A_1 \cos \omega_d t - e^{-Pt} A_1 \omega_d \sin \omega_d t - p e^{-Pt} A_2 \sin \omega_d t + e^{-Pt} A_2$$

$$y(t = 0) = A_1 = y_0$$

$$v(t = 0) = -pA_1 + A_2 \omega_d = -p y_0 + A_2 \omega_d = v_0$$

Thus: $A_1 = y_0, \quad A_2 = \dfrac{v_0 + p y_0}{\omega_d}$

b.

$$A = \sqrt{A_1^2 + A_2^2} = \sqrt{y_0^2 + \frac{p^2 Y_0^2}{\omega_d^2}} = y_0 \sqrt{1 + \frac{p^2}{\omega_d^2}}$$

$$\tan \phi = -\frac{A_2}{A_1} = -\frac{p}{\omega_d}$$

3.4

a. See fig.3.3.4; with rule (3.4): $f_0 = \dfrac{1}{2\pi}\sqrt{\dfrac{1}{LC}} = 800$

from this: $C = 100$ nF.

b. $Q = \sqrt{\dfrac{bm}{R^2}} = \sqrt{\dfrac{L}{CR^2}} = 2000$

See fig.3.4.1.c.

With rule (3.25): $\ln r = -\dfrac{\gamma}{2Q}$, $\gamma = \dfrac{2\pi f_0}{f_s} = 0.2$

From this: $r = 0.99995$, $c = 2\cdot\cos\gamma = 1.9597$
Thus: $c\cdot r = 1.9596$ and $-r^2 = -0.9999$.

3.5

a. $Q = 750/150 = 5$ and $Q = 1800/150 = 12$
b. $t_d = 1/p = 1/(B\cdot\pi) = 2.122$ ms.

3.6

a. When the frequency of the undamped vibration f_d is equal to 0.

b. See page 78: $\dfrac{R^2}{4m^2} = \dfrac{b}{m}$ or $\dfrac{bm}{2}$ $(= Q^2) = \frac{1}{4}$; thus $Q = \frac{1}{2}$

c. Yes; if $\gamma = 0$ (thus $c = 1$).

3.7

$$B = \sqrt{B_1^2 + B_2^2} = \sqrt{x_0^2\left(1 + \frac{\cos^2\gamma}{\sin^2\gamma}\right)} = x_0\sqrt{\frac{\sin^2\gamma + \cos^2\gamma}{\sin^2\gamma}} = \frac{x_0}{\sin\gamma}$$

$$\phi = -\tan^{-1}\frac{\cos\gamma}{\sin\gamma}$$

According to (2.40): $\dfrac{\cos \gamma}{\sin \gamma} = \dfrac{\sin(90° - \gamma)}{\cos(90° - \gamma)} = \tan(90° - \gamma)$

Thus: $\phi = -\tan^{-1}\{\tan(90° - \gamma)\} = \gamma - 90°$

Together: $y_k = \dfrac{x_0}{\sin \gamma} \cos(\gamma k + \gamma - 90°) = \dfrac{x_0}{\sin \gamma} \sin(\gamma k + \gamma)$

3.8 With rule (3.3): $f_0 = \dfrac{34000}{2\pi}\sqrt{\dfrac{2}{2.5 \cdot 125}} = 432.9\,\text{Hz}$

3.9 $\theta = \tan^{-1} \dfrac{R}{\dfrac{1}{\omega C} - \omega L}$ $C = \dfrac{1}{\omega\sqrt{R^2 + (\dfrac{1}{\omega C} - \omega L)^2}}$

Chapter 4

4.1

	decimal	octal	hexadecimal	binary
a.	2748	5274	ABC	0000101010111100
b.	-300	7324	FED4	1111111011010100
c.	592	1120	250	0000001001010000
d.	-592	76660	FDB0	1111110110110000

4.2

	sum	product
a.	1110001_2	010001100010_2
b.	710_8	57554_8
c.	$1D0_{16}$	$436F_{16}$

4.3 a. 7126_8 b. 67737_8 c. 161625_8

4.4 a. 0.00256 V   b. 0.2615 V c. 0.4146 V

4.5 $(72 + 4.77 - 20\log 3) = 67.23\,\text{dB}$

4.6 2250 V/s

4.7 $y(t) = 5\cdot\sin 2\pi ft$, $dy/dt = 5\cdot 2\pi f\cdot\cos 2\pi ft$, $10\pi f = 2250$ V/s thus $f = 71.62$ Hz.

4.8 $a_4 = -0.178,\ b_4 = -1.681,\ C_4 = 1.691,\ \phi_4 = 83°$

4.9

$$C_n = \frac{8}{(2n-1)^2\pi^2} \quad \text{thus} \quad y_{RMS} = \sqrt{\frac{1}{2}\frac{64}{\pi^4}\left(1 + \frac{1}{81} + \frac{1}{625} + \dots\right)} = 0.577$$

4.10

$$a_0 = \frac{2}{T}\left\{\int_0^{\frac{1}{2}T}(+1)\,dt + \int_{\frac{1}{2}T}^{T}(-1)\,dt\right\} = \frac{2}{T}\{\tfrac{1}{2}T - 0 - (T - \tfrac{1}{2}T)\} = 0$$

$$a_n = \frac{2}{T}\left\{\int_0^{\frac{1}{2}T}\cos 2\pi nft\,dt + \int_{\frac{1}{2}T}^{T}-\cos 2\pi nft\,dt\right\}$$

$$= \frac{2}{2\pi nft}\left(\sin 2\pi nft\,\Big|_0^{\frac{1}{2}T} - \sin 2\pi nft\,\Big|_{\frac{1}{2}T}^{T}\right)$$

With $f = 1/T$ this becomes:

$$a_n = \frac{2}{2\pi n}\{\sin 2\pi n\tfrac{1}{T}\tfrac{1}{2}T - 0 - (\sin 2\pi n\tfrac{1}{T}\tfrac{1}{2}T)\}$$

$$= \frac{2}{2\pi n}(\sin \pi n - 0 - \sin 2\pi n + \sin \pi n) = 0$$

$$b_n = \frac{2}{T}\left\{\int_0^{\frac{1}{2}T}\sin 2\pi nft\,dt + \int_{\frac{1}{2}T}^{T}-\sin 2\pi nft\,dt\right\}$$

$$= \frac{2}{2\pi nfT}\left(-\cos 2\pi nft\,\Big|_1^{\frac{1}{2}T} + \cos 2\pi nft\,\Big|_{\frac{1}{2}T}^{T}\right) =$$

$$= \frac{2}{2\pi n}\{-(\cos 2\pi n\tfrac{1}{T}\tfrac{1}{2}T - 1) + (\cos 2\pi n\tfrac{1}{T}T - \cos 2\pi n\tfrac{1}{T}\tfrac{1}{2}T)\} =$$

$$= \frac{2}{2\pi n}(-\cos \pi n + 1 + \cos 2\pi n - \cos \pi n) = \frac{1}{\pi n}(2 - 2\cos \pi n)$$

4.11 $f_o = 20000/512 = 39.0625$ Hz

a. $q = 0.4, f = 15.4 \cdot 39.0625 = 601.6$ Hz, $A = 720 \cdot \pi \cdot 0.4/\sin(\pi \cdot 0.4) = 951.3$

b. $750 = 19.2 \cdot 39.0625$. The value of q thus is 0.2 and the largest amplitude coefficients are found at C_{19} and C_{20}. The value of C follows from: $550 \cdot \sin(0.2 \cdot \pi) = C_{19} \cdot 0.2 \cdot \pi$ thus $C_{19} = 514.519$.
From C_{19} and q now follows $C_{20} = 128.63$.

4.12 The function rule is: $y(t) = \sin \pi t/T$

$$a_0 = \frac{2}{T} \int_0^T \sin \frac{\pi t}{T} \, dt = \frac{2}{T} \frac{-T}{\pi} \cos \frac{\pi t}{T} \Big|_0^T = \frac{4}{\pi}$$

$$a_n = \frac{2}{T} \int_0^T \sin \frac{\pi t}{T} \cos 2\pi n \frac{t}{T} \, dt$$

With rule (2.46) this becomes:

$$a_n = \frac{1}{T} \int_0^T \sin(1 + 2n) \frac{\pi t}{T} \, dt + \frac{1}{T} \int_0^T \sin(1 - 2n) \frac{\pi t}{T} \, dt = \frac{2}{(1 + 2n)\pi} + \frac{2}{(1 - 2n)\pi} =$$

From a similar calculation of b_n follows: $b_n = 0$.

4.13

a.   125 Hz

b.   The 8-th, 16-th, 24-th etc. harmonics, thus all multiples of 1000 Hz.

c. $C_1 = (2A/\pi)\sin(\pi/8)$; $C = (2A/4\pi)\sin(4\pi/8)$, $_1C/C = 4 \cdot \sin(\pi/8) = 1.531$. This corresponds to 3.698 dB.

4.14

$$a(f) = 2A \int_{t_0 - \frac{1}{2}W}^{t_0 + \frac{1}{2}W} \cos 2\pi ft \, dt = 2A \frac{\sin 2\pi ft}{2\pi f} \Big|_{t_0 - \frac{1}{2}W}^{t_0 + \frac{1}{2}W} = \frac{2A}{\pi f} \cos 2\pi ft_0 \sin \pi fW$$

$$b(f) = 2A \int\limits_{t_0-\frac{1}{2}W}^{t_0+\frac{1}{2}W} \sin 2\pi ft \, dt = 2A \frac{-\cos 2\pi ft}{2\pi f} \Bigg|_{t_0-\frac{1}{2}W}^{t_0+\frac{1}{2}W} = \frac{2A}{\pi f} \sin 2\pi ft_0 \sin \pi fW$$

$$C(f) = \sqrt{a^2(f) + b^2(f)} = \frac{2A}{\pi f} \sin \pi fW$$

$$\tan \phi(f) = -\frac{\sin 2\pi ft_0}{\cos 2\pi ft_0} = -\tan 2\pi ft_0, \quad \text{thus}$$

$$\phi(f) = -2\pi ft_0, \quad \text{and} \quad f_p = -\frac{1}{2\pi} \frac{d\phi}{df} = -\frac{1}{2\pi} -2\pi t_0 = t_0$$

4.15 The first two harmonics are below half the sampling frequency; the third harmonic (21 kHz) generates 19 kHz, the fourth harmonic (28 kHz) generates 12 kHz, the fifth harmonic (35 kHz) generates 5 kHz.

4.16 The attenuation by a zero-order hold filter is described by: $\dfrac{\sin \dfrac{\pi f}{f_s}}{\dfrac{\pi f}{f_s}}$

(See fig.4.4.5). When $f = \frac{1}{2}f_s$ the attenuation that should be corrected is $2/\pi = -3.92$ dB, and with $f_s' = 4 \cdot f_s$: $(\sin \pi/8)/(\pi/8) = -0.22$ dB.

4.17 $a_1 = x_0 + k \cdot x_1 - k \cdot x_3 - x_4 - k \cdot x_5 + k \cdot x_7$
$b_1 = k \cdot x_1 + x_2 + k \cdot x_3 - k \cdot x_5 - x_6 + k \cdot x_7$
$b_3 = k \cdot x_1 - x_2 + k \cdot x_3 - k \cdot x_5 + x_6 - k \cdot x_7$

4.18

$$a_o = \frac{2}{T} \int\limits_{-\frac{1}{2}W}^{\frac{1}{2}W} A \, dt = \frac{2A}{T} t \Bigg|_{-\frac{1}{2}W}^{\frac{1}{2}W} = \frac{2AW}{T}$$

$$a_n = \frac{2}{T} \int_{-\frac{1}{2}W}^{\frac{1}{2}W} A \cos 2\pi \frac{n}{T} t = \frac{2A}{T} \left. \frac{\sin 2\pi \frac{n}{T} t}{2\pi \frac{n}{T}} \right|_{-\frac{1}{2}W}^{\frac{1}{2}W} = \frac{A}{\pi n} \left(\sin 2\pi \frac{n}{T} \cdot \frac{1}{2} W - \sin 2 \right.$$

$$b_n = \frac{2}{T} \int_{-\frac{1}{2}W}^{\frac{1}{2}W} A \sin 2\pi \frac{n}{T} t = \frac{2A}{T} \left. \frac{-\cos 2\pi \frac{n}{T} t}{2\pi \frac{n}{T}} \right|_{-\frac{1}{2}W}^{\frac{1}{2}W} = \frac{A}{\pi n} \left(-\cos 2\pi \frac{n}{T} \cdot \frac{1}{2} W + c \right.$$

Chapter 5

5.1 FIR; with a transversal filter.

5.2 With linear distortion, because $h(k) \neq \delta(k)$.

5.3 Yes, because $h(k)$ is symmetrical.

5.4 $y(k) = \frac{1}{2}x(k) + x(k{-}1) + \frac{1}{2}x(k{-}2)$

5.5 $y(k) = \frac{1}{2}\sin sk + \sin s(k{-}1) + \frac{1}{2}\sin s(k{-}2)$
     $= \sin s(k{-}1) \cdot \cos s + \sin s(k{-}1) = \sin s(k{-}1)(1 + \cos s)$

5.6 Low-pass filter.

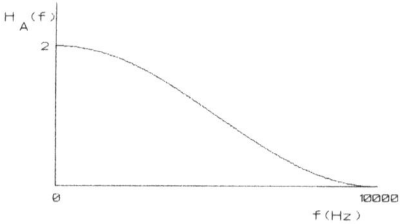

5.7 Use $s = 2\pi f/f_s$:
     At 3500 Hz $s = 1.100$ and $1 + \cos s$: 1.454
     At 7000 Hz $s = 2.199$ and $1 + \cos s$: 0.412
     The proportion is 10.9 dB.

5.8 50 µs, 0.314 rad $= 18°$, linear.

5.9 $\sqrt{\dfrac{1}{9} + \dfrac{1}{25} + \dfrac{1}{49} + \ldots + \dfrac{1}{225}}$ = 0.45 thus 45%.

5.10 The presence filter, because this is a system with waveshape distortion.

5.11 At the resonance frequency, because there the steepness $d\phi/df$ is maximal.

5.12

a. $y(k) = \dfrac{x(k) + x(k-1) + \ldots + x(k-20)}{21}$

b. The impulse response has the shape of a single impulse with pulse-width $W = 21/10000$ s. The amplitude response is thus described by $\sin \pi fW/\pi fW$. The filter is a low-pass with cut-off frequency (better: first zero point): $10000/21 = 476$ Hz.

c. Yes, the impulse response is symmetrical around $h(k=10)$ or $t_o = 1$ ms; this is therefore the time delay.

d. The sample frequency should be above twice the highest signal frequency = $2 \cdot 476$ Hz; 1000 Hz is a good choice. The easiest way to achieve this reduction is to skip 9 clock pulse before calculating a new y(k)-value.

5.13 From the superposition principle it follows that
if $x \rightarrow y$
then $x+x+x+\ldots$ (*n* times) $\rightarrow y+y+y+\ldots$ (n times) or $nx \rightarrow ny$
Suppose $x' = (n/m) \cdot x$ then $mx' = m \cdot (n/m) \cdot x = nx$
And as $mx' \rightarrow my'$ we can conclude $nx \rightarrow my'$
This means $my' = ny$ or $y' = (n/m) \cdot y$

Chapter 6

6.1 The total load is 1 kΩ parallel to 9 kΩ = $(1 \cdot 9)/(1+9) = 0.9$ kΩ.
Input voltage: $0.9/(4+0.9) = 0.184$ volt.
The pre-amplifier delivers $900/4900^2 = 37.5$ µW; $0.184^2/1000 = 33.7$ µW in the resistance of 1 kΩ and $0.184^2/9000 = 3.8$ µW in the other one.

6.2 $(N_s/N_p)^2 = 0.9/4$ thus $N_s/N_p = 0.474$. The output voltage without load resistance is 0.474 volt, with load resistance it is half that value. The power is $0.474^2/(4·900) = 62.5$ μW. From this power $0.237^2/1000 = 56.25$ μW in the resistance of 2 kΩ, and $0.237^2/9000 = 6.25$ μW in the other one.

6.3 $1.6 = \dfrac{1}{1+x} \cdot 2$, thus $1.6 + 1.6x = 2$, $x = 250 \, \Omega$

6.4

a. The total load resistance consists of the series circuit of $(1-\alpha)R$ and the parallel resistance of αR and R_B thus

$$R_L = \frac{R_B \cdot \alpha R}{R_B + \alpha R} + (1 - \alpha)R$$

R_L varies between $(R_B·R)/(R_B+R)$ if $\alpha=1$ and R if $\alpha=0$.

b. We can determine the output resistance by connecting a certain resistance R_x, with such a value that the output voltage is reduced by a factor 2. This R_x is then equal to the output resistance. Let us indicate the 'upper' resistance of the potentiometer with R_1 and the 'lower' one with R_2.
The total output resistance is now the sum of R_A and the contribution of the potentiometer. To find this contribution we may assume $R_A = 0$.

Half the value of the unloaded voltage is: $\dfrac{1}{2} \dfrac{R_2}{R_1 + R_2}$

and this is equal to: $\dfrac{\dfrac{R_x \cdot R_2}{R_x + R_2}}{\dfrac{R_x \cdot R_2}{R_x + R_2} + R_1}$

If we work out this equation we find the following result: $R_x = \dfrac{R_1 \cdot R_2}{R_1 + R_2}$

The output resistance due to the potentiometer is equal to that of the parallel circuit of the upper and lower section of the potentiometer and is thus dependent upon its setting.

6.5 The factor β is $4/(1+4) = 0.8$

Open loop gain A: 80 dB,

closed loop gain: $A' = \dfrac{10000}{1 + 0.8 \cdot 10000} = 1.2498 = 1.937\,\text{dB}$

This is almost equal to $1/\beta = 1.25$ (= 1.938 dB). The gain factor has been reduced from 10000 to 1.25; this reduction corresponds to 78 dB, the difference between the open loop and closed loop gain.

To determine the output resistance we connect a load resistance with such a value R_x that the output voltage is reduced by a factor 2. Assume an input voltage of 1 volt. The output voltage without load is then 1.25 V, and with load 0.625 V. The voltage across R_{out} is: $10000(1 - 0.8 \cdot 0.625) = 5000$ V, the output current is $5000/1000 = 5$ Ampere. This current also flows through R_x , across which a voltage of 0.625 V exists. From this it follows that $R_x = 0.625/5 = 0.125\ \Omega$. The reduction factor is again $1.25 \cdot 10^{-4}$.

6.6 The length of the trajectory of the reflected sound is $2 \cdot (\sqrt{(1.75^2 + 2.50^2)} = 6.10$ m. The difference in distance is 1.10 meter, the time delay $\tau = 3.24$ ms. The circuit of fig.6.3.1a can be used when q is given the value -0.9. This means that in the table below fig.6.3.3 the maxima change into minima and vice versa.

Maxima at multiples of $1/\tau$: 309 Hz, 618 Hz etc;

minima at $1/2\tau$, $3/2\tau$,... : 154.5 Hz, 463.6 Hz etc.

Peak/valley proportion: $(1+0.9)/(1-0.9) = 19 = 25.6$ dB.

6.7

$$A \sin(\omega t + \phi) = \sin \omega t + qA \sin\{\omega(t - \tau) + \phi\}$$

$$= \sin \omega t + qA \sin(\omega t + \phi) \cdot \cos \omega \tau - qA \cos(\omega t + \phi) \cdot \sin \omega \tau$$

$$qA \sin \omega \tau \cdot \cos(\omega t + \phi) + A(1 - \cos \omega \tau) \cdot \sin(\omega t + \phi) = \sin \omega t$$

The left-hand expression can be written as: $B \sin(\omega t + \phi + \gamma)$ with

$$B = \sqrt{q^2 A^2 \sin^2 \omega \tau + A^2(1 - \cos \omega \tau)^2}, \quad \tan \gamma = \dfrac{qA \sin \omega \tau}{A(1 - q \cos \omega \tau)}$$

$B \sin(\omega t + \phi + \gamma) = B \sin \omega t \cdot \cos(\phi + \gamma) + B \cos \omega t \cdot \sin(\phi + \gamma) = \sin \omega t$

$\{B \cos(\phi + \gamma) - 1\} \sin \omega t + B \sin(\phi + \gamma) \cdot \cos \omega t = 0$

This is true if $\phi = -\gamma$ and $B = 1$. From this:

$$A\sqrt{q^2 \sin^2 \omega \tau + (1 - \cos \omega \tau)^2} = 1 \quad \text{or} \quad A = \frac{1}{\sqrt{1 - 2q \cos \omega \tau + q^2}}$$

6.8   The first equation is: $\quad V_y + RC\dfrac{dV_y}{dt} = V_x$

The step response is found by giving V_x the value V_o. In that case the solution is:

$$V_y = V_0\left(1 - e^{\frac{-t}{RC}}\right)$$

Verification: substitute V_y and $\dfrac{dV_y}{dt} = \dfrac{V_0}{RC}e^{\frac{-t}{RC}}$:

$$V_o = RC\frac{V_0}{RC}e^{\frac{-t}{RC}} + V_0 - V_0 e^{\frac{-t}{RC}}$$

For the sine response we substitute the same V_x and V_y as was done with the differentiating network of fig.6.3.7:

$\quad RC \cdot \omega B \cos(\omega t - \phi) + B \sin(\omega t - \phi) = \sin \omega t$

$\quad D \cos(\omega t - \phi + \gamma) = \sin \omega t$

with

$$D = \sqrt{B^2\omega^2 R^2 C^2 + B^2} = B\sqrt{1 + \omega^2 R^2 C^2}, \quad \tan\phi = -\frac{B}{B\omega RC} = -\frac{1}{\omega RC}$$

$\quad D \cos \omega t \cdot \cos(\gamma - \phi) - D \sin \omega t \cdot \sin(\gamma - \phi) = \sin \omega t$

This is true if $\gamma - \phi = -\frac{1}{2}\pi$ and $D = 1$ or $1 = B\sqrt{1 + \omega^2 R^2 C^2}$

With this we know $B \,(= H_A(f))$ and $\phi \,(= \Delta_\phi(f))$:

$$H_A(f) = \frac{1}{\sqrt{1 + 4\pi^2 f^2 R^2 C^2}}, \quad \Delta_\phi(f) = \gamma + \frac{1}{2}\pi = \arctan - \frac{1}{2\pi f R C} + \frac{1}{2}\pi$$

6.9 $2\pi f = 1/RC$; with $f = 20$ Hz and $R = 5$ kΩ: $C = 1.59$ μF.

6.10 Same formula as in 6.9; with $f = 1000$ Hz and e.g. $R = 5$ kΩ we find $C = 31.9$ nF.

6.11

					mean	
.001	.193	.585	.350	.823	.391	
.174	.710	.304	.091	.147	.285	
.989	.119	.009	.532	.602	.450	
.166	.451	.057	.783	.520	.395	
.876	.956	.539	.462	.862	.739	
.780	.997	.611	.266	.840	.699	
.376	.677	.009	.276	.588	.385	
.838	.485	.744	.458	.744	.654	
.599	.735	.572	.152	.425	.497	Between 0.0 and 0.2 : 0
.517	.752	.169	.492	.700	.526	0.2 and 0.4 : 5
.148	.142	.693	.427	.967	.475	0.4 and 0.6 : 10
.153	.822	.191	.817	.156	.428	0.6 and 0.8 : 5
.732	.280	.682	.722	.123	.508	0.8 and 1.0 : 0
.835	.517	.426	.949	.550	.655	
.472	.847	.456	.983	.739	.699	
.196	.839	.501	.027	.573	.427	
.531	.843	.658	.842	.110	.597	
.314	.286	.140	.835	.600	.435	
.253	.002	.806	.211	.553	.365	
.114	.752	.543	.437	.696	.509	

6.12

a. $f_o = 1000$ Hz, amplitude $= 1$

    $f_o + f_m = 1200$ Hz, amplitude $= \frac{1}{2}m = 0.1$

    $f_o - f_m = 800$ Hz, amplitude $= \frac{1}{2}m = 0.1$

b. Simplify the function rule to $\sin(a + m \cdot \cos b)$. This can be worked out to:

$$\sin a \cdot \cos(m \cdot \cos b) + \cos a \cdot \sin(m \cdot \cos b).$$

    Use rule (6.23) and (2.42):

$$\cos(m \cos b) = \cos\left(m \sin\left(b + \tfrac{1}{2}\pi\right)\right) = J_0(m) + \{-J_2(m)\cos 2b + J_4(m)\cos 4b - \ldots\}$$

$$\sin(m \cos b) = \sin\left(m \sin\left(b + \tfrac{1}{2}\pi\right)\right) = 2\{J_1(m)\cos b - J_3(m)\cos 3b + J_5(m)\cos 5b - \ldots$$

    With this we find:

$$\sin(a + m \cos b) =$$

$$= J_0(m)\sin a - [J_2(m)\{\sin(a + 2b) + \sin(a - 2b)\} + J_4(m)\{\sin(a + 4b)$$

$$+ [J_1(m)\{\cos(a + b) + \cos(a - b)\} - J_3(m)\{\cos(a + 3b) + \cos(a - 3b)\}$$

    For the given function rule $y(t) = \sin(2\pi f_c t + m \cdot \cos 2\pi f_m t)$ we find:

$$y(t) = \sin(2\pi f_c t + m \cos 2\pi f_m t)$$

$$= J_0(m)\sin 2\pi f_c t +$$

$$+ J_1(m)\cos 2\pi(f_c + f_m)t + J_1(m)\cos 2\pi(f_c - f)m)t +$$

$$- J_2(m)\sin 2\pi(f_c + 2f_m)t - J_2(m)\sin 2\pi(f_c - 2f_m)t +$$

$$- J_3(m)\cos 2\pi(f_c + 3f_m)t - J_3(m)\cos 2\pi(f_c - 3f_m)t +$$

$$+ \ldots.$$

c. $f_c = 1000$ Hz, amplitude $= J_0(0.2) \approx 1$

    $f_o + f_m = 1200$ Hz, amplitude $= J_1(0.2) \approx 0.1$

    $f_o - f_m = 800$ Hz, amplitude $= J_1(0.2) \approx 0.1$

    Higher and lower frequency components are not significant.

d. The only difference is a phase shift of $90°$ between the central components of the AM and the FM signal.

6.13

frequencies		amplitudes ($J_n(10)$)
	940	0.246
740	1140	0.043
540	1340	0.255
340	1540	0.058
140	1740	0.220
(-) 60	1940	0.234
(-) 260	2140	0.014
(-) 460	2340	0.217
(-) 660	2540	0.318
(-) 860	2740	0.292
(-)1060	2940	0.207
(-)1260	3140	0.123
(-)1460	3340	0.063
(-)1660	3540	0.029
(-)1860	3740	0.012
(-)2060	2940	0.005
(-)2260	3140	0.002

Chapter 7

7.1

a.

$$\overline{R}(\tau) = \frac{1}{T} \int_0^T \sin 2\pi ft \cdot \sin 2\pi f(t + \tau)\, dt$$

$$= \frac{1}{T} \int_0^T \{\tfrac{1}{2} \cos 2\pi f\tau - \tfrac{1}{2} \cos(4\pi ft + 2\pi f\tau)\}\, dt$$

$$= \frac{1}{2T} \{\cos 2\pi f\tau \cdot t \Big|_0^T - \frac{\sin(4\pi ft + 2\pi f\tau)}{4\pi f} \Big|_0^T\} = \tfrac{1}{2} \cos 2\pi f\tau$$

The conclusion that R is sinusoidal also follows from rule (7.6).

b. The power spectrum of a square wave is equal to the amplitude spectrum of a triangular wave. According to (7.6) R also has a triangular shape.

7.2 $0.918 = a_1 \qquad + \qquad 0.918 \cdot a_2$

$0.690 = 0.918 \cdot a_1 + \qquad a_2$

From this: $a_1 = 1.807$ and $a_2 = -0.969$

7.3 $h(0) = 1$, $h(1) = -1$; $y(k) = x(k) \cdot h(0) + x(k-1) \cdot h(1) = x(k) - x(k-1)$

If $x(k) = \sin s \cdot k$: $y(k) = \sin s \cdot k - \sin s(k-1) = 2 \cdot \sin \frac{1}{2}s \cdot \cos(sk - \frac{1}{2})$

Thus $H_A(f) = 2 \cdot \sin \pi f/f_s \approx 2\pi f/f_s$ if $\pi f \ll f_s$.

7.4 $4 \cdot \cos^3 x - 3 \cdot \cos x = (4 \cdot \cos^2 x - 3) \cdot \cos x = (2 + 2 \cdot \cos 2x - 3) \cdot \cos x =$

$= 2 \cdot \cos 2x \cdot \cos x - \cos x = \cos 3x + \cos x - \cos x = \cos 3x$

7.5

a. This will exhibit a peak with direct sound and strong reflections. It will decrease with reverberation.

b. By introducing linear distortion that changes the waveshape of the signals. For this purpose a filter could be used with a straight, horizontal amplitude response and a non-linear phase response.

INDEX